Operation of Complex Water Systems

Operation, Planning, and Analysis of Already Developed Water Systems

W0042753

NATO ASI Series

Advanced Science Institutes Series

A series presenting the results of activities sponsored by the NATO Science Committee, which aims at the dissemination of advanced scientific and technological knowledge, with a view to strengthening links between scientific communities

The series is published by an international board of publishers in conjunction with NATO Scientific Affairs Division

A	**Life Sciences**	Plenum Publishing Corporation
B	**Physics**	London and New York
C	**Mathematical and Physical Sciences**	D. Reidel Publishing Company Dordrecht and Boston
D	**Behavioural and Social Sciences**	Martinus Nijhoff Publishers Boston/The Hague/Dordrecht/Lancaster
E	**Applied Sciences**	
F	**Computer and Systems Sciences**	Springer Verlag Berlin/Heidelberg/New York
G	**Ecological Sciences**	

Series E: Applied Sciences – No. 58

Operation of Complex Water Systems

Operation, Planning and Analysis of Already Developed Water Systems

edited by

Emanuele Guggino
Professor and Director of the Institute
for Hydraulics, Hydrology, and Water
Management at the University of Catania
Catania, Sicily, Italy

Giuseppe Rossi
Professor at the Institute for Hydraulics,
Hydrology, and Water Management at the
University of Catania
Catana, Sicily, Italy

David Hendricks
Professor of Civil Engineering
at Colorado State University
Fort Collins, Colorado, USA

1983 **Martinus Nijhoff Publishers**
A member of the Kluwer Academic Publishers Group
Boston / The Hague / Dordrecht / Lancaster
Published in cooperation with NATO Scientific Affairs Division

Proceedings of the NATO Advanced Study Institute on "Operation of Complex Water Systems", Erice, Sicily, May 23 - June 2, 1981

Library of Congress Cataloging in Publication Data

NATO Advanced Study Institute on "Operation of
    Complex Water Systems" (1981 : Erice, Italy)
    Operation of complex water systems.

    (NATO advanced study institutes series. Series E,
Applied sciences ; v. 58)
    1. Water supply engineering--Congresses.
I. Guggino, Emanuele. II. Rossi, Giuseppe.
III. Hendricks, David W. IV. North Atlantic Treaty
Organization. V. Title. VI. Series.
TD201.N37 1981   628.1   82-22529

ISBN-13: 978-94-009-6809-7   e-ISBN-13: 978-94-009-6807-3
DOI: 10.1007/978-94-009-6807-3

Distributors for the United States and Canada: Kluwer Boston, Inc., 190 Old Derby Street, Hingham, MA 02043, USA

Distributors for all other countries: Kluwer Academic Publishers Group, Distribution Center, P.O. Box 322, 3300 AH Dordrecht, The Netherlands

CONTENTS

PREFACE

Most water systems in the industrial regions of the world are already developed. At the same time they are highly complex. This is true with respect to physical configuration, managment, operation, political goals, environmental interactions, etc. Thus the basic systems are already in place. This realization is the starting point for any new water developments and for operation.

From this we conclude that whatever we do to meet new exigencies requires an understanding of the presently in-place complex water systems. Their operation is the important thing. And how can we adjust their operation to meet the new demands upon the system?

This book deals with complex water systems and their operation. Some chapters are highly theoretical while others are rooted in practical applications. How can we analyze the operation of a complex water system and determine how its performance can be improved? Several chapters on mathematical analysis give approaches involving different aspects of this problem. But operation also has political, management, and physical aspects. These problems are addressed in chapters by managers who operate such systems.

The main theme of all chapters is how to deal with the different aspects of a complex water system, already in place. We feel the book, in dealing with this question could be a start for new theoretical premises in water planning.

The book has been developed from the papers presented at the NATO Advanced Study Institute entitled, "Operation of Complex Water Systems," held at Erice, Sicily, May 23-June 2, 1981. The institute was organized by the School of Water Resources Management of the Centre Ettore Majorana, an enterprise headquartered at the Istituto di Idraulica, Idrologia, e Gestione delle Acque and the Istituto di Idraulica Agraria, Universita di Catania. The book is one of the significant accomplishemnts of the first ten years of three-way cooperation in the area of water management between the University of Catania, FORMEZ, and Colorado State University.

Each chapter was edited during a process of continuing refinement to change the style from conference paper to book chapter. Thus the book is more than a collection of seminar papers, it is a document intended for communication with others in the field, and as a beginning of professional thought about operation of already developed complex water systems.

Emanuel Guggino
Giuseppe Rossi
David Hendricks

ACKNOWLEDGMENTS

This book was developed from the proceedings of the NATO Advanced Study Institute, Operation of Complex Water Systems, held 23 May to 2 June, 1981 at the Ettore Majorana Centre for Scientific Culture, Erice, Sicily, directed by Antonio Zichichi. The NATO Institute was conceived during two year period which involved the participation of the editors together with Professor Vujica Yevjevich, and Professor Guido Calenda. Dr. Mario DiLullo, Director of the ASI Program, provided continuing support in helping the development of the institute. The NATO ASI Program provided the major financial support.

Other sponsors for the institute include Centro Formazione e Studi per il Mezzogiorno, Cassa per il Mezzogiorno, Ministero dei Lavori Pubblici, Ministero della Pubblica Isstruzione, Ministero per la Ricerca Scientifica e Tecnologica, and Regione Siciliana. Dott. Giulio Centemero was instrumental in providing the FORMEZ support.

Several Italian professional organizations were involved. They are: ANIAI, Associazione Nazionale Ingegneri ed Architetti Italiani, ANDIS Associazione Nazionale di Ingegneri Sanitaria, AII, Associazione Italiana Idrotecnica.

We appreciate the support of the following persons:

Dott. Gulio Centemero, Manager of Special Projects, FORMEZ, Roma
Dott. Alberto DeMaio, Director, FORMEZ, Roma
Dott. Ing. Gabriele Di Palma, Ministry of Public Works, Roma
Honorable Franco Nicolazzi, Minister of Public Works, Roma.
Dott. Giovanni Torregrossa, Advisor to the Government for Public Works, Roma
Prof. Antonio Zichichi, Director, Centre Ettore Majorana, Erice
Dott. Sergio Zoppi, President of FORMEZ, Roma

PARTICIPANTS AND AUTHORS

BELGIUM

Dr. Vladimir Mandl*
Commission of European
  Communities
Rue de la Loi 200, B
1049 BRUXELLES

Dr. H. Vrijhof
Commission of European
  Communities
Rue de la Loi 200,B
1049 BRUXELLES

CANADA

Dr. G. A. Fuller*, Head
Regional Water Systems
  Engineering
University of Regina
REGINA SASKATCHEWAN S4S OA2

FRANCE

Prof Yves Emsellem*
CE.FI.GRE.
Sophia Antipolis
B.P. 15
VALBONNE (Alpes Maritimes)

GERMANY

Dr. Ricardo Harboe*
Ruhr Universität Bochum
Lehrstuhl für Wasserwirtschaft
  und Umwelttechnik I
Postfach 102148
4630 BOCHUM - QUERENBURG

ITALY

Dott. Giuseppe Angelo
Ente Acquedotti Siciliani
Via Torino
PARTANNA

*Author

Prof. Salvatore Enrico Battiato*
Istituto Finanze
Facoltà di Economia e Commercio
Universita di Catania
Corso Italia, 55
CATANIA

Prof. Ing. Marcello Benedini*
Instituto Ricerca sulle Acque
Via Reno, 1
ROMA

Dott. Domenico Bertucci
FORMEZ
Via Salaria, 229
ROMA

Dott. Ing. Giuseppe Boscarino
Consorzio Lisimelie
Via Siracusa Belvedere
SIRACUSA

Prof. Ing. Guido Calenda
Istituto Costruzioni Idrauliche
Facoltà di Ingegneria
Via Eudossiana, 18
ROMA

Dott. Amedeo Calenzo
FORMEZ
Via Salaria, 229
ROMA

Dott. Giulio Centemero
FORMEZ
Via Salaria, 229
ROMA

Dr.ssa Lucia De Anna
FORMEZ
Via Salaria, 229
ROMA

Dott. Alberto De Maio, Direttore
FORMEZ
Via Salaria, 229
ROMA

Dott. Ing. Gabriele Di Palma
Ministero dei Lavori Pubblici
P.le Porta Pia
ROMA

Dott. Ing. Maurizio Di Stefano
Scuola di Ingegneria Aerospaziale
Istituto di Aerodinamica
Via Eudossiana, 16
ROMA

Prof. Ing. Costantino Fassò*
Istituto di Idraulica e
    Costruzioni Idrauliche
Facoltà di Ingegneria
V.le Merello, 92
09100 - CAGLIARI

Dott. Ing. Massimo Ferraresi
IDROSER
Via Alessandrini, 13
BOLOGNA

Prof. Ing. Mario Gallati
Istituto di Idraulica
Facoltà di Ingegneria
Piazza Leonardo da Vinci
PAVIA

Prof. Emilio Giardina*
Istituto Finanze
Facoltà di Economia e Commercio
C.so Italia, 55
CATANIA

Prof. Ing. Mario Gramignani
Istituto di Idraulica Idrologia
    e Gestione delle Acque
Facoltà di Ingegneria
Viale Andrea Doria, 6
CATANIA

Prof. Ing. Emanuele Guggino*
Istituto di Idraulica Idrologia
    e Gestione delle Acque
Facoltà di Ingegneria
Viale Andrea Doria, 6
CATANIA

Prof. Ing. Salvatore Indelicato*
    Direttore
Istituto di Idraulica Agraria
Facoltà di Scienze Agrarie
Via Valdisavoia, 5
CATANIA

Dr.ssa Lisa Kholchtchevnikova
Dagh Watson S.p.a.
Piazza Amendola, 3
20149 Milano

Prof. Paolo Leon
A.R.P.E.S.
Via XX Settembre, 98
ROMA

Prof. Ing. Gianmarco Margaritora
Istituto di Costruzioni
    Idrauliche
Facoltà di Ingegneria
Via Eudossiana, 18
00184 ROMA

Prof. Benedetto Matarazzo
Facoltà Economia e Commercio
Corso Italia, 55
CATANIA

Dott. Ing. Mario Rosario Mazzola
Istituto di Idraulica
Facoltà di Ingegneria
Viale delle Scienze
PALERMO

Dr.ssa Vanna Messora
Via Roncobonoldo, 20
46020 PALIDANO (MN)

Dott. Ing. Carlo Modica
Istituto di Idraulica Idrologia
  e Gestione delle Acque
Facoltà di Ingegneria
Viale Andrea Doria, 6
CATANIA

Prof. Alberto Petaccia
Istituto di Costruzioni
  Idrauliche
Facoltà di Ingegneria
Via Eudossiana, 18
00184 ROMA

Dott. Ing. Bartolomeo Reitano
Istituto di Idraulica Idrologia
  e Gestione delle Acque
Facoltà di Ingegneria
Viale Andrea Doria, 6
CATANIA

Prof. Francesco Rizzo
Istituto di Idraulica Idrologia
  e Gestione delle Acque
Facoltà di Ingegneria
Viale Andrea Doria, 6
CATANIA

Dott. Giuseppe Rossetto
FORMEZ
Via Salaria, 229
ROMA

Prof. Ing Giuseppe Rossi*
Istituto di Idraulica, Idrologia,
  e Gestione delle Acque
Universita di Catania
Viale Andrea Doria, 6
95125 CATANIA

Dott. Giovanni Sarnataro
Ripartizione VI - Divisione 6
Cassa per il Mezzogiorno
Piazza Kennedy, 20
ROMA-EUR

Dott. Ing. Paolo Serraglini
Ripartizone VI - Divisione 6
Cassa per il Mezzogiorno
Piazza Kennedy, 20
ROMA

Dott. Ing. Augusto Sbraccia
Capo Divisione
Progetto Speciale 30
Cassa per il Mezzogiorno
Piazza Kennedy, 20
ROMA

Dott. Giovanni Torregrossa
Consigliere di Stato
Ministero Lavori Pubblici
P.le Porta Pia
ROMA

Dott. Ing. Roberto Viviani
Capo Ufficio
Progetto Speciale 30
Cassa per il Mezzogiorno
Piazza Kennedy, 20
ROMA

Geom. Francesco Vasque
Studio Boscarino
Consorzio Lisimelie
Via Siracusa Belvedere
SIRACUSA

PORTUGAL

Dr. Luis Veiga da Cunha*
Head of Hydrology and River
  Hydraulic Division
Laboratòrio Nacional de
  Engenharia Civil
Avenida do Brasil, 101
1799 LISBOA CODEX

Dr.ssa Vitòria Mira Da Silva
Secretaria de Estado do
Ordenamento e Ambiente
Presidência do Conselho de
  Ministros
Rua Prof. Gomes Teixeira
1300 LISBOA

Dr. Ing. Antonio Goncalves
  Henriques
Laboratòrio Nacional de
  Engenharia Civil
Av. do Brasil, 101
1799 LISBOA

Prof. Luis A. Santos Pereira*
Instituto Superior de Agronomia
Technical University of Lisbon
Tapada da Ayuda
1399 LISBOA

ROUMANIA

Dr. Ing. Mihaela Jonescu
Istitutul Politehnic
Traian Vuia
TIMISOARA

SPAIN

Dr. Amable Sanchez
Servicio Geologico del
  Ministerio de Obras Publicas
Avenida de Portugal, 81
MADRID - 11

TURKEY

Dr. Hilmi Dogan Altinbilek
King Abdulaziz University
Civil Engineering Department
P.O. Box 9027 JEDDAH

Prof. Dr. Mehmeticik Bayazit
Teknit Universite
Insaat Fakültesi
INSTANBUL

Dr. Ibrahim Gürer*
Hacettepe Universitesi
Verbilimleri Enstitüsü
BEYTEPE - ANKARA

Dr. Eng. Ferhat Turkman
Dept. of Civil Engineering
Civil Engineering Faculty
Ege University
IZMIR

UNITED KINGDOM

Dr. Asit K. Biswas*
International Water Resources
Association
76 Woodstock Close
OXFORD OX2 8DD

Dr. Eng. R. P. Jones
Control Theory Centre
Department of Engineering
University of Warwick
CONVENTRY CVA 7AL

Dr. Andrew Spink
Department of Civil Engineering
University of Birmingham
P.O. Box 363
BIRMINGHAM B15 2TT

Dr. Eng. K. Vijayaratnam
164, Clensham Lane
SUTTON, SURREY SMI 2NG

Dr. Geoffrey Conybeare Williams
  Chief Executive
South West Water Authority
3.5. Barnfield Road
EXETER EX1 1RE
DEVON

U.S.A.

Dr. Neil Cline*
Manager
Orange County Water District
10500 Ellis Avenue
P.O. Box 8300
Fountain Valleys
CALIFORNIA 92708

Dr. Fulvio Croce*
Hydro-Triad Ltd
12687, W. Cedar Dr.
LAKEWOOD, COLORADO 80228

Prof. Lucien Duckstein*
The University of Arizona
Systems and Industrial
  Engineering Department
TUCSON, ARIZONA 85721

Dr. Lloyd Fowler*
Manager
Goleta County Water District
P.O. Box 788
GOLETA, CALIFORNIA   93116

Prof. Warren A. Hall*
Civil Engineering Department
Colorado State University
FORT COLLINS, COLORADO   80523

Prof. David W. Hendricks*
Civil Engineering Department
Colorado State University
FORT COLLINS, COLORADO   80523

Prof. José D. Salas*
Civil Engineering Department
Colorado State University
FORT COLLINS, COLORADO   80523

Prof. Vujica Yevjevich*
Director
International Water Resources
  Institute
George Washington University
2000 L Street N.W. Suite 301
WASHINGTON D.C.   20037

Prof. Norman Wengert*
Department of Political Science
Colorado State University
FORT COLLINS, COLORADO   80523

YUGOSLAVIA

Dr. Eng. Milan Andjelić
Institute Mihailo Pupin
11000 BELGRADE, VOLGINA, 15

Dr. Eng. Boris Berakovic
Elektroprojekt
37 Proleterskih Brigada
41000 ZAGREB

Dr. Eng. Branko Karan
Mihailo Pupin Institute
11000 BELGRADE, VOLGINA, 15

Dr. Eng. Milden Petricec
Institute Za Elektroprivredu
37 Proleterskih Brigada
41000 ZAGREB

Dr. Vladimir Todorović
Institute Mihailo Pupin
11000 BELGRADE, VOLGINA, 15

BIOGRAPHICAL NOTES

Emanuele Guggino is Professor and Director of the Institute for Hydraulics, Hydrology, and Water Management at the University of Catania in Sicily. He is also principal in his own engineering consulting firm in Palermo, having a long history of important water projects. For more than two decades he has been one of the major leaders involved in the economic development of Sicily and southern Italy, recognizing water mangement as one of the key factors.

Guiseppe Rossi is Professor at the Institute for Hydraulics, Hydrology, and Water Management, University of Catania. He has been in charge of numerous basic and applied water research projects at the Institute in the areas of flood hydrology, systems analysis, and water planning, working in collaboration with Italian water planning agencies. His teaching at the Institute is in hydrology and water planning. He has worked on numerous consulting assignments with firms in Sicily, for the national government, and internationally.

David Hendricks is Professor of Civil Engineering at Colorado State University. He began his career with the California Department of Water Resources and since has had twenty-five years' experience in irrigation, water planning, water quality modeling, environmental assessment, water reuse, and sanitary engineering. He has been principal investigator for some twenty government-funded research projects, and has had consulting assignments for various government agencies, consulting firms, and international organizations.

The three editors have had a collective experience of nearly eighty years as engineers in the water business. They have had continuing associations with each other on previous projects and in activities which have dated back several years.

The authors also are experienced water professionals. They have expertise in a variety of discipline backgrounds including law, political science, agronomy, sociology, and engineering, and come from academia, consulting firms, government agencies, and local and regional water agencies. Many have associated with each other and with the editors from previous joint activities.

Introduction: THE SYSTEMS APPROACH TO WATER MANAGEMENT
Giuseppe Rossi

1. INTRODUCTION

One of the major characteristics of the post-industrial age is complexity. Social, economic and political phenomena are not restricted within isolated systems; they each mutually interact causing secondary impacts which often are more important than primary effects. The finite size of our world increases this level of complexity, as decisions made within one system rebound and impact upon others. This idea, outlined in the 1978 report "Science and Rebirth in Europe" to the Commission of the European Communities by Dr. André Danzin, President of the European Research and Development Committee (1), was a central theme in a proposal to develop a political orientation for scientific and applied research in Europe. Dr. Danzin suggested two approaches for "complexity control:" (1) reduce the nonessential complexity, (2) increase the effectiveness of services. With this perspective the methods of "system analysis" have a role. We must recognize, however, the difficulty of having these ideas used by decision makers. In another part Danzin states that a complex system needs a device for internal control. System analysis is now applied to solve an increasing number of applied problems. But of equal importance the "systems approach" has become an item of cultural debate.

2. THE SYSTEMS APPROACH

The difficulty of finding a definition for "system" which is universally accepted by experts from various fields is well known (2). It is generally accepted however, that a system is an assembly of interconnected parts to be considered as a whole rather than an agglomerate of elements, and whose essential characteristics are feedback and a structure oriented to achieve a common objective.

Despite the differences in meaning, the concept of system and the methods proposed to study systems (system analysis, system

---

(1) Danzin A., Scienza e rinascita dell'Europa. Scienza e Technica 1979. Mondadori, Milano, 1979.

(2) Machol, R., Methodology of system engineering in "System Engineering Handbook," McGraw Hill, New York, 1965.

engineering, system theory) have become increasingly widespread during the last thirty years. There has been no lack of opposition and suspicion, however. Very often "the instinctive reaction of the scientific community to a new discipline is to say that if it is really new then it is not science, and if it is really science, then it is not new" (3). In the case of "system analysis" which not only claimed to be a new development of a well tried and tested conceptual framework, but also introduced new scientific criteria, the most common reaction was to relegate the innovations out of the scientific orthodoxy.

One of the fathers of a general system theory, Bertalanffy (4), summed up the emerging findings from biological research originating prior to World War II, which expanded the conceptual outline of classical physics. According to Bertalanffy, up to only a short time ago, the field of science, seen as the attempt to discover the laws describing phenomena and to forecast their future behavior, was identified with theoretical physics. Classical physics, however, basically dealt with problems with only a few variables: in the simplest case unidirectional causalty, i.e., the relationships between one cause and effect. It found adequate solutions to these problems (e.g., mechanics to explain the movement of the celestial bodies). However in many fields of science (physics, biology, behavioral and social science) new problems have arisen, with organized complexity, a high number of variables, far more complex structures than the simple cause-effect relationship, and needing new conceptual instruments, or rather an enlargement of the science itself. As a matter of fact, every science can be considered a model of reality, i.e., a conceptual structure capable of describing some aspects of reality. Hence the growing conviction that the mechanistic methodology of classical physics, apart from the undeniable merits, is not the only model for dealing with reality. Modern biology has helped to break the "monopoly" of a single, wholly comprehensive model of reality by stressing the need to take into account the concepts of organization, intention and finality, which were considered illusory or metaphysical by classical physics. The main role played by biology in creating and developing the system theory is based on the analysis of the living organism as an organized body evolving to ever more complex states, and which needs to be treated as an open system, capable of receiving material containing free energy, in order to compensate the increase of entropy caused by irreversable processes within the system itself.

---

(3) Agazzi, E., I sistemi tra scienza e filosofia, S.E.I., Torino, 1978

(4) Bertalannffy, von L., "General System Theory," Braziller, New York, 1968.

The system theory, however, does not only derive from biology: it is a result of parallel developments in many disciplines, e.g., automatic control and cybernetic procedures (to study complex artificial systems with feedback), contributions form information theory, economic analysis, and management science, and the latest developments in the decision making theory to analyze rational decisions within human organizations.

The variety of scientific fields from which "system analysis" derives, plus the variety of phenomena which system concepts and methods are used to describe, forecast the control have rapidly given rise to a consequence which at first sight may seem paradoxical. The system theory, which proposed an inter-disciplinary point of view as opposed to a widespread tendency towards excessive specialization, has itself come to be a highly specialized discipline.

Yet this is less paradoxical than may seem. Above all, because interdisciplinarity does not eliminate disciplines but presupposes them: a work method can automatically be considered interdisciplinary when a typical problem from one discipline can be dealt with using theoretical and technical means from other disciplines. And the stimulus not to bury oneself within the limits of one's own discipline, but to consider the globality of the system does not eliminate the need to deepen and develop concepts and techniques to find a solution.

Second, the relationship between the system theory and both natural and social sciences is analogous to the role of applied mathematics in the various disciplines.

Like mathematics, also the system theory is aiming to develop its own content matter and its own logical structure which do not specifically belong to any other discipline, despite their being applicable to many sectors of science. In this respect the system theory has a significant and advantageous difference (5). The system approach tries to develop methods aimed at solving problems rather than "adjusting" the problems to the methods to be applied, i.e., it tends to respect the original formulation without assuming highly simplifying hypotheses, which do make the problem manageable while introducing profound distortions in the meantime. The tools used to solve the problem in a sense become secondary; as regards the nature of the problem, they may be not only mathe-matical, but can also be of a combination of mathematical, heuristic, experimental or other kinds of aspects.

However, the system theory recognizes that most problems contain, besides the component depending on the specific context

(5) Klir, G., Applied General Systems Research. Plenum Press, New York, 1978.

of the system examined, also a component which is in some way independent. This latter can be treated with a considerable saving in time and work by applying a suitable model for that category of processes, even though it may have been developed in a completely different kind of discipline. To go deeper into the models suited to different process categories in different science thus becomes the field of research of system theory in its strict sense.

Up to now we have examined the disciplines where system analysis initiated and has been developed. But now let's examine which classes of problems where system analysis can be applied significantly. A first level is the application to the various discipline in which there is a practical objective, the need for solution of a particular problem, or the need to evaluate the alternatives open to the decision maker. Here, too, belong the growing applications of system engineering to problems of large-scale system: communication system, urban systems, transport systems and large-scale water systems (6). There is, also a second level in which the system theory is beginning to be recognized: the philosophy of science or epistemology. Here, systems theory is being considered as a cognitive methodology unifying different sciences, or rather as an element crossing the divide between scientific and humanistic culture (7).

Today the position has been consolidated in the scientific culture, that the concept of system can be viewed as a new "paradigm," i.e., a set of new basic principles capable of attracting steady support and offering the opportunities for the development of specific applications to actual problems. Kuhn (8), in The Structure of Scientific Revolutions, has challenged the premise that scientific progress is the steady accumulation of knowledge, theoretical development, and research. He has pointed out the importance of the "scientific revolutions" which upset accepted principles and establish a new paradigm. The systems approach is becoming such a paradigm. It allows to address the problem of complexity. It is becoming a useful tool for the

---

(6) An excellent presentation of very significant applications in the last two categories of civil engineering problems is provided for example by De Neufville, R., Marks, D., Systems planning and design, Prentice-Hall, Inc. Englewood Cliffs, 1974

(7) This position emerged especially at the Convention of the Society for General System Research held at the University of Maryland (USA) in 1973. (cfr. Dechert, C. R., Sistemi, Paradigimi, sociotà, France, Angeli, Milano, 1978).

(8) Khun, T. S., "The structure of scientific revolution," University of Chicago Press, 2nd Ed., Chicago, 1970.

physical and biological sciences in the analysis of natural systems, just as it is now accepted in the engineering sciences and the social sciences in applications to human oriented systems.

Finally, in the most recent development a new need has arisen: to apply the systems approach as an overall model to solve problems which are at the same time scientific, techno- logical, economic, social, and human. The first overall models, which are set up by the Club of Rome and developed by M.I.T. from 1971 onwards, were based on the conviction that the world is a system which can be represented by a model. These overall models were useful, because minds were drawn to sit up and notice the seriousness of the situation, and the need to do something about it. However, they also raised a storm of controversy, rightly I feel, since errors and omissions in describing the system probably distorted future forecasts.

Today there is a tendency to limit analysis to less vast systems, but at the same time to examine more thoroughly the points of view from which to study the systems considered. Thus the position taken up by Checkland at a recent meeting (9) is significant. In affirming the need to "rethink the system approach," he maintained that most of the past applications of system engineering were directed towards well structured systems with fairly clear goals, and in which time and effort were to be spent in finding the optimum solution for the prefixed goal. He added that the new frontier in applying the system approach is represented by problems in which the goals to be reached are themselves problematical (e.g., problems relating to urbanization, to the delicate balance between industrialization and environ- mental conservation, risk levels for nuclear power stations, etc.). Rather than find a so called optimum solution for these problems, the system approach can help to explore the perceptions of a situation, i.e., it can help to understand the problem.

3. THE SYSTEMS APPROACH TO WATER MANAGEMENT

There is a long tradition of applying systems approach to water resource systems, in particular at the planning level. But despite the two decades of experience with the systems approach in water resources two main attitudes still remain. These are: prejudicial mistrust, and blind enthusiasm. The prejudicial mistrust about the possiblity of improving the decisions through the systems engineering is based upon various factors. These

(9) Checkland, P. B., Rethinking a systems approach, Meeting on "Systems Analysis in Urban Policy Making," organized through the Systems Science Programme of NATO, New College Oxford, Sept. 1980.

include: (1) a suspicious attitude toward the logical processes and the mathematical techniques which are viewed as academic exercises without links with the reality; (2) the lack of sufficient data, which would lead to doubt of the results even if powerful computers are employed; (3) the preoccupation that the spreading of the computer science could lead to neglect of the value of the professional experience and common sense, and that the concentration of the decision making in the hands of few experts would limit democratic participation.

On the other side a blind enthusiasm toward the new methodologies derives from the tendency, which is largely established in university circles, to believe that the current procedures are irrational and lack the strict logic of theoretical solutions. It is significant that the same persons, who two decades ago participated in the Harvard Water Resources Program and tried to demonstrate the convenience of the systems approach and techniques, today tend to criticize the worship of optimization and warn against the fragility of the so called optimal solutions which may be derived from simplified hypotheses (10). A more balanced attitude is needed. It should be open to accept the value of the systems approach as an aid in addressing the problems of complex systems, but tempered by the realization that the procedures should be designed to help decision makers and not replace them.

The management of water resource systems has evolved substantially in the last decades. The reasons for this include the development of engineering techniques generated by the increase in water demands, an increasing attention to the environmental problems and to the environmental empact of the water related facilities and the evolution of laws and institutions which govern our society. Amont major changes, at least the following should be remembered (11):

(1) to substitute single-source facilities to supply a single use sector with multisource and multipurpose facility systems;

---

(10) Fiering, M. B., Reservoir planning and operation. In "Stochastic Approaches to Water Resources" by H. W. Shen (editor), Fort Collins, 1976.

(11) A more detailed presentation of the recent trends in water management is given by Guggiono, E., Hendricks, D., and Reitano, B., (editors) "Conjunctive use of multiple source of water" F. P. M. Catania, 1980. In particular, the evolution of water systems is treated by Vevjevich, V., Conjunctive water use, and an examination of methods is in Rossi, G., Metodologie di approccio all'uso congiunto di risorse idriche superficiali sotterranee e non convenzionali.

(2) to abandon the concept of planning, designing and operating
     the water supply facilities as entities which are separate
     and independent from the pollution and flood control
     facilities;

(3) to consider using "demand limiting measures" (e.g., recycle
     in the same use sector, reuse of treated wastewater,
     consumption-increasing billing rates, etc.) as an alternative
     to constructing new facilities (e.g., new reservoirs);

(4) to evaluate the alternatives by multiobjective criteria
     rather than a single economic criterion (usually minimum cost
     or maximum benefit);

(5) to make the water law a tool to use water resources within
     the perspective of social goals (including protection of the
     environment) rather than to satisfy the needs of the
     individual users (with little regard for the environment).

We can briefly state that water resource management is becoming
increasingly complex. Above all it is a quantitative complexity,
owing to the huge size of the water plant systems which are also
made up of formerly independent subsystems. It is also a
qualitative complexity, however, which does not only derive from
structural interventions (engineering works) but also from
nonstructural measures (economic and legal tools).

    Three groups of consideration seem particularly relevant.
First of all the water laws are generally inadequate: for example
they do not include in an unitary context the problems of water
supply, flood control and pollution control, that often occur in a
single complex water system: they frequently cause constraints
regarding new possibilities offered by technological development
(e.g., water reuse) or the new needs arising from conjunctive
operation (e.g., flexibility of water rights). An appropriate
reform of the water law could produce major benefits in water
management.

    Second, it is necessary to reform a management structure with
too many organizations sharing responsibility in the various
supply sectors (municipal, agricultural, industrial) and even in
the same supply sector itself. This can be achieved by
coordinating the agencies regarding their functions and
territorial jurisdiction. The coordination of this management
structure appears a priority requisite in order to effectively
operate a complex water system, as the British experience has
already shown.

    Finally the use of system engineering techniques and
methodologies, and in general the system approach is required by

the growing complexity of the water systems in order to improve the decisions taken by the management throughout every single phase of water system planning, designing, construction, operation and control.

This use seems to be particularly urgent with reference to operational problems. As long as there was only a low percentage of water resources used, competition among the various uses was limited and systems were fairly simple, the lack of interest in operational problems could be justified. Now, however, the situation has greatly changed and it has become ever more necessary to use the appropriate tools to solve the technical, organizational and institutional problems in order to ensure a more efficient operation of the present or still developing complex water resource systems.

4. APPLICATION TO OPERATION OF WATER SYSTEMS

The value of the systems approach in handling complex systems and the increasing importance of operation in the management of complex water systems could lead to an easy consensus about the helpfulness of system approach to the solution of the operation problems of water systems. Most of the methodologies and their actual implementation relate, however, to planning problems of the water systems. Then, the analysis of the actual decision process in operation of water systems, which is affected significantly by the legal constraints and by the organization of the management structure, seems to confirm the need of orienting the efforts toward the development of appropriate system approach procedures which could be employed at the operation stage while being acceptable by the decision makers.

Then, the following needs arise:

(1) to define which particular sector of system approach should be developed in order to provide a tool for facing success-fully the operation problems;

(2) to define the lines of activity which could help to bridge the gap between the practitioners of water system operation and the specialists in the related areas (water resource management, system engineering, decision thoery).

About the first point, it is a general view that system approach, or at least its part oriented to the engineering and management problems, include three main aspects:

(1) definition of objectives and formulation of measures of effectiveness, i.e., the quantitative indices for evaluating the degree of attainment of the objectives;

(2) modeling of the systems, i.e., construction of models describing system, inputs, outputs and control and feedback mechanism;

(3) evaluation of the alternatives by the application of search techniques for the identification of the most convenient range of solutions in order to facilitate the choices in the decision-making process.

In particular a few requisites seems to be related to the first point:

- the presence of multiple objectives is a common situation in the actual operation of many water systems;

- a complete description of the systems from the operation point of view cannot neglect the particular features of the input data (e.g., hydrological and demand forecasts) and the relevant role of the control and feedback mechanisms;

- the choice of the techniques for evaluation of alternatives (optimization and/or simulation) should take into account the sequential nature of the operation decisions.

The above example of requisites give some insight even into the second point.

It is advisable that the contribution of a large range of expertise about the specific methodologies of system approach and decision theory be integrated with the contribution of other aspects which often were neglected (e.g., legal, institutional and environmental) and with the contribution of the personnel involved in the operation of the actual systems. In order to overcome such gap, the comparison of the various aspects should give equal value to the experiences of those persons who manage the solution techniques but lack in information about the real systems and to those who know the problems but ignore tha appropriate tools.

5. PRESENTATION OF THE NASI

In order to face the methodological and the substantial problems which delay a larger diffusion of the systems approach in the operation of water systems, the "School of Water Resource Management," in continuation of the line of activity which was initiated since 1971, dedicates its 12th course, in form of the NATO Advanced Study Institute, to "operation of Complex Water Systems." The objectives of the course were:

(1) to discuss the broadest applications of the systems approach to the oepration of water resource systems including engineering, the socioeconomic, institutional and educational implications;

(2) to provide an international forum to bring together those concerned with the solution of the main problems of the operation of complex water resources in the developed as well as in the developing countries;

(3) to establish and reinforce the necessary conditions for a speedy transfer of useful criteria and methods arising from scientific research to the operation of water resource systems.

The institute theme was developed through twenty-one lectures, eight Case Studies and three Pannels.

The Institute was for scholars, researchers and experts from public and private agencies operating in the sector of water resource management, coming from both technical and socioeconomic studies. Particular attention was paid to the following participants:

- experts from universities and research organizations (both public and private) with authority in the following fields: hydrology, water resources, mathematics, computer science, economics, law, etc.;

- decision makers from complex water resource system management agencies (engineeris responsible for operation, administrators, etc.).

Participants from eight NATO countries plus Spain and Yugoslavia took part in the Institute.

Many observers from public agencies in Italy attended the course. The inaugural meeting was opened by Mr. F. Nicolazzi, the Minister of Public Works, and Mr. S. Zoppi, the President of the FORMEZ (Agency for Education and Studies in the South of Italy).

The papers duscussed at the Institute, and here edited and published, can be divided into the following four main groups:

(1) An introduction, including three lectures, presented the characteristics of complex water resource systems and analyzed the main operational problems and the ways of improving it. C. Fassò reviewed the possible definitions of a complex system (large-scale systems, multiobjective systems, natural systems as opposed to artificial systems)

and examined their applicability to the water systems. He also presented the basic criteria of some techniques which have been proposed for large-scale and multiobjective systems. S. Indelicato discussed the kinds of alternatives open to water system operators and showed the specific objectives and criteria with reference to three different situations involving the balance between resources and demands within each system (i.e., water resources superior, equal or inferior to demand). Finally L. Fowler reviewed several strategies to improve the operation of existing complex water systems (i.e., from reducing consumption and losses, to increasing recycled and reused water, to getting a more widespread coordination in using surface and ground-waters). He also stressed that the key factors in introducing such methods are operator training and institutional arrangements.

(2) In another group of lectures the applications of the tools of systems analysis to the problems of water management were discussed. J. Salas and W. Hall dealt with three fundamental principles and standard techniques to disaggregate and aggregate water systems. Moreover they examined in detail the problem of determining the optimum operation of a system of reservoirs by using the concept of "equivalent reservoir." L. Duckstein showed how some recent techniques of the system theory (e.g., the catastrophe theory) are used to study the structure of complex water resource systems and the operation of subsystems (in particular those devoted to water quality control). V. Yevjevich discussed the definition of the objective function for the case when economic may be considered the main objective (and a trade-off can be looked for among the cost benefits, on the one hand, and amoung the risks and uncertainties on the other), and the case when several objectives must be considered using the multicriteria optimization technique. The other lectures dealt with specific models, developed ato help in the solution of operational problems of various categories of water systems. They included reservoir systems (R. Harboe); groundwater (Y. Emsellem); quality of water courses (D. Hendricks); and pollution in aquifers, particularly coastal aquifers (M. Benedini).

(3) A third group of lectures dealt more specifically with the decision-making processes in the operation of water resource systems. They included hydrological forecasting methods (V. Yevjevich); demand forecasting (L. Cunha); idenfication of operational rules for multipurpose reservoirs systems (J. Salas and W. Hall); application of decision-theory methods in the form of an Information Response System, either in the case of perfect or imperfect information, or in the case of optimum or nonoptimum response (L. Duckstein);

and the analysis of decision constraints including those not quantified (N. Wengert).

(4) The fourth group of lectures was devoted to the institutional aspects of the operation of water systems (N. Wengert and L. Cunha); to the legal persepctives of pollution control (G. Torregrossa); to the environmental implications (A. Biswas); to the criteria of sharing operational costs (A. Williams); and to the educational requirements related to a greater awareness of the complexity of water systems (E. Guggino).

Besides these lectures which gave a state-of-the-art view of the main aspects of the operation of water systems, some case studies were discussed concerning different physical and socio-economic situations in various countries; U.K. (A. Williams); FRG (R. Harboe); Italy (G. Rossi, E. Guggino, S. Indelicato, E. Giardina, S. E. Battiato, F. Croce); USA (N. Cline, L. Fowler); The Rhine, International River (V. Mandl and H. Vrijhof); Turkey (I. Gurer); Portugal (L. Periera); and California (N. Cline).

Case studies have a special role in this book. The editors believe that case studies provide the link between theory and practice, and will encourage us to make use of the latest scientific developments in modeling water systems, in carrying out the decision-making process, and in other relevant aspects of operation. The mixture of case studies and theory attempts to find an appropriate mixture of actual experiences of water managers and the latest theory.

While the topic of this book may appear to be very specific at first glance, the theme, which we should not miss, is the application of the systems approach and attitude. Also we should realize that our problem is one of the most vital in human organization, i.e, the management of water resources. My wish is that this book lead to new developments in various countries. Also we should hope that the methods of the systems approach and decision theory will be viewed in the proper perspective. They are not instruments able to produce magic solutions, but they are advanced tools to be used to aid in the development and operation of human systems. Just as we adapt nature and organize society to fulfill the human needs, we should try to orient the technology toward improving human welfare.

PART I: OPERATION OF COMPLEX WATER SYSTEMS

1. **CHARACTERISTICS OF COMPLEX WATER SYSTEMS**
   Costantino A. Fasso
2. **PRACTICAL ASPECTS OF WATER RESOURCE SYSTEM OPERATION**
   Salvatore Indelicato
3. **IMPROVED METHODS OF WATER SYSTEM OPERATION**
   Lloyd C. Fowler

1. CHARACTERISTICS OF COMPLEX WATER SYSTEMS

Costantino A. Fassò
University of Cagliari, Italy

This chapter introduces the theme of the book, Complex Water Systems. Because it is such a broad topic, I shall deal with their characteristics in general terms, concentrating more on problems rather than on solutions. Specific case problems on the topic are outlined in later chapters along with their specific solutions. So in this chapter I will focus on theoretical concepts related to analysis of complex water systems.

1. DEFINITIONS

The widely accepted definition of a "system" is that it is a set of elements mutually interrelated. And, as the biologists say, the system functions as a whole in a holistic manner.

The emphasis in the preceding definition is on the interrelation between the elements or parts of the system. We cannot say we understand a system if we know all the characteristics of its isolated parts; we must also know the relations between them, a concept which is often summarized in the apparently paradoxical expression "the whole is more than the sum of parts". In other words, the characteristics of complex system, as compared to those of the elements, appear as new or emergent, so that the behavior of the system may be derived from the behavior of the parts only if we can identify all of the parts contained in the system and the relations between them. Also, it can be said that while a simple arithmetic sum can be added to gradually, a system, as the total of its parts with its interrelations, has an instant identity. These statements stress the difference between the mechanistic conception of classical physical sciences, whose tendency has been towards resolution of phenomena into independent elements and casual chains by-passing interrelations, and the modern idea of a system typical in the fields of biology, psychology, and sociology.

The general, meanings of the adjective "complex," to be found in a dictionary, are "composed of two or more parts; not simple, complicated, intricate, involved, knotty." These explanations and synonyms give us an idea of "complex." But certainly it is not precise enough to distinguish between complex and non-complex systems. We realize also that, since a system is necessarily composed of at least two parts, any system could be complex by the dictionary. So we had better leave dictionaries aside and pass to the technical literature, in which we can find at least three types of systems qualified as "complex":

   i)   large-scale
  ii)   multi-objective
 iii)   natural

In the following sections I shall review the characteristics of these three types of complex systems. The mathematical characteristics are emphasized for the first two. In Section 5, I will examine their applicability in describing complex water systems.

2. LARGE-SCALE SYSTEMS

A large-scale system is a system in which some parameter N is presumably "large." This parameter N may denote the number of the subsystems (or "components") of which it is made up, or the number of variables that are used in its mathematical representation, or the number of constraints, or it may have other interpretations.

It should be pointed out that the problems of large-scale systems are not just the problems of small-scale systems magnified: new problems are also created by the large dimensions of the system. One of these peculiar problems, for instance, is how the system will perform when one or more of its many components fail, viz., the problem of failures. Another problem is the possible decentralization of the control of the system, which leads to a controller network operating on a large number of relatively autonomous subsystems. Thus questions arise about the number of components of the controlling system and their coordination, which is a challenging topic in control theory (Drennick, 1981). A third problem is the increasing role of uncertainties in large-scale systems. The consequence is that they can no longer be treated as first-order perturbations.

Now, limiting this review to those problems already existing in small-scale systems, which are magnified in the large-scale ones, we should note that increasing the scale may easily lead to dimensions so large that the solutions cannot be reached even with the aid of powerful computers. A lot of mathematical curiosities support this statment. Let us consider, for instance, a directed graph of N points, as outlined by Von Bertalanffy (1969). Suppose that between one point and another one arrow may exist, or may not exist, which means two possibilities for each pair of points and leads to $2^{N(N-1)}$ different ways of connecting the N points. If N = 2, the possible ways are 4, if N = 3 they are 64, if N = 4 they are 4096, if N = 5 they are 1,048,576, and so on. If N = 17, the number of possible ways is 7.59×10^{81}, exceeding the estimated number of particles in the universe (10^{80})!

In their well-known book Hall and Dracup (10) discuss the very simple system in which 20 controllable variables could each take on values of over a range of 100 discrete units. To such a curiosity a more "hydraulic" interpretation is given by Hall himself (9) by applying it to the problem of deciding what amount of water, up to 100 distinguishable units, should be allocated to each of twenty agricultural districts. Computations are carried out as follows, assuming, for the sake of simplicity, that only whole units of water (for example 1,2 ..... 100 million cubic meters per year) can be distributed. Since water could be allocated to each single district in 100 different ways, the number of different possibilities for all 20 districts will be:

$$100^{20} = 10^{40} .$$

The magnitude of this number is such that the choice of the optimum solution by direct comparison of the alternatives possible is physically unfeasible, even if a very fast computer is employed, for instance one which requires only one thousandth of a second to evaluate the goal impact of one of the alternatives. As a matter of fact, the total time required for the evaluation of all the alternatives would be:

$$10^{40} \times 0.001s = 10^{37}s = 3 \times 10^{30} \text{ years,}$$

namely, one hundred billion-billion times the age of the earth, according to the latest geological estimations. Of course, in practice most of the alternatives could be discarded by elementary and logical arguments and the optimum solution could be found very rapidly by examining the benefits of allocating water to each district (unless serious interactions between different agricultural areas need to be taken into account). That is why I have called this a hydraulic curiosity.

What I want to emphasize is that although computers have enabled men to solve difficult mathematical problems, even in those cases where analytical methods fail and exact solutions do not exist, the capability of solving numerical problems by means of computers has theoretical limits, no matter how much the hardware and software of computation are improved.

Table 1.   Classification of Mathematical Problems and Their Ease of Solution by Analytical Methods (after Franks, 1972)

Equation	Linear Equations			Nonlinear Equations		
	One Equation	Several Equations	Many Equations	One Equation	Several Equations	Many Equations
Algebraic	Trivial	Easy	Essentially Impossible	Very difficult	Very difficult	Impossible
Ordinary differential	Easy	Difficult	Essentially impossible	Very difficult	Impossible	Impossible
Partial differential	Difficult	Essentially impossible	Impossible	Impossible	Impossible	Impossible

To illustrate, consider Table 1. The heavy line represents the border between the mathematical problems which can be solved analytically and those which are difficult to solve, or are not solvable at all. It is apparent that the entire field of non-linear equations, both algebraic and differential, can be categorized as essentially nonsolvable. Also in the case of linear equations, problems involving many equations (which of course means hundreds or thousands) are essentially impossible to solve. Such a situation also arises with smaller numbers of linear equations if they are partial differential. The problems depicted in the right-hand section of the table can be solved by computers with a high degree of approximation. Computer methods, however, cannot deal with problems requiring an "immense" number of steps, i.e., not infinite, but very large numbers like those we have encountered in the two preceding examples.

Actually, in many systems even of a moderate number of components, but with strong interactions, such immense numbers appear, as soon as exponentials, factorials and other explosively increasing functions are involved (fortunately they are not likely to appear in water systems). In recent years a mathematical "complexity theory" has been developing, which aims at discovering the hard or intractable problems, i.e. not actually unsolvable problems but those which are too complex to be solved within a reasonable period of time.

3. MULTI-OBJECTIVE SYSTEMS

Multi-objective analysis applies to those complex systems where there are several objectives which may be conflicting and noncommensurable. A typical example is offered by water resources problems in which economic efficiency (measured in monetary units) and environmental quality (measured in units of pollutant concentration) are to be maximized.

Until 20 years ago or so these types of problems were drastically simplified by selecting only one objective (usually economic efficiency) and relegating the others among the constraints. But in recent years, with the increasing importance attached to nonpecuniary objectives, multiple objective analysis has been developed and applied to a wide variety of problems, including those of water resources.

Multi-objective problems differ from those involving a single objective in that they require the optimization of a vector, instead of a scalar function. If there are n objective functions of N decision variables, we have to minimize (or maximize) the n functions

$$f_i(\underline{x}), \quad i = 1, 2 \ldots, n$$

of the N-dimensional vector, $\underline{x}$, subject to the m constraints

$$g_k(\underline{x}) \leq 0, \quad k = 1, 2 \ldots, m .$$

By introducing a vector $\underline{f}(\underline{x})$ having the n functions f_i as scalar components, and a vector $\underline{g}(\underline{x})$ having the m functions g_k as scalar components, we may write synthetically:

$$\text{MIN } \underline{f}(\underline{x})$$

subject to the constraints $\underline{g}(\underline{x}) \leq \underline{0}$

where,

$\underline{x} \ \varepsilon \ R^N$ is the decision vector

$\underline{f}: \ R^N \rightarrow R^n$ is the objective vectorial function

$\underline{g}: \ R^N \rightarrow R^m$ is the constraint vector

$\underline{0} \ \varepsilon \ R^m$ is a vector having all elements equal to zero.

A feasible set, T, for the decision vector, viz:

$$\underline{x}; \quad T = \{\underline{x} \mid \underline{g}(\underline{x}) \leq \underline{0}\}$$

is determined by the constraints, $\underline{g}(\underline{x}) \leq \underline{0}$, and a corresponding feasible set, S, for the objective function vector,

$$S = \{\underline{f}(\underline{x}) \mid \underline{x} \ \varepsilon \ T\}$$

follows from the fact that a unique value $\underline{f}(\underline{x})$ exists for each vector $\underline{x} \ \varepsilon \ T$.

Thus the multiple objective problem can be defined either in the decision space as,

$$\text{MIN } \underline{f}(\underline{x})$$

subject to $\underline{x} \ \varepsilon \ T$,

or in the functional space as,

$$\text{MIN } \underline{f}(\underline{x})$$

subject to $\underline{f}(\underline{x}) \ \varepsilon \ S$

An illustrative example is given in Figure 1, where the sets S and T are shown for the following problem, with N = 2, n = 2, m = 4:

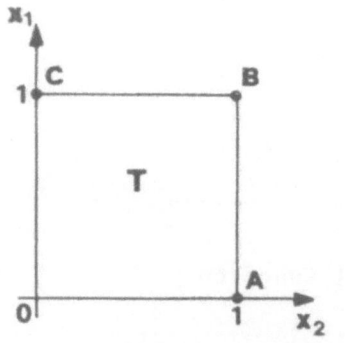

Decision Space

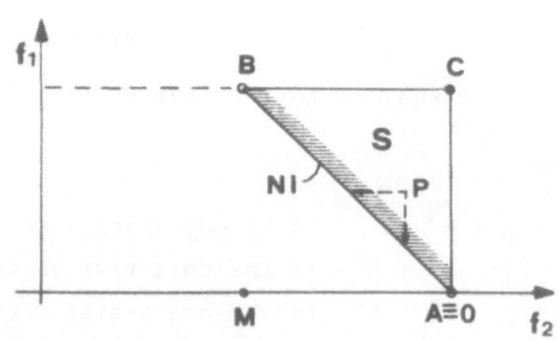

Functional Space

Figure 1. Decision and Functional Spaces Illustrating a Two-Decision Problem

$$f_1(\underline{x}) = x_1$$

$$f_2(\underline{x}) = 2 - x_1 x_2$$

$$\text{MIN } \underline{f}(x)$$

subject to $x_1 \geq 0$

$$x_2 \geq 0$$

$$1 - x_1 \geq 0$$

$$1 - x_2 \geq 0 .$$

It appears from this example that if we define an "optimal solution" as one which attains the minimum value of all the objectives at the same time, there are cases where such an optimal solution does not exist.

In the example, in fact, the optimal solution should correspond to MIN $f_1 = f_1(0) = 0$, MIN $F_2 = f_2(1,1) = 1$, which are two minima which cannot be attained simultaneously. In other words, such a solution is represented in the functional space by point M, in which $f(0,1) \varepsilon$ S.

In the frequent cases where a situation like this arises, the concept of "noninferior solutions" is introduced. A noninferior solution is defined as one in which no one objective can be decreased unless at least one of the other objectives is increased. It is the classical concept of "Pareto optimality." In the example of Figure 1, starting from any point $P \varepsilon$ S and moving parallel to each axis in the negative direction, one of the scalar objective functions remains constant, whilst the other decreases until the boundary of S is reached: therefore the set of noninferior solutions NI coincides with this boundary, namely with the line AB, and it corresponds in the decision space to the solutions:

$$x_2 = 1$$

$$0 \leq x_1 \leq 1$$

The fact that all $f(\underline{x}) \varepsilon$ NI lie on the boundary of S is a general property of noninferior solutions.

The determination of the NI set is not sufficient, since the decision-maker has to choose one of the noninferior solutions. Thus he has to apply some preference criterion in order to select the "best" or preferable among the NI solutions. Some authors, such as Belenson and Kapur (1), call this solution, which is optimal only in terms of a particular set of value judgements, the "best-compromise solution."

4. NATURAL SYSTEMS

Some authors, when speaking of "artificial systems vis-a-vis complex systems" identify complex systems with "nonartificial," viz., "natural systems." For instance, Gérardin (6) says: "as soon as we leave the domain of artificial systems, whether they are simple or strongly complicated, to move into that of complex systems, it is imperative to emphasize what characterizes the system in a functional manner." Examples of complex systems according to such a definition include: a living cell, and a social organization.

This concept of a complex system implies the following characteristics:

1) organization; when an element is drastically altered or a new relationship appears or disappears, the system may lose its identity;

2) internal coherence; the system must evolve in such a way that its structural identity is maintained;

3) resilience, which enables the system to recover from or adjust to changes induced by the environment;

4) a certain degree of autonomy with respect to the environment;

5) teleology; a given final state of the system can be reached starting from different initial stages so that the final state should be interpreted as an almost natural goal of the system.

5. COMPLEX WATER SYSTEMS

Since water systems interact with natural systems such as the meteorological and hydrological, as well as with social environment, we could say that a complex water system is depicted best by a system of the third type, namely the natural system. Such a broad concept for water systems planning or operation is inappropriate, however, due to the fact that in water systems the artificial components always play an essential role.

Therefore either of the other two definitions of complex systems mentioned in Section 1 can be applied. As a matter of fact, water systems are often large-scale so that definition i) is applicable in this case. Obviously, this is especially true when large catchment areas are involved and the resources are to be utilized for multiple purposes and/or in a conjunctive way. For instance, in the case of the "Progetto 14" of the "Cassa per il Mezzogiorno," which is aimed at planning the utilizaton of the water resources in two regions of Southern Italy, Puglia and Basilicata, the system to be studied includes 7 reservoirs, 8 diversion weirs, 11 agricultural areas, 6 industrial areas, 4 groups of users for municipal supply and 2 hydroelectric plants. As a further index of its complexity, it has 2637 constraints, even though simplifications were introduced, a "large" figure indeed.

Even if this project has been approximately optimized with respect to only one objective, viz., the maximization of the cumulated net benefit of the four categories of water users, I have already pointed out that the current tendency of water systems analysts is towards the multi-objective approach. Therefore it is necessary to consider a water system as "complex" even according to the type ii model of Section 1.

One well-known characteristic, which is always present even in the small-scale and single-objective water systems, is the number and variety of knowledge aspects which have to be involved in the study. This idea is illustrated in Figure 2, by A.D. Hall (9), which depicts the three major "dimensions" of systems engineering.

According to the description by Sage (12), the time dimension includes the systems engineering phases of a project extending from the initial conception of an idea through system retirement. The logic dimension deals with the steps involved in each of the systems engineering phases. The knowledge dimension refers to specialized knowledge from various professions and disciplines. The seven steps of systems engineering listed on the time axis are carried out in an iterative fashion, i.e., it is possible to go back to refine and improve the results of any lower-numbered step as a consequence of the results of any higher-numbered step. In each of the seven phases of systems engineering it is necessary to conduct each of the seven steps.

In the knowledge dimension I have listed seven disciplines, which seem to me more strictly related to water systems. But this is by no means an exhaustive list and their subdivisions can also vary to some extent.

The names of the disciplines listed in Figure 2 are sufficient to define the fields of expertise of the respective

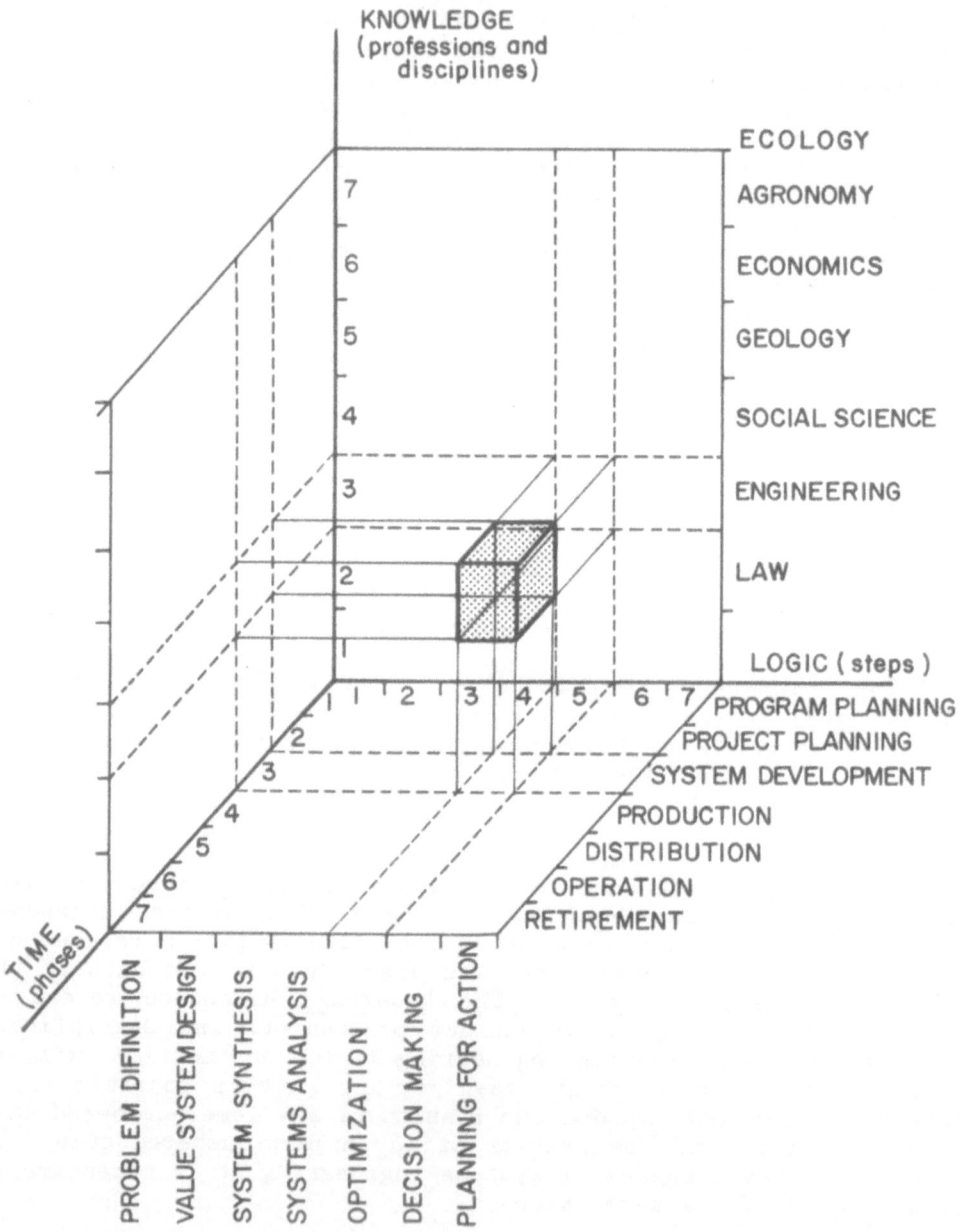

Figure 2.   Multimorphological Box for Systems Engineering
(adapted from Hall, 1969)

subjects. But the word "ecology" warrants further remarks. I intended to refer, with a term that is not strictly correct, to the studies to be conducted in order to verify the compatibility of the reservoirs and of the utilization plants with the adjacent physical, faunistic and vegetative environments.

Now I will recall those techniques of analysis for the large-scale and multi-objective model types which can be applied to water resources problems. But because the task is large, the discussion is limited to basic ideas.

5.1 Large-Scale Water Systems

Of the analysis techniques for large-scale water systems, probably the most widely used is "scale reduction," sometimes also called "model simplification," "equivalencing" or "aggregation." This technique seeks out simpler systems of smaller scale which can be mastered more easily than the original one, even though they possess the same fundamental characteristics.

Figure 3 gives the general schematic representation of a large-scale water system suggested by Hall and Dracup (1970), used for the purpose of decomposing the total system into independent subsystems. Since total independence obviously can not be achieved, the authors suggest selecting system decomposition points (SDP in the figure) which result in the minimum number of interactions between the parts. The authors' basic concept is to disaggregate the complex system into two principal subsystems: "the supply subsystem," consisting of the storage and other water regulation and collection systems; and the "distribution-and-use subsystem," consisting of all distribution aqueducts, intermediate regulatory storage, and the use and final disposal of the water by reuse, reclamation or direct discharge outside the system.

In Figure 3 only one supply subsystem, e.g., surface water A , groundwater E , saline water F and reused water G , and one distribution-and-use subsystem, e.g., direct inter-reservoir uses D , distribution system B and wastewater collection system C , are represented. But it is obvious that in real cases several of the features, represented here only once, will be repetitive.

The optimal design of the complex system must be achieved through a number of iterations for which Hall and Dracup suggest the following succession:

14

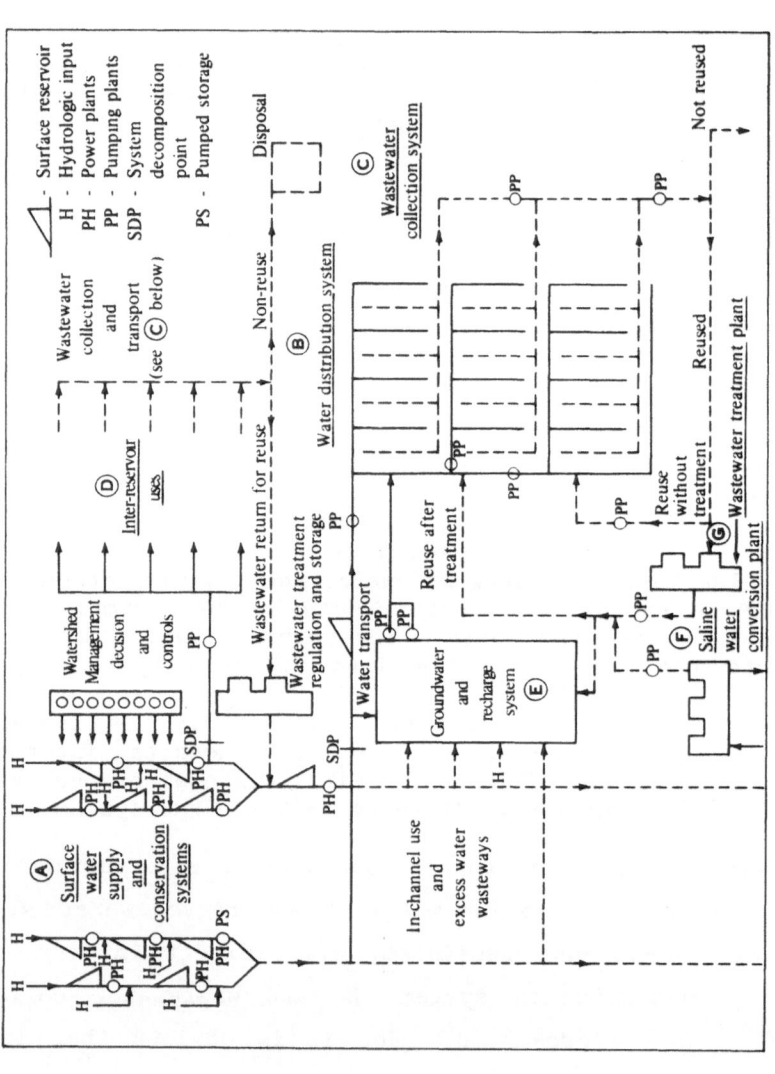

Figure 3. Schematic Representation of Complex Water Resources Systems Development Systems (Hall and Dracup, 1970)

i) Reconnaissance and tentative configuration of the land and water resources of a region, tentatively delimited on the basis of geographic and hydrographic locations,

ii) distribution-and-use subsystem design, analyzing the uses and determining the net value of the delivery water to each of the terminal points of the tentative distribution system and leading to output as an optimal function of input,

iii) tentative supply system specifications as a function of water output; this analysis allocates the storage capacity of the various reservoir sites in such a manner so as to minimize the cost of storage as a function of the total storage requirements,

iv) optimization of operations for planning purposes leading to the optimum output for the given preliminary design of the supply system,

v) adjustment of the preliminary design by iteration of step iv until satifactory design has been obtained, and

vi) final review, including the readjustment of the region boundaries, also taking into account the economic, legal, social and political aspects.

A procedure in some way opposite to scale reduction, is that of enlarging the scale by increasing the number of components. This takes advantage of the fact that, under certain conditions, the mathematical analysis of a system can be simplified by executing the limit of $N \rightarrow \infty$. Several other procedures exist, but as far as I know, no applications to water systems have been developed so far.

5.2 Multi-Objective Water Systems

In the field of multi-objective problems several methods have been proposed. Some have already been successfully applied to water systems, while others have proved less suitable. Among the former is the "surrogate worth trade-off" method, elaborated by Haimes, Hall and Freedman (1974, 1975), which enables the decision-maker to choose the best solution through a systematic comparison of objectives, two at a time. In addition, the comparison is performed in the functional space, S, which usually is much smaller and easier to work with than the decision space, T. This is becuase usually the number of objective functions, n, is much smaller than the number, N, of decision variables. Of course, the information collected is later transformed into the decision space. Moreover, the decision-maker's task can be greatly simplified because many of the noninferior solutions can be eliminated previously, based on his knowledge and preferences.

Two other promising methods have been applied to water systems: the Multiple Objective Optimization Theory (MOOT) and Multiple Attribute Utility Theory (MAUT). The essential features of MOOT can be briefly illustrated for the case, based upon an example by De Wispelare and Sage (3) of a water system having two inputs (e.g., capital C and Labour L) and two outputs (e.g., a--water for agriculture, and b--hydropower).

We assume that all agricultural water and energy will be distributed between two consumers with the aim of maximizing their utilities u_1 and u_2. The problem is to find the amounts C_a and C_b of capital and those L_a and L_b of labour to be allocated for producing the outputs a and b respectively. Since the maximum levels of capital $\overline{C}$ and labor $\overline{L}$ to be supplied to the system are fixed inputs, the production functions which determine the amount of water and energy produced are:

$$a = f_i(C_a, L_a),$$

$$b = f_j(C_b, L_b),$$

with the constraints:

$$\overline{C} = C_a + C_b,$$

$$\overline{L} = L_a + L_b .$$

Since the MOOT is particularly suitable for confronting different design alternatives, we introduce in our example two alternative actions (A_1, A_2), each characterized by a set of two production functions. If a_1 and b_1 are the volume of water and the amount of power allocated to the first consumer and a_2 and b_2 the corresponding quantities allocated to the second consumer, for each alternative we have the new constraints

$$a = a_1 + a_2,$$

$$b = b_1 + b_2,$$

and two objective functions, to be maximized:

$$u_1 = g_1(a_1, b_1),$$

$$u_2 = g_2(a_2, b_2).$$

Each one of the objective functions is first optimized, maintaining the value of the other objective function constant.

This first step of the optimization problem can be formulated as follows:

if $\bar{u}_2$ indicates a specific level of utility for consumer 2, then

maximize $\quad u_1$,

subject to: $\quad u_2 = \bar{u}_2$,

$$\bar{C} = C_a + C_b,$$

$$\bar{L} = L_a + L_b,$$

$$a = f_i(C_a, L_a) = a_1 + a_2,$$

$$b = f_j(C_b, L_b) = b_1 + b_2 .$$

Taking into account the constraints, by means of Lagrange multipliers, and writing the usual optimization conditions, a set of three differential equations is found which represent efficient consumption, production and product mix, respectively:

$$\frac{\partial u_1/\partial a_1}{\partial u_1/\partial b_1} = \frac{\partial u_2/\partial a_2}{\partial u_2/\partial b_2} , \qquad \frac{\partial a/\partial L}{\partial a/\partial C} = \frac{\partial b/\partial L}{\partial b/\partial C} , \qquad \frac{\partial u_1/\partial a_1}{\partial u_1/\partial b_1} = \frac{\partial b/\partial C}{\partial a/\partial C} .$$

For each alternative this computation, repeated with many $\bar{u}_2$ leads to a curve such as Figure 4a, which represents solutions which are Pareto optimal for each alternative. As a second step, a scalar social choice function (SCF) is constructed which is used to select the best solution from among the Pareto optimal ones. For instance for an SCF of the form:

$$\hat{Y} = \alpha u_1 + (1 - \alpha)u_2 , \tag{1}$$

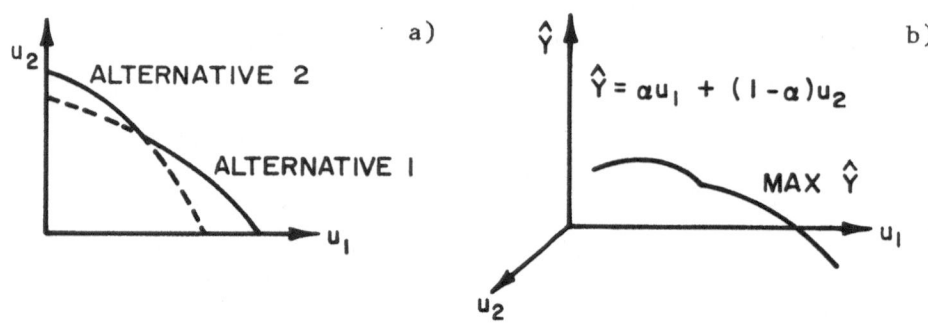

Figure 4. Illustrative Example of an Application of MOOT
to a Two-Consumer, Two-Alternative Problem
(De Wispelare and Sage, 1981)

the result is found as in Figure 4b.

The MAUT method differs from the MOOT method in that, first
of all it assumes that society is concerned with equity, viz.,
with the fact that all members (the two consumers in the preceding
example) receive the same level of satisfaction; moreover it
introduces from the beginning of the mathematical optimization
procedure the "attributes" or "objective indices" (to some extent
analogous to the coefficients α and $1-\alpha$ of Eq.(1)), that
measure the degree to which the planning objectives are met in any
solution.

After having introduced an SCF of the form

$$U = h(u_1, u_2)$$

the two assumptions lead to a fourth differential equation

$$\frac{\partial u_1/\partial a_1}{\partial u_2/\partial a_2} = \frac{\partial U/\partial u_2}{\partial U/\partial u_1} \, ,$$

in addition to the three of MOOT. Such an equation, which, in the
right-hand member, includes the marginal effects on U by the
utility of consumer 1 and 2 respectively, selects the operating
point among the Pareto optimal solutions. An interesting
application of MAUT to the selection between five development
alternatives for the Tisza River basin in Hungary has been
developed by Keeney and Wood (1977).

According to De Wispelare and Sage (3), both MOOT and MAUT "are capable of producing either identical or strategically equivalent policies when applied at the same level to a common decision situation." The same authors emphasize that MOOT is adept at solving physically oriented parameter refinement and optimization problems, while MAUT is well suited for a wide range of evaluation application including those with behavioral aspects and qualitative elements as well as uncertainties. De Wispelare and Sage also propose an interesting approach realizing the combination of both methods in modelling and resolving large-scale decision situations.

6. THE WORLD OF COMPLEX WATER SYSTEMS

With this brief introduction to some of the mathematical characteristics of complex water systems, and approaches to analysis, the stage may be set to appreciate the kind of problem with which we are dealing. The ensuing chapters broaden the perception of complex water systems with further examples of their characteristics, how they may be analyzed, and specific case studies. Both theory and practice are described.

REFERENCES

1. Belenson, S. M., and K. C. Kapur, "An algorithm for solving multi-criterion linear programming problems with examples", Operations Research Quarterly, Vol. 24, No. 1, 1973, pp. 65-77.

2. Bertalanffy, Von, L., "General system theory," George Braziller, New York, 1969.

3. De Wispelare, A. R. and Sage, A. P., "On combined multiple objective optimization theory and multiple attribute utility theory for evaluation and choicemaking," Large-Scale Systems, Vol. 2, No. 1, 1981.

4. Drenick, R. F., "Large-scale system theory in the 1980's," Large-Scale Systems, Vol. 2, No. 1, 1981.

5. Franks, R. G. E., "Modeling and Simulation in Chemical Engineering," Wiley-Interscience, New York, 1972.

6. Gérardin, L. A., "How complex is complex?," Large Scale Systems, Vol. 2, No. 1, 1981.

7. Haimes, Y. Y. and Hall, W. A., "Multiobjectives in water resource systems analysis: the Surrogate Worth Trade-Off Method," Water Resources Research, Vol. 10, No. 4, 1974.

8. Haimes, Y. Y., Hall, W. A., Freedman, H. T., "Multi-objective optimization in water resources systems," Elsevier Scientific Publishing Co. Amsterdam, 1975.

9. Hall, A. D., "Three-dimensional morphology of systems engineering," IEEE G-SSC Transactions, Vol. 5, No. 2, 1969.

10. Hall, W. A. and Dracup, J. A., Water Resources Systems Engineering, McGraw-Hill Book Company, New York, 1970.

11. Keeney, R. L. and Wood, E. F., "An illustrative example of the use of multiattribute utility theory for water resource planning," Water Resources Research, Vol. 13, No. 4, 1977.

12. Sage, A. P., Methodology for Large-Scale Systems, McGraw-Hill Book Company, New York, 1977.

2. PRACTICAL ASPECTS OF WATER RESOURCE SYSTEM OPERATION

Salvatore Indelicato
University of Catania, Catania, Italy

1. INTRODUCTION

The operation of a water system is defined here as a set of practices designed to meet demands with the available water resources. In operation, resource allocation and water demands have been defined already. With this constraint various combinations of sources can be used to satisfy the water demands and other goals.

Each of these alternatives has associated with it different results in terms of demand fulfilment and costs. Thus, operation is unique for each situation, depending upon the results desired.

In this chapter I will review practical aspects of water resource system operation which will include:

1) identification of "alternatives," e.g., the kinds of choices to be made, the variables to quantify, and the relationships between these variables;

2) discussion of goals,

3) determination of the water resource system characteristics which condition the operating pattern, and which need to be known for decision making.

In addition, there is concern also about data collection, projection of water demand, and public education of the users regarding waste.

2. OPERATION PROBLEMS

The first area of focus in the operation of a water resource system is upstream from the distribution network. System components include facilities for intake, storage, and conveyance. The operation variable of concern is the water quantity delivered from each source to each of the users during a given time interval.

In interconnected systems the amount of water delivered from each source for the different uses must be less than or equal to the water available. This is, of course, a constraint on the system, which is true for any given time interval. During years in which the total water supply exceeds the total demands, the yearly amount of water delivered to each user from all connected sources is equal to annual demands. In this case, the operation decisions require dividing the water demands among the connected sources. When the water supply is not adequate to cover demands, the operators must make decisions about how to spread the shortage in time and to allocate the deficiency among the users, i.e., the global demands must equal the resource available.

For any given time period, each individual decision about the distribution of the available water supply will influence both the global amount of supply and the operation costs. Such decisions will affect also the total economic damages caused by shortages, and the distribution of such damage among the users. This is due to the unique characteristics of each water source and the users, and to the kinds of linkages between sources and users which involve various types of costs and benefits.

By introducing new variables and decisions, other kinds of operating actions also affect the extent of water damands and their time distribution. These are interventions related to the billing system (mixed systems, changes in rates according to time of year, amount of use, etc.), or of distribution characteristics (flow, hydraulic loads, etc.).

3. GOALS AND CRITERIA IN OPERATION

The goals in operation may differ from one situation to another. In addition, they may not be fully achievable, i.e., some will be given priority over others. This is true especially where system operation is governed by several agencies, each having its own particular interests. Thus operational goals may be different and even conflicting. But rather than evaluate priorities and approaches for achieving several objectives at once, decisions often are based on compromise or the intervention of external authority. As a rule, however, the specific operation objectives are set in relation to the particular context of the resource-demand equation, i.e., whether the system's resources are more than, sufficient for or less than the demand.

3.1 The Supply Surplus Case

The surplus of resources with respect to demand is a frequent situation, though usually it is not permanent. It comes about during the fairly long period in which water uses are established. For example, in a new irrigation district many years go by before the irrigation potential is fully developed.

In such circumstances the operation objective should be to provide incentives for more widespread water use so as to improve the economic and social return from the investments. Such operation involves a atrategy which minimizes controls and water distribution constraints. It translates to the maximum adjustment to user demands in terms of schedules, flow deliveries, etc., and the nonuse of rates in relation to the volumes used. In practice, this kind of operation strategy must be used with caution in order to avoid wasteful habits. Such habits are difficult to change when the period of surplus has ended.

3.2 The Supply-Demand Balance Case

In a supply-demand balance situation the objective should be to maximize the economy of operation. Thus the demands for each time interval will have to be met so that costs, e.g., pumping, treatment, etc., will be minimum, or alternatively, that net benefits will be maximum.

In practice, this objective rarely leads to definite and unequivocal decisions. This is true especially when costs and benefits are expressed only in terms of probability. This kind of situation arises, for example, when there is an alternative between diversion from a watercourse or a groundwater basin and drawing water from a reservoir with hydroelectric power production. When such system is managed by various agencies without a single cost-sharing mechanism, the above objective is viewed from the perspective of each agency, since each administrator decides his managerial policy on the basis of least cost for his customers.

To reduce these problems and to maximize the utility of the water resource system only a single management agency should be set up. This agency should coordinate and reconcile the needs of different users and also administer a cost distribution mechanism.

3.3 The Water Shortage Case

Periods of water resource shortage may occur frequently in a fully utilized water resource system, especially when naturally variable surface waters are the main supply source. Operation is particularly critical during these periods of shortage. The predominant operating objective during such shortages should be to minimize the overall damages resulting from the reduced ability to fulfill demand. Operation based upon this objective will minimize the problems of having inequalities among different users. At the same time there should be a compensation mechanism to handle inequalities; these are, however, difficult to handle when negotiating.

In practice, the allocation of the water resources among the different users, and therefore also the division of the water deficit, is effected according to the prevailing standards of each country. Examples include priority usage of municipal use over agricultural and industrial uses, and temporal priority. Another approach is to utilize a percent reduction as regards to the resources normally allotted to each use.

With respect to time-sharing, one can apply the foreseeable damage minimization criterion to each user. Another criterion which is occasionally convenient is to use one part of the system's water resources to irrigate annual crops. These are sown or planted in a part of the year in which the extent of the available water resources has been evaluated already. In deficit years, one can substitute dry crops, and thus reduce the damage caused by reduction in water resource availability.

4. SPECIAL CONSIDERATIONS IN OPERATION

A fundamental consideration for satisfactory operation of a water resource system is the contact between the manager and the water users. Operating choices are conditioned, not so much by a knowledge of the water demands already made, as by anticipating the subsequent demands. Furthermore, during periods of shortage one must know the amount of damage each user is likely to incur.

The user needs to be satisfied on one hand and the damages borne on the other, depend on the user's own choice, as well as on external elements which are often difficult to foresee. The latter include such diverse aspects as demand trends for products and probabilistic effects, e.g., weather and climate. By knowing the influence such data may have on resources and the distribution of costs, the users tend to manage their own choices so as to derive maximum benefit either for themselves or for the consumers of their products. Therefore, it is advisable for the operating organization in charge to acquire the necessary data first hand and to manage for the overall good.

In addition, the characteristic rapid adjustment in user behavior when faced with a water reduction usually will reduce the extent of damages. This is true particularly with respect to agriculture, as noted previously. Such response, however, requires a constant flow of information concerning the users and their adjusted behavior.

Because of this inherent complexity in the overall system, it would be best for the whole water resource system to be managed as a single unit. This means a single organization should have the operating authority. Such organization should also promote the widest possible user involvement, and provide the necessary technical assistance.

5. CONCLUSIONS

In summary, it is possible to outline principles on how a water resource system should be operated. They are enumerated as follows:

1) Ensure coordination with respect to operating decisions. This requires one managing body for the whole system.

2) Share operating costs and damage compensation costs during water shortages.

3) Define clear objectives for water system operation to facilitate decision making.

4) Integrate water system operation with an information service in order to mesh the projected water availability with demands.

5) Provide technical assistance to water users to help increase water use efficiency.

Effective management of the system operation, based upon these principles, will go a long way toward squeezing more productivity and benefits from the available water resources, and at reduced costs.

3. IMPROVED METHODS OF WATER SYSTEM OPERATION

Lloyd C. Fowler
Goleta Water District, Santa Barbara, California

1. INTRODUCTION

Water supply systems are commonly planned, designed, constructed, and operated to provide a firm yield each year. The
firm yield or the safe yield is the amount of water available from
the water supply project every year including those years of a
critically dry period. The critically dry period may or may not
be the most severe drought of record. Nevertheless the intent is
to supply a certain amount of water from the project over a long
period of years regardless of the hydrologic conditions existing
during those years.

When the water needs of the area being served begin to exceed
or are projected to exceed the firm supply from the project,
planning is usually initiated to develop a new project to supplement the existing water supplies. This is the conventional
approach and is demonstrated by practice throughout the world.

In recent years it has been shown that improved water management of existing systems can extend the water supply to serve new
uses without the addition of new projects. This reduces the
problems engendered by the potentially adverse effects some water
supply projects have on the environment. The basic concept is one
of living within the available water supply by improved operation
and management techniques.

2. IMPROVED OPERATION AND MANAGEMENT

Methods of improved operation and management include the
elimination of waste of water, reductions in water use through the

practice of water conservation measures, recycling and reuse of water supplies, coordinated operation of surface and underground water systems, better training of operators, and improved institutional arrangements. The characteristics of each are outlined in the following paragraphs.

2.1 Elimination of Waste of Water

Losses in water supply systems commonly amount to 10 to 50 percent of the available supply. It is clear that the elimination of this water loss would allow the existing water supply to meet expanding needs.

There are many opportunities to prevent the loss of water from a water supply system. The obvious ones are to look for and eliminate leaks from water distribution pipelines and canals. Seepage losses from canals can be large and can be eliminated by many different types of linings as long as these are maintained and kept in good repair. Overflows from storage reservoirs in the distribution system and wastage from the ends of the distributary system can be minimized, if not entirely eliminated, by proper system operation. In irrigation systems this requires carefully scheduling water deliveries. This can be helped by the use of a control system. Automatic valves and other devices in piped distribution systems together with skillful operation will eliminate the overflows from system storage reservoirs.

2.2 Reduction in Water Use

Careful water management by the user can be an effective means to extend the community water supply. The water user needs to be informed as to the best means of utilizing water. Old habits are difficult to change but it is possible to develop wise water using habits by education and by appropriate incentives such as changed water pricing measures. Declining block rate pricing (the more water used the cheaper becomes the unit price), should be eliminated. A scheme of lifeline water rates, minimum payment for a minimum amount of needed water, and uniform unit rate schedules for amounts over the minimum will encourage reasonable water use habits. Increasing block rate pricing, wherein the unit price charged for water increases as the amount of water used increases, does the most to assure careful consideration of water use. The latter type of pricing structure achieves notable reductions in water waste and maintains these water savings as long as the pricing structure is in effect.

Efficiency in water use can be achieved by regulating water using devices in urban communities. Building codes should be developed and enforced that require the installation of water savings devices and prohibit the use of devices with excessive water uses. Agriculture can be encouraged to use the most efficient water application schemes. The water savings that can be accomplished through the use of drip irrigation systems are well known. Industry can reduce its water use by improved production techniques and product control. Probably the greatest opportunity for saving water in industry comes from recycling water supplies in cooling and production areas. The savings of water through water conservation techniques in homes, industry and agriculture makes available water for new uses. It also can be considered to extend the life of existing water supply projects should the project yield have been over estimated. Water conservation is an economical means of extending a water supply.

2.3 Reclamation and Reuse

Wastewaters from urban and agricultureal areas can be reclaimed and made available for reuse. Recycling in industry was mentioned before as a means of conserving water and extending the community water supply. Urban wastewaters can be reclaimed and used to provide for irrigation of parks, landscaping, and agriculture providing the wastewater has been properly treated to meet public health standards. Such reclaimed waters can be used for irrigation in lieu of presently used potable water supplies. These released potable waters can then be made available for such uses as may require a potable supply. Similar use and reuse of water in agriculture can be a means of expanding the water supply. Care must be taken, however, that the normal increments of minerals dissolved in the water during each use do not accumulate excessively so as to adversely affect downstream water use. As with any water supply the quality of the reclaimed water must be compatible with the type of crop and soil. Provided that these quality factors are taken into consideration it is possible to reclaim wastewaters for reuse not only in agriculture but also in industry and urban environments.

2.4 Coordinated Operation

Most surface water supply systems involving storage reservoirs are operated on a safe yield basis. Unfortunately water supplies derived from groundwater basins are generally not operated on the basis of a safe yield or any other reasonable basis. In many areas the groundwater safe yield is rather limited; therefore, there is a tendency to overdraft the groundwater basin. In those areas where both surface water systems and groundwater systems are or can be made available there is the reasonable expectation the the safe yields of both systems can be increased by coordinated operation of the two. This coordinated operation is frequently referred to as conjunctive use.

The advantages of a surface water system are generally one of elevation, that is, water can be stored at a higher elevation than the service area. In addition floodwaters that would ordinarily move rapidly through the stream system can be stored for use during times when the surface flows of the streams are not adequate to meet the needs of the area. Surface reservoirs on streams are also susceptible to multipurpose uses increasing their desirability. Surface reservoirs do have the disadvantage of high cost, flooding of valleys, and evaporation losses.

The advantages of a groundwater supply system are that the source of the water can be tapped by wells located in the service area reducing the size of the distribution system. In addition the underground storage available is usually large, free from evaporation losses, and relatively inexpensive to develop. Groundwater basins are subject to water quality degradation from overdraft conditions that may bring about sea water intrusion or intrusions of conate brines. In addition overdraft conditions can bring about land surface subsidence. Groundwater basins are also susceptible to pollution by surface spills of chemicals and leachates from dump sites. Groundwater basins must be protected, as should surface streams, against the disposal of chemical wastes of all types that may pollute the water.

Where surface water systems and groundwater systems can be developed in conjunction with each other improved water supply availability can result. For example, it is possible to operate surface water reservoirs on a maximum yield basis rather than a safe yield basis if the waters from the storage reservoirs can be placed in underground storage. This type of operation calls for the surface reservoirs to store the floodwaters when available and release these floodwaters into groundwater basin recharge areas before the next floodwater flow occurs. This type of operation is especially suitable where surface runoff is seasonal in nature. It does require that groundwater basin recharge areas be available and capable of recharging the surface storage capacity before the next season runoff. The waters stored underground are recovered through wells which pump directly into the distribution system.

The yield from groundwater basins can be increased by improving the recharge characteristics of the area. Surface waters that tend to flow through the recharge areas before being fully recharged can be diverted to offstream storage where the normal silt content can be removed. These waters are then released to recharge areas to increase the storage underground. Special canal type recharge facilities can be constructed to maintain flowing streams over longer paths in recharge areas to increase the total recharge. The flowing water in canal recharge systems reduces the problems of surface clogging of the recharge areas normally found with standing water in recharge basins. For maximum efficiency, recharge waters in standing basins should be as close to zero turbidity as possible. It is also possible to enter a stream system after the initial flood peaks that carry the maximum turbidity have passed and construct stream material berms or dams. This ponding of low turbidity stream water increases the amount of recharge over that which would normally occur. These are some examples of improved operation that can increase the yield of groundwater basins without large capital costs for additional facilities.

The coordinated use of surface and groundwater systems can result in increased total water supply. An example of such an opportunity which would improve the yield of a surface water project is described in Chapter 26. In this instance a surface water storage reservoir operated on a safe yield basis could have its yield increased by almost 10 percent without major increases in costs of the works utliized. The reservoir fills and spills on the average of about once in every 3.3 years but there may be periods of 10 to 15 years in which no spill occurs. These spill-waters occur for generally short periods of time and in rather large volumes. Spillwaters are diverted through the operation facilities which are only partially used during the spill period because the water needs of the service area are low. The transferred spillwaters pass through a water treatment plant which normally provides water for urban purposes. After treatment the water is released into the distribution system where it is allowed to flow into wells connected to the distribution system. These wells are normally used to produce groundwater to help meet the water supply needs of the area. During the low demand times of the year these production wells can be utilized as injection wells placing newly captured spillwaters in underground storage. The normal production wells are supplemented by unused production wells that act solely as injection wells. The unused production wells are ones that encountered problems during production and their use had to be ceased; however, the wells serve adequately as recharge wells or injection wells increasing the total system recharge capability.

2.5 Operator Training

One of the key factors to utilization of improved methods of water system operation is additional training of system operators. The opportunities for elimination of waste of water, for water savings through water conservation, for reclamation and recycling, and for the coordinated operation of surface and groundwater supplies require knowledgeable operators. Improved hydrologic forcasting can help in selecting the best method to use in operating any water supply system. This, too, requires intelligent operation to utilize the information that can be made available. The education of water system operators will be aided

by the developement of responsive educational institutions and
water system agencies which will allow imporved operations to be
implemented.

2.6 Institutional Arrangements

Institutions at all levels of government must operate in a
cooperative and coordinated fashion to enable optimum operation of
water supply systems. Too often it is the institutional problems
which interfere with system operations and prevent operators from
maximizing the yields from existing water systems. Local water
supply agencies can be most responsive to the needs of the area
they serve but at the same time they can stand in the way of
regional operation which could improve the total water supply. In
many instances improved operation of water supply systems call for
the transfer of water from one basin to another or from one watershed to another. The lack of regional authority which can appro-
priately handle such interbasin transfers prevents full utilization of the available water supplies. Local water supply
agencies working within cooperative regional water supply direc-
torates can achieve maximization of the water supply potential in
the area.

Good water management should be total water management. This
means that an institution, the water management agency, should
encompass the entire water supply basin area, both surface and
underground, that is to be managed; it should have the authority
to regulate water use; it should have the authority to finance
water supply projects and to impose reasonable water pricing
schemes; and it should have the authority to integrate local and
imported surface water supplies with the groundwater supply.
Among these authorities are the implied duties to handle water
rights, water quality protection, and environmental concerns.

3. CONCLUSIONS

The operation of most water supply systems can be improved to
increase the availability of water supply. Improved operation is
one means to live within the water supplies currently available
without having to make major capital investments in new water
supply facilities with all of the attendant environmental
problems. It is generally better and more economical to improve
operational methods than it is to try and develop new projects.
Improved operator training can pay large dividends in this regard.
It is reasonable to expect that such training can be institutionalized through water users organizations and with guidance from
centralized authorities. The present methods of water system
operations can be improved to achieve the concept that better
operations will make a little water go a long way.

PART II: SYSTEMS ANALYSIS METHODOLOGIES

4. DISAGGREGATION AND AGGREGATION OF WATER SYSTEMS

Jose D. Salas and Warren A. Hall
Colorado State University
Fort Collins, Colorado
and
R. A. Smith
Interamerican Center for Water and Land Development
Merida, Venezuela

The paper outlines fundamental principles of disaggregation-aggregation of systems and reviews the standard mathematical and physical approaches to the problem together with other considerations such as alternative subobjectives and judgemental optimization which have proved useful in disaggregation-aggregation models for decision making. A structural basis for a relatively general hierarchical decomposition model for both planning and operations of complex water resource system is discussed. Also an aggregation-disaggregation approach based on the concept of equivalent reservoir is presented to find the operation of a multireservoir system. Finally the Valencia-Schaake statistical disaggregation model originally proposed for data generation in hydrology is applied in an operational context. Results indicate that such techniques are promising for simplifying the analysis of the operation of multireservoir systems.

1. SIMPLIFYING COMPLEX SYSTEMS

Water resources systems are inherently complex, whether considered in a planning sense or operational sense. For example, the system in Northern California involving the Trinity, Sacramento, Feather and American Rivers and the related irrigation projects would require on the order of 500,000 decision variables involving nothing more than capacities for water flow, water storage and energy production in order to describe it as a two purpose (water and energy), single objective (economic) system. If these magnitudes are coupled with the many additional purposes plus a dozen or so additional noncommensurate objectives, the result is clearly impossible to treat as a total system with any known optimization methods.

It is for this reason that we are concerned with disaggregation, even for a relatively simple, one reservoir system. The historical approach to complex systems has always been to break them down into simple subsystems which can be treated logically. Indeed all mathematical and judgemental optimization methods themselves are in reality disaggregation models which allow us to treat subsets logically, then logically proceed to other subsets.

For example, linear programming deals successively with basic feasible solutions as subsystems of the larger problem after first proving that only such subsystems need be considered out of the infinitude of subsystems that could be analyzed. Dynamic programming deals successively with two relatively independent subsets which are then combined into single optimal subunits. Similarly with all other procedures.

For most water resources systems even these standard procedures for analyzing sequences of subsystems become burdened with excessive computation and additional decomposition logics must be sought, whether mathematical or judgemental optimization or some combination thereof is to be used.

Although the need for disaggregation of water resources systems is apparent, it is by no means obvious what form the decomposition should take to be useful, let alone best. A procedure useful in one case may easily be useless or perhaps erroneous in another. Even more critical to the selection of the appropriate disaggregation model is the common requirement for subsequent aggregation of the disaggregated components back into a complete system before any use can be made of the analysis.

The general optimization procedures mentioned above are examples of disaggregation where subsequent aggregation is not desired. Here the objective is to find a particular subset out of the many possible which is expected to be "best." For this reason the disaggregation model not only seeks to identify subsystems but more importantly, to find logical reasons why large numbers of such subsets cannot possibly contain the optimum set, hence these can be eliminated from any subsequent analysis. In this case we deal with mutually exclusive subsets, since we seek and use only one of the many.

For physical systems such as those required for water resources development and management, however, our complexity is due to physical subsystems, a substantial number of which must be retained in the ultimate system rather than be rejected. In this sense the complex physical system consists of both the decision subsets which can be reduced by logical means and of physical subsets which will always remain essential interacting subunits of the total system. The complex physical subsystem thus deals first of all with subsets which are not mutually exclusive but rather mutually interdependent and only second with mutually exclusive decision sets.

With respect to the latter, elimination of large numbers of mutually exclusive decision sets is an objective of their disaggregation as before. However, because of the necessary interactions between the physical components of the system, it is no longer sufficient to optimize each physical element of the system independently (the result of most mathematical optimization methods), but rather to optimize each sub-element as a function of all of the significant interactions between that component and all other components and between that element and the environment. In this way the reaggregation can be optimized.

2. FUNDAMENTAL PRINCIPLES OF DISAGGREGATION-AGGREGATION

Although the complex physical systems requiring disaggregation-aggregation are often rather different in form, there are, nevertheless, three principles which can be used to guide the process. These are:

(1) The optimum of the union of a set of nonindependent subsystems is not equal to the union of the optima of the individual subsystems.

(2) In order to determine the optimum combination of a set of nonindependent subsystems, each subsystem must be optimized for all combinations of feasible values of the signigicant interactions between it and all other subsystems in the total system.

(3) The boundaries used to partition the various subsystems must be selected in such a manner as to minimize the significant interactions between the subsystems. This will simplify the subsequent aggregation.

The first principle dictates the necessity of aggregation following disaggregation. Only when the subsystems contribute to objectives as independent entities can one optimize the policies for each individual component and be assured that the combination will be optimal. In effect the interactions between non-independent elements act as additional constraints, the level of which is determined by the policies or sets of decisions proposed for all the other elements. These values become implicit constraint levels.

The second principle requires the identification of all such implicit constraints and the probable range of values each might assume. When these are identified, each subunit can then be optimized as a function of the values of these implicit constraints over the range anticipated. Obviously the greater the number of such significant interactions and the greater the range of values each might take, the more difficult the optimal aggregation analysis will be. Likewise the more difficult and time consuming the subsystem analysis will be.

This brings us to the third principle, since the primary objective of disaggregation-aggregation is to produce a computationally feasible and hopefully effective analysis. By minimizing the number of interactions the number of implicit constraints required for any subsystem will also be a minimum.

3. OTHER CONSIDERATIONS OF DISAGGREGATION-AGGREGATION

The foregoing basic principles provide excellent guides to the process of disaggregation-aggregation of complex systems for purposes of analysis, particularly optimization analysis. However, there are many other considerations which are useful in specific situations. The two most general approaches to the problem of disaggregation-aggregation can be described under the general headings of "mathematical disaggregation" and "physical disaggregation." Commonly both procedures will be used, each to a degree depending on circumstances.

Mathematical disaggregation is based on constructing a suitable mathematical model representing the entire system to be considered. In general, the interactions between the subsystems will appear in the set of internal constraints which limit the set of feasible decision sets. The mathematical disaggregation approach uses this set of constraints to identify the potential subsystem boundaries by noting the structure of the set of equations (or inequalities) which describe these limiting interactions. If all significant interactions are so described, the application of the third principle of decomposition becomes a relatively simple task.

For example consider the set of continuity constraints which govern the availability of water from a river-reservoir subsystem. These are often written as:

$$s_{i+1} - s_i + x_i + E_i(s_i) \leq y_i , \quad i = 1, 2, \ldots n$$

where s_i is the volume of water in storage at the beginning of the ith period of time, x_i is the volume of water released for purposes of the system during period i, E_i is the evaporation volume lost during period i, y_i is the inflow volume during period i, and n is the total number of time periods.

Note that the inequality for $i = 1$ contains only one variable with a subscript 2 (i.e., s_{i+1}) and none with higher subscripts. The inequality for $i = 2$ contains no subscript lower than 2 and only one subscript 3 larger than 2. Thus the total interconnection (interaction) between any two subsets of such equations, the first consisting of consecutive values of $i = 1, 2 \ldots, k$ and the second consisting of values of $i = k + 1$, $k + 2, \ldots n$, is expressed by the magnitude of the variable s_{k+1}. It is the only variable which appears in both subsets. Thus if $i = 1, 2 \ldots 500$ and this proved to be excessively large for one computation, the problem could be subdivided into two sets $i = 1, 2 \ldots 250$ and $i = 251, 252, \ldots 500$. The first problem would then be solved for $i = 1, 2 \ldots 250$ with the magnitude of the variable s_{251} given as a parametric constant over its range. This means that each subsystem must be solved for a suitable number of magnitudes of s_{251} to represent the range of expected variation.

For example, ten such values might be satisfactory for the required accuracy. In such a case, 20 solutions of the smaller but computationally feasible problem of 250 constraints each, replaces the computationally infeasible problem (at least so presumed in this illustration) of 500 constraints. If we had two reservoirs instead of one, then there would be 1000 such constraining inequalities, 500 for each site, with only two storage values appearing in both of two subsets, thus defining a simple disaggregation mode. In this case, again assuming parametrization with 10 magnitudes for each, the two 500 subunits would require 200 subsystem solutions to be equivalent. However, again if the larger system is computationally infeasible and the two smaller systems are feasible the gain is obvious.

Note that the second problem (two reservoirs) could be divided into any number of subsystems as desired. Depending on the nature of the objective function, the limit of this process treats each i as a separate problem as a function of the succeeding pair of s_{i+1} or the preceding pair s_{i-1}. This limit corresponds to the dynamic programming with two state variables. Thus in one sense, dynamic programming (DP) is essentially a disaggregation model for the set of interaction constraints. Where only one variable appears in more than one constraint we have the standard "one dimensional" or one state variable DP, regardless of the functional form of each of the i constraints (or the objective). If two variables share this dubious honor, the problem can be disaggregated into a two state variable (two stage variables) dynamic programming problem. In general if there are m "carryover" variables from i to the next we have an m dimensional dynamic programming problem.

Another example of mathematical decomposition is Monte-Carlo analysis. It is basically a mathematical decomposition which disaggregates the very complex problem of all relevant joint probabilities of various numbers of events in sequence. These events are not completely independent nor completely dependent. In water resources we are not only concerned with the probabilities of (for example) the magnitude of the streamflow on any given day, hour, month or minute, but rather on the probability that these events (correlated in time) will result in a wide variety of adverse sequences of flows when taken one at a time, two at a time, three at a time, four, five ... ten, twelve ... fifty ... one hundred, etc. If the interval concerned in as large as one month, then we are concerned in the possible outcomes of monthly streamflow events taken 600 at a time, 599 at a time, 598, etc., on down to 12, 11, 10, 9 ... 3, 2, 1. We can disaggregate the problem of dealing with all the complexly interrelated probability density functions involved for each level of each possible outcomes for all these groups of sequences into a set of sequences, month by month, the total of which represents one equally likely sample of the many combinations of sequences which could occur over a total of n years. Each subsequence in any one of these sequences is a "subsystem" on this mathematical decomposition which "interacts" with its neighbors through correlations. By dealing with the significant correlations, it is possible to create a sequence of as many of these "subsystems" which should be an "equally likely sequence," hence contains an appropriate sample of the kinds of sequences with shorter durations which might prove adverse or favorable as the case might be. By generating a number of these equally likely sequences we can derive some probability statements about the impacts of events in sequences short or long. There are many other examples of mathematically based disaggregation models, but the above is illustrative of the approach.

A second approach might be termed physical disaggregation. It differs from mathematical disaggregation only in the point at which one begins. In physical decomposition the complex system is first divided into subunits on the basis of the physical or functional characteristics, then the mathematical models are prepared, rather than vice versa. If both procedures are correctly executed, the results will be identical. Certain simplifications are commonly made in either the physical or mathematical modeling which will cause the result to differ unless the simplifications are identical. One cannot say a priori which of the two approaches is best for any particular problem until most of the analysis has been completed (but not necessarily the computations). However, there often are some specific advantages for the physical desaggregation approach for optimization problems in water resources where multiple objectives are involved.

Physical disaggregation itself may have a number of basic approaches. One of the more useful of these for water resources is the identification of subsystems by their functions or purposes within the larger system. Frequently this allows disaggregation of objectives and objective functions as well as of the interactions. This is particularly useful for multiple-objective systems such as are usually encountered in water resourses planning or management. Some objectives dominate certain subsystems but are essentially negligble for others. Frequently this eliminates serious problems of attempting to commensurate fundamentally different (noncommensurate) objectives at all levels. Even where it does not eliminate the problem, the number of such different objectives which must be treated for any one functional subsystem will usually be substantially less. Frequently the functional subdivisions may correspond to geographic locations but this is not always true nor necessary.

As an oversimplified example, consider an automobile as a system (the driver must be included in most applications). Here virtually every part has a function to perform. Some parts, such as the spark plugs are essentailly identical but located at different points on the engine. Functionally, however, they are the same. Physically they interact through the fixed firing order of the cylinders. All are subsystems of the engine subsystem whose function is to provide a source of torque and power. Thus it is obviously possible to consider the spark plugs in terms of their contribution to the efficient production of torque and power (through the intermediary of other subsystems of the engine subsystem) without the necessity of considering all of the objectives of an automobile and its potential uses. Obviously the range of torque, power and efficiency outputs must be sufficiently broad so that the best magnitudes of these can ultimately be considered for their contributions to the real objectives of the auto. Note that, except for some psychological objectives, neither torque nor power are objectives in and of themselves. They become subsystem objectives only because the achievement of the true objectives of the auto (whatever these may be) appear to require a driving force (torque) and power for their accomplishment.

At the same time, in considering the design of the spark plugs it would be awkward, if not downright foolish to attempt to evaluate each design alternative on the basis of the potential future owners desire to drive from Los Angeles to New York to take a new job. It would obviously be more effective to deal directly with the effect of spark plug design directly on the torque, power and efficiency of the internal combustion engine, itself a subsystem intended to convert the chemical energy of the fuel into mechanical energy with appropriate characteristics. Had we begun the problem with mathematical modeling of the system, the objective function would contain neither torque nor energy. Rather these essential requirements would be buried in the rather large set of constraints which defines the interactions of all of the elements of the auto with ecah other and with the final desired objectives involving ride, comfort, speed, economy, acceleration, appearance, etc., as well as destinations, time limits, etc.

This trivial, but nevertheless instructive example illustrates most of the principles of function-based physical decomposition and its advantages. Most subsystems have a limited number of purposes in the form of outputs to other components of the system for further processing into the actual desired objectives of the complete system. The objectives in this case are the productions of the maximum level of the most desirable combination of these outputs at minimum "cost" whether expressed in terms of resources used or in the usual economic sense of cost.

4. A PROPOSED DISAGGREGATION MODEL FOR WATER RESOURCES SYSTEMS

In this section we present an open-ended, hierarchial disaggregation model of commonly encountered water resources systems. It is hierarchial in the sense that the primary functional elements of the system are themselves composed of functional subelements. Each of these is also further composed of functional subelements. Thus the imbedded subsystems form a hierarchial set. It is open-ended because this disaggregation model always leads to elements which may, under some circumstances require further decomposition in order to accomplish the purposes of the total system. On the other hand, any further hierarchial disaggregation can be eliminated as soon as it is clear that the essential characteristics of the system can be properly represented by a particular level of detail, or where the behavior of these subsystem elements passes from the direct or indirect control of the decision-maker(s) for whom the analysis is being conducted.

The fundamental function of all water resources developemnt and management systems (physical and institutional) is to provide a relatively high degree of assurance that water will be available for various purposes when needed, where needed and with the appropriate quality and other characteristics, with some known level of reliability of that availability, and/or to provide assurance that it will not occur where and when it is not desired or with undesirable quality characteristics, again with a known level of reliability. The terms assurance and reliability in the previous statement are not redundant expressions. Here reliability represents the known and accepted levels of risk that the quantities and qualities will not occur. Assurance represents the credibility that that level of reliability will in fact be maintained.

The necessity for assurance of the reliability arises from the fact that most of the outputs of water systems have value which is determined more by the reliability (and its assurance) than on the quantities nominally producible. For example, lightning represents a sizable quantitiy of high voltage electricity. Yet nowhere is it utilized as an energy source because (1) it is unreliable in both time, place and quality (voltage) and (2) it is virtually impossible to convert it into a reliable source. In fact, in most instances it is a destructive potential, with systems required to prevent the adverse effects of its being where it is not wanted, when it is not wanted.

Water is a similar resource. It has no value unless it is available with an appropriate reliability for the purposes intended, where needed, when needed in appropriate quality. Its cost is often a minor component of the costs associated with its use. At the same time it is usually an absolutely essential resource for that use within the designed limits. Indeed, if water were in fact capable of being properly evaluated adequately in economic or monetary units per unit volume (as they are customarily stated), water resources systems would not be required. Few if any water supply systems actually produce additional water. Most reduce to total volume of freshwater by evaporation and other system losses.

These background comments focus attention for disaggregation on the system processes which convert the unsatisfactory time-place distribution of water as it naturally occurs into a more satisfactory temporal and spacial distribution. We will presume that our system begins with precipitation, the set of stochastic events which deliver the liquid water condensed from the atmosphere to the surface of the ground. We could move back one step and initiate our system with the water vapor transport system which brings the essentially pure water to the area in the first place, if anyone so desired, but the concepts presented in this disaggregation model would be basically unchanged. The potentially feasible alternatives might change but not the concepts.

The primary interaction between any and all possible subsystems of this total system (converting unsatisfactory conditions to more satisfactory ones) is clearly the flow of water. From a functional point of view there are five major functions or procedures involved in this flow as it progresses from the original precipitation, serves the actual use, and the residual is ultimately disposed. The first function is basically one of concentration of the diffuse sources into streamflow or groundwater flow. This will be referred to as the Watershed Subsystem. The second function is one of storage to convert the unsatisfactory unreliable timing of flow arrivals at the storage location into more satisfactory, more reliable outflow sequences over time. This will be referred to as the Flow Regulation Subsystem. The most common significant interactions between the two are flows of water sediment and dissolved solids sequentially varying over time. By utilizing the point of inflow (conceptual more than physical) as the boundary between them the number of such significant interactions is minimized.

The third major function is one of distributing the water to the points of use in the geographic sense. Again the sequence of flows of water constitute the significant interactions. It may be distributed using only the existing stream channels, only using off channel aqueducts or both. Any channel so used becomes by its function a part of the Water Distribution Subsystem. Note that under this definition the penstocks to hydroelectric facilities are technically a part of the distribution system. This subsystem accomplishes the desired assurance of the reliability of water at the point of use in geographic space.

The fourth major function is the actual use of the water. The interaction once again is the flow sequence for water. For the first time we move from expenditure of other resources to one of utilization of the water resource (Water Use Subsystem). Note that under this definition recreational use of a reservoir surface is a part of the use subsystem not the flow regulation subsystem. Its distribution system however has zero dimensions. Hydroelectric energy production is also a part of the subsystem if the definition is strictly held. However, because of the difficulty of uncoupling such "use" subsystems from the reservoir management decisions our basic principle of minimization of interactions argues that they should be treated as part of the flow regulation subsystem.

The fifth major function (all too often overlooked) is the removal, reclamation, and disposal or availability for reuse of the liquid water and its quality constituents after use is completed (Water Reclamation-Disposal Subsystem). Again a flow of water (and its quality constituents) constitutes a major significant interaction.

Presentation of the complete detailed analysis of the secondary, tertiary and high order subsystems in the hierarchial structure of all five primary subsystems is probably not warranted for the purposes of this paper. These levels differ considerably with specific locations, problems, objectives, etc., hence it would be impossible to present a complete, all-inclusive case. For illustrative purposes, the flow regulation subsystem will be developed further since it is often the principal focus of water management, particularly "real time" operations.

As indicated above the function of this subsystem is to convert the stochastic, unreliable arrivals of streamflow quantities over time, commonly out-of-phase with requirements at the potential storage sites (as a group), into a more reliable time series reasonably compatible with the time distribution of requirements. It accomplishes this function by the use of storage. This consists in concept of filling a storage volume during high flow periods so that it can be released at a later time for beneficial purposes. The latter is clearly related to some assured "reasonably reliable" level of service which might be represented by some volume level per year, X. The planning objective associated with this subsystem must be represented as a minimum by a quantity to be maximized (X) and a quantity to be minimized (cost). Furthermore, the previous comments regarding the primary importance of temporal reliablility require that the volume X be distributed over the year in a manner compatible with the proposed uses and acceptable to the proposed users.

Table 1 summarizes the characteristic functions of the primary subsystems. Each of these primary subsystems will normally be further subdivided into additional subsystems, depending on the actual physical system. For example, the water-shed subsystem normally has two basic functional elements, the "catchment surface" and the "channel collection system." The former acts primarily to partition the precipitation into direct runoff and infiltration-interception. It then usualy acts as an evaporation surface. The function of the channel collection system is self-evident. It can also be further subdivided into its main stem, principal tributaries, etc., as a continuation of the hierarchial subsystem model to whatever degree is necessary.

Table 1. Water Subsystems and Their Functions

Primary Subsystem	Water Resources	Other Relevant Processes
1. Watershed Subsystem	a. Partition of precipitation into evapotranspiration and liquid water runoff (surface and subsurface). b. Aggregation of diffuse quantities into concentrated volumes and their transport to subsystem 2.	a. Erosion sediment transport and deposition. b. Solution and transport of various soluble minerals. c. Other economic activities (e.g., forestry, agriculture, grazing, mining, urban development, recreation, wildlife management, etc.).
2. Flow Regulation Subsystem	a. Transformation of stochastic unrealiable time sequences of flow into more deterministic and reliable flow sequences over time.	a. Uses of the water resource directly in the subsystem (e.g., navigation, recreation, hydropower production, etc.). b. Evaporation and other system losses. c. Sediment deposition.
3. Water Distribution Subsystem	a. Transformation of flows at locations not needed or desired to locations where needed and desired (spatial reliability).	a. Seepage and evaporation losses. b. Hydropower production and/or use, other potential uses.
4. Water Use Subsystem	a. Utilization of water (consumptive and/or nonconsumptive) to accomplish desired basic objectives.	a. Virtually all other aspects of the socio-economic-political system. b. Quality degradation.
5. Water Reclamation-Disposal Subsystem	a. Removal or neutralization of adverse quality constituents in liquid water. b. Removal and disposal (including return for reuse) of liquid water residual.	

The flow regulation subsystem usually consists of a number of reservoirs at various points on the streams. Each of these ultimately represents an individual subsystem. However, it is usually necessary or at least desirable to consider these in combinations (series or parallel). Using position on the "channel collection" subsystem as a guide, going down each branch, consider as a series subsystem all reservoirs on that branch until a tributary branch containing at least one reservoir is reached. Continue this process until no more upper branches with reservoirs in series remain.

Some of these series elements will contain only one reservoir, others several reservoirs (or sites for reservoirs) in series. However, all such subsystems thus defined now constitute elements in parallel combination with at least one other such element. By combining adjacent parallel subsets on the same stem the next level of the hierarchy is defined. These are then combined with downstream series units as before for the next hierarchy, continuing in this manner until the entire stream flow regulation system is in one equivalent system. Figure 1 illustrates such a hierarchial structure for this subsystem.

The distribution subsystem usually consists of a branching noncross-connected set of aqueducts, mains, submains, laterals, sublaterals, etc., until the geographic point of actual use is reached. The wastewater collection, reclamation and disposal subsystem normally consists of a branched, main cross-connected system of open or closed drains, leading to waste treatment systems and/or point of ultimate disposal. The hierarchy of subsystems here is similar to that for the distribution system.

Utilizing this hierarchical subsystem procedure to any desired degree, permits the analysis for each subsystem (and sub-subsystem) in terms of its functional contribution to the whole. The primary interactions between any two subsystems at any level of the hierarchy consists of a flow of water into and/or out of each unit thereof. Each such unit either modifies the time distribution or the spatial distribution of the water to serve the basic requirement of providing the necessary degree of reliablity for water to be present when needed, where needed, in the appropriate quantity and quality for the purpose intended, or for water not to be present when not needd, where not needed or with inappropriate quality characteristics (including the usual surface and subsurface drainage requirements).

5. OPERATION OF A MULTIRESERVOIR SYSTEM BY THE EQUIVALENT RESERVOIR CONCEPT

The problem of the optimal operation of a multireservoir system can be simplified by transforming the problem into three

48

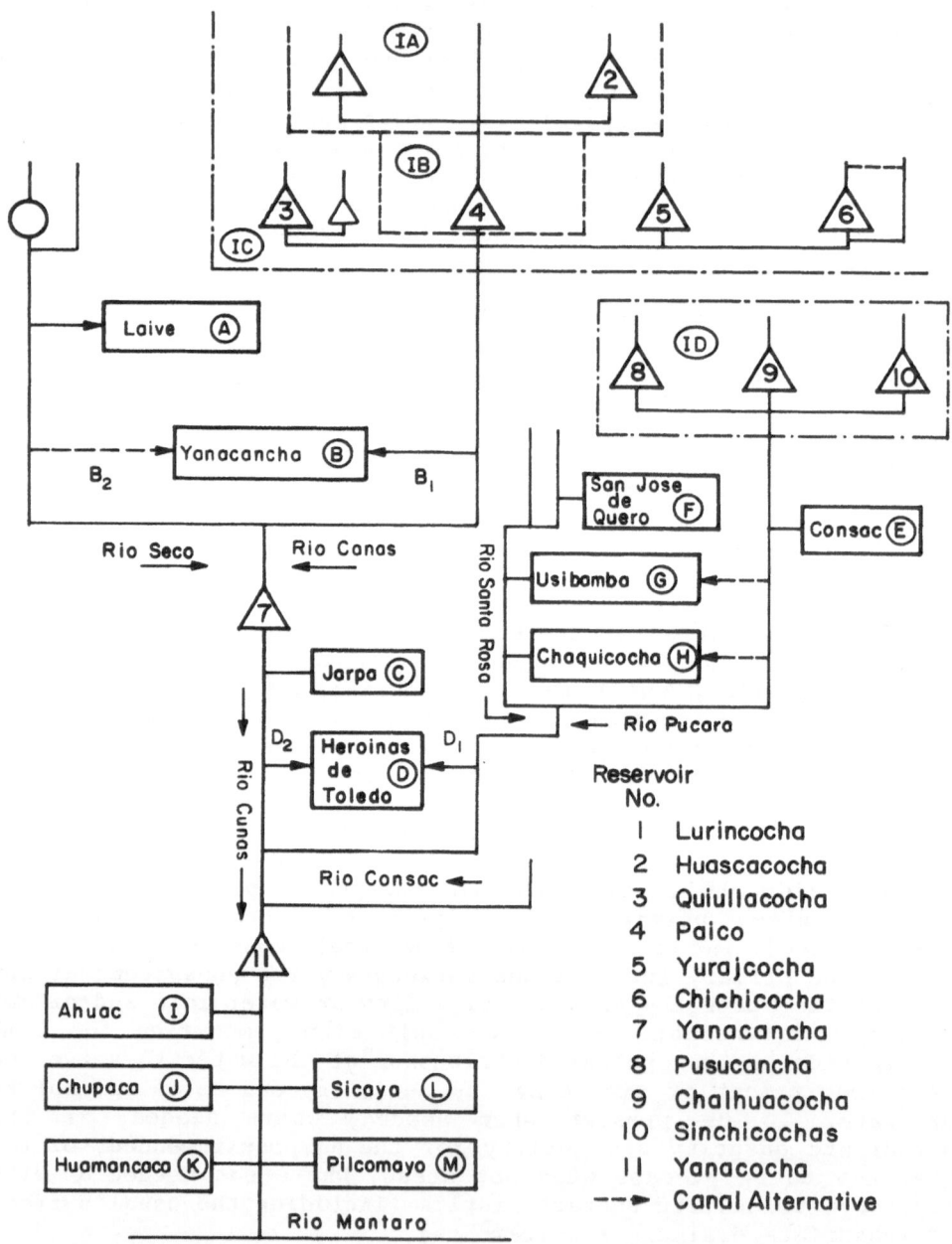

Figure 1.   Hierarchical Structure of a Flow Regulation System

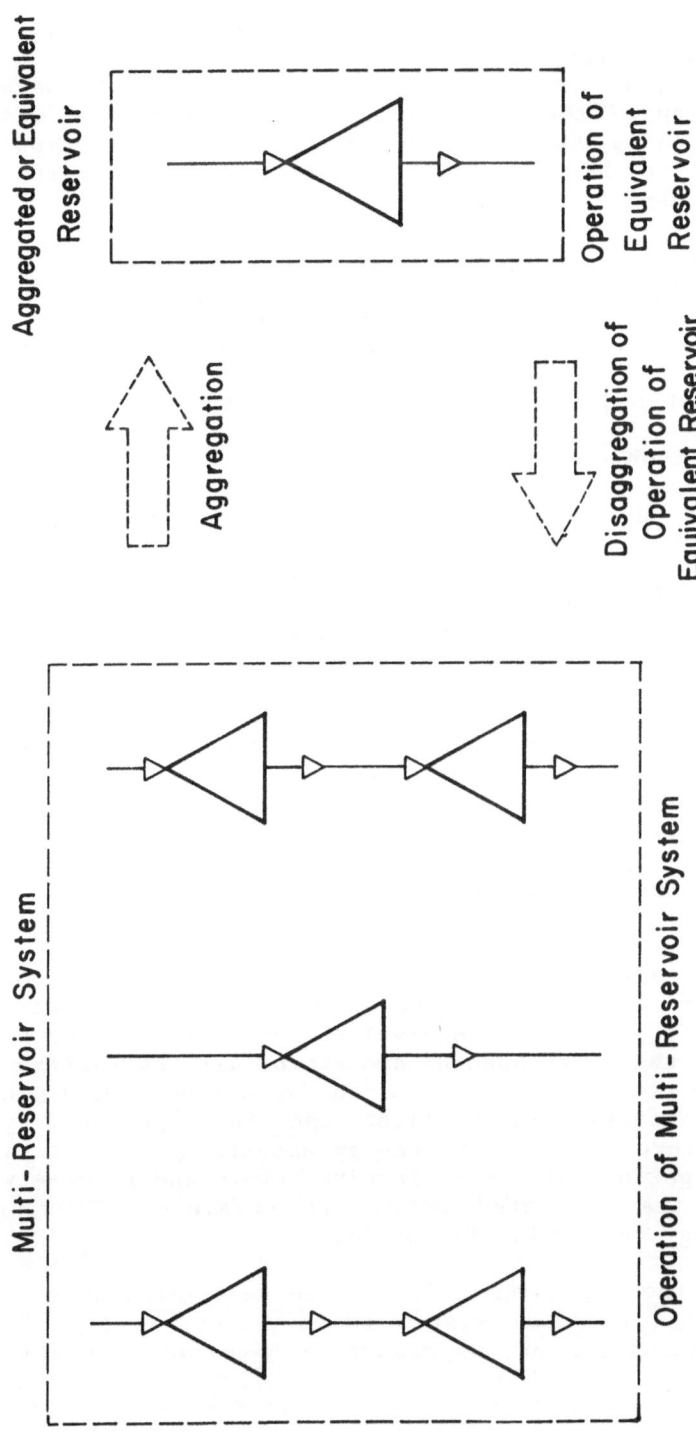

Figure 2. Schematic Representation of the Aggregation of a Multireservoir
System into an Equivalent Reservoir and the Disaggregation of
the Operation of such Equivalent Reservoir into the Operation
of the Original Multireservoir System

subproblems. The first is to determine a single equivalent reservoir representing the overall multireservoir system. The second is to determine the optimal operation of such single reservoir. The third is to disaggregate such optimal operation into the operation of the original multireservoir system. Figure 2 shows schematically the foregoing process. The main purpose of this section is to describe how such aggregation-disaggregation process can be made.

Attempts to use aggregation procedures for solving multi-reservoir systems began in 1951 with the work of Morlat (3). Since then extensive studies have been made directed to better approaching the three subproblems mentioned above. Literature on the subject can be found in Smith (4). In fact most of the rest of this paper is based on Smith's recent studies.

5.1 Definition of the Equivalent Reservoir System

In general an equivalent reservoir or the aggregated representation of a multireservoir system is a single reservoir with a capacity and inputs that will produce the same water outputs (in the form of water releases, hydropower generation, etc.) as would be produced by the original multireservoir system.

Let us take the example of a 5 reservoir water supply system such as the one schematically shown in Figure 3. The equivalent reservoir input, storage and outputs (system releases, evaporation and spills) will be simply the summation of the corresponding variables of the indivdual reservoirs, considering of course, local inflows and outflows in the system and the usual mass balance equations. Figure 3 shows the schematic representation of a 5 reservoir system where the input and output variables are defined occurring during a time period t except that $v_t^{(i)}$ represents the storage volumes at the beginning of time period t. Likewise maximum and minimum bounds for storage volumes and releases for the aggregated or equivalent reservoir will be simply the summation of the corresponding amounts of all reservoirs. The reservoir for a hydropower system can be defined in terms of energy or water amounts. In the first case, the aggregated system will receive, store and release energy amounts (1) while in the second the aggregated system will receive, store and release water quantities and the generated energy is obtained by using an appropriate energy convertion factor (2).

In general the variables defining the aggregated system will depend whether the original system is in series, in parallel or both. For instance, let us assume we have two reservoirs in

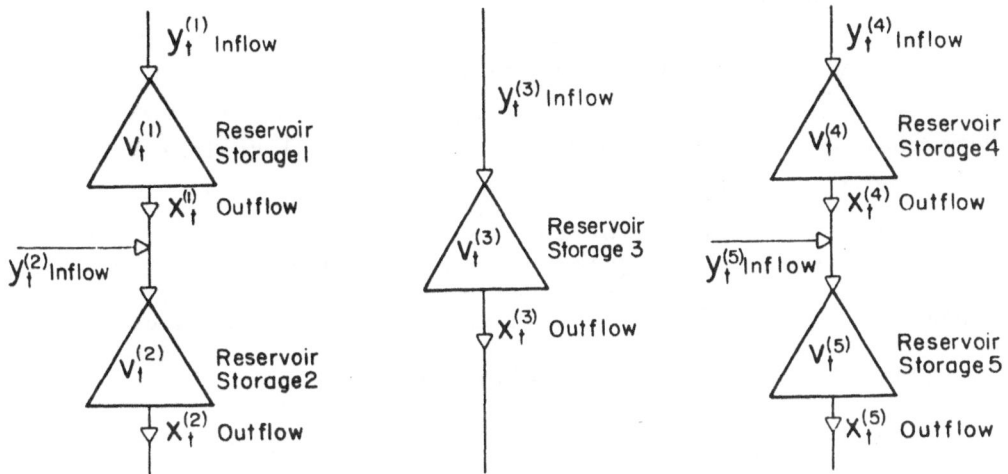

Figure 3.   Five Reservoir System in Series
and in Parallel

parallel where $v_t^{(1)}$ and $v_t^{(2)}$ are the volumes at the beginning of time t, and $h^{(1)}$ and $h^{(2)}$ are the corresponding energy convertion factors. The aggregated energy storage v_t may be determined by

$$V_t = \int_0^{v_t^{(1)}} h^{(1)} \, dv^{(1)} + \int_0^{v_t^{(2)}} h^{(2)} \, dv^{(2)}$$

while the aggregated energy inflow Y_t is

$$Y_t = h_t^{(1)} y_t^{(1)} + h_t^{(2)} y_t^{(2)}$$

On the other hand for reservoirs in series, the energy generated downstream by releases from upstream reservoirs must be considered.   For instance, for 2 reservoirs in series the aggregated energy storage V_t becomes

$$V_t = \int_0^{v_t^{(1)}} h^{(1)} \, dv^{(1)} + v_t^{(1)} h_t^{(2)} + \int_0^{v_t^{(2)}} h^{(2)} \, dv^{(2)}$$

while the aggregated energy inflows Y_t is

$$Y_t = y_t^{(1)} \ [h_t^{(1)} + h_t^{(2)}] + y_t^{(2)} \ h_t^{(2)}$$

Note that in the above equations the individual variables such as $x_t^{(i)}$, $y_t^{(i)}$ and $v_t^{(i)}$ are given in water amounts while the corresponding variables for the aggregated reservoir are given in energy amounts. The aggregated releases, spills and evaporation as well as the maximum and minimum storage and release capabailities may be also determined by appropriate calculations.

5.2 Optimal Operation of the Equivalent Reservoir

Once the original multireservoir system has been transformed into an equivalent reservoir system, the operation of such equivalent reservoir may be determined by a standard optimization technique or in general by an appropriate algorithm. For instance, if D_t represents the aggregated demand (water or energy) during time period t and X_t is the actual reservoir release the optimization problem for the operation of the equivalent reservoir may be stated as

$$\text{Min } \{ \sum_1^T (D_t - X_t)^k \}$$

subject to: bounds in storage $\quad V_{min} < V_t < V_{max}$

bounds in releases $\quad X_{min} < X_t < X_{max}$

and the mass balance equation

$$V_{t+1} = V_t + Y_t - X_t - E_t - L_t$$

where T is the operation horizon, k is an exponent which may be equal to 2, E_t and L_t are the aggregated evaporation and aggregated spill during the time period t and the rest of the variables are as defined previously. This problem may be solved by say dynamic programming.

5.3 Disaggregation of the Operation of the Equivalent Reservoir

Generally speaking it is very difficult to end up with an equivalent reservoir which will incorporate all the limitations and constraints that the multireservoir system has. Thus it may

be stated that the optimal operation of the equivalent reservoir
represents an upper bound for the yield attainable by the original
multireservoir. In other words, the optimal operation of the
multireservoir will produce at most the yield of the optimally
operated equivalent reservoir. Therefore, the search for an
operational policy to disaggregate the optimal operation of the
equivalent reservoir into the operation of the multireservoir
system, can be tested by comparing the relative yield of the
multireservoir system (relative to the upper bound yield) operated
by alternative disaggregation policies.

The disaggregation of the optimal policy can be defined as an
optimization problem whose objective function is to minimize the
differences between the aggregated yield and the one produced by
the multireservoir system. Thus if U_t is the yield of the
equivalent reservoir and $u_t^{(i)}$, $i = 1,...,N$ are the
corresponding yields of the individual reservoirs then the
optimization problem can be stated as

$$\text{Min} \left\{ U_t - \sum_1^N u_t^{(i)} \right\}$$

subject to: bounds in storage for each reservoir, bounds in
releases for each reservoir, mass balance equations for each
reservoir and the upper limit in yield production. Several
techniques such as LP, DP and QP were tested by Smith (4) for
solving the foregoing optimization problem.

5.4 Application

The case of the North Platte River multireservoir system
located in Wyoming and Nebraska, USA is used as an example of the
aggregation-disaggreagtion approach described above. The primary
purpose of the system is for irrigation with hydropower, flood
control, recreation and wild life preservation as secondary
purposes. The schematic representation of the system is shown in
Figure 4.

The equivalent reservoir was defined as explained in section
5.2 and the optimal operation of such reservoir was obtained by
dynamic programming. The optimal reservoir contents for the
period 1960-1965 are shown in Figure 5. The disaggregation of
this optimal policy was determined by linear programming. It led
to a capacity reduction of the equivalent reservoir of about 2
percent in order to have the same yield for both the equivalent
reservoir and the multireservoir system. Figure 6 shows the
operation of the Seminoe Reservoir.

54

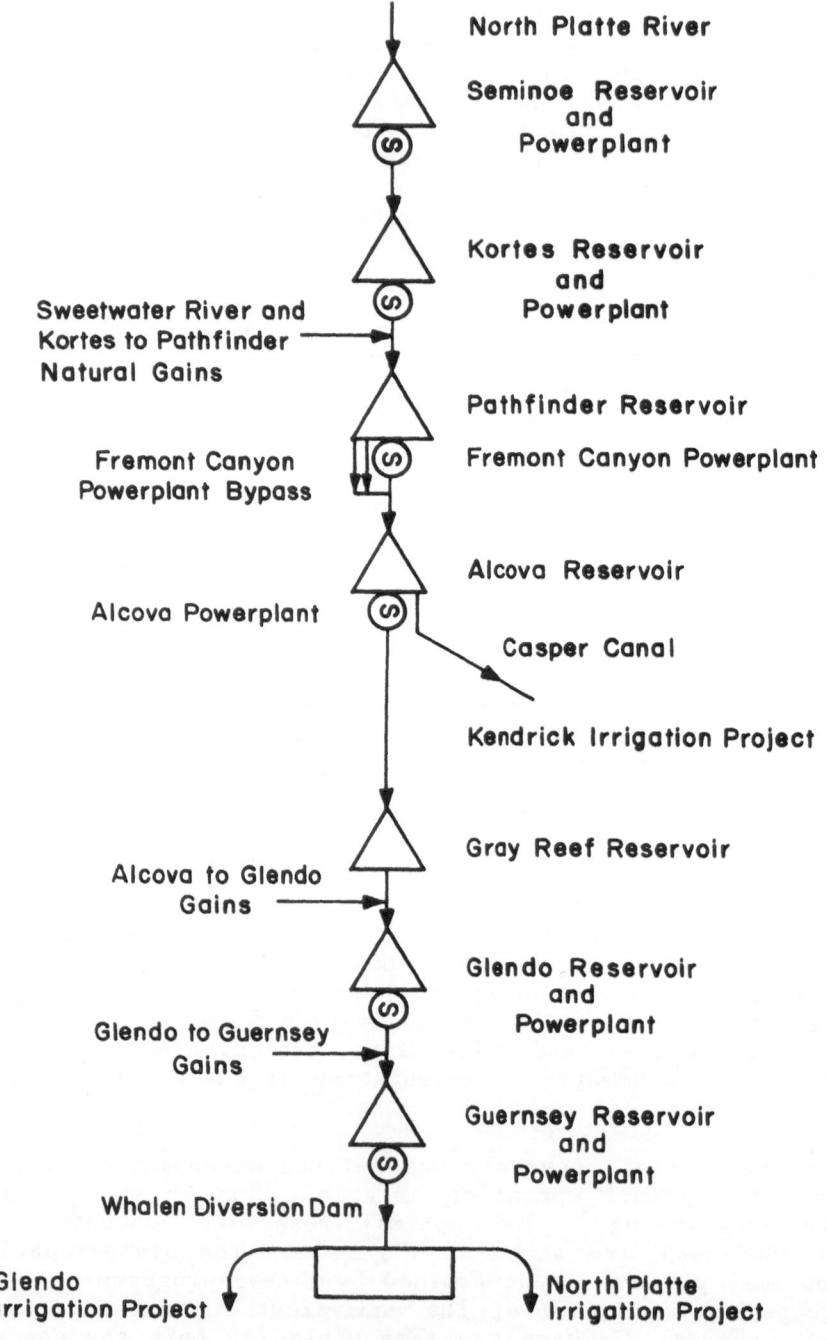

Figure 4. Schematic Representation of the North Platte Multireservoir System (4)

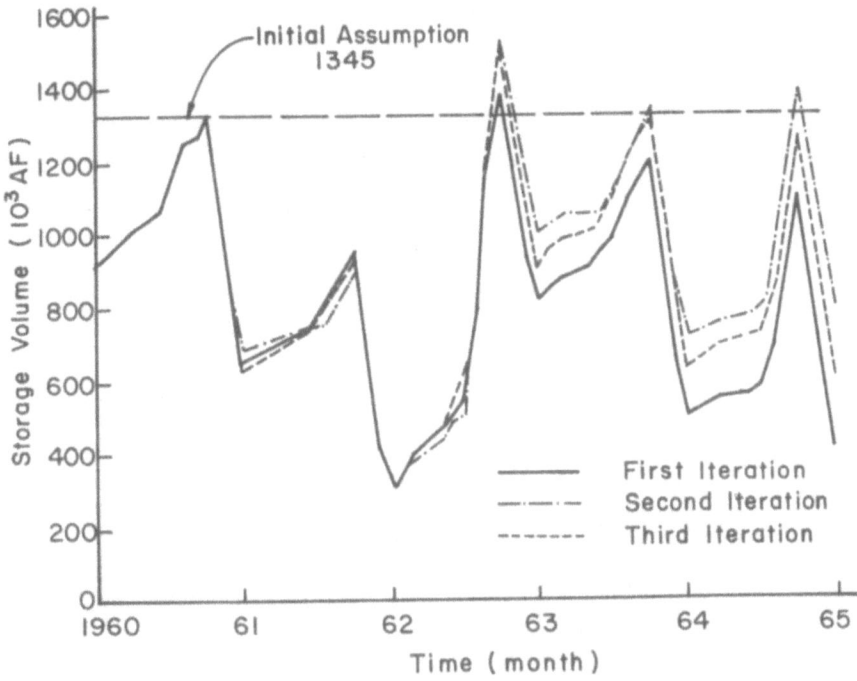

Figure 5. Optimal Operation of the Equivalent Reservoir
of the North Platte Multireservoir System with
a 2 percent Capacity Reduction (Smith, 1981)

6. OPERATION OF A MULTIRESERVOIR SYSTEM BY STATISTICAL DISSAGREGATION

A statistical disaggregation model was proposed by Valencia
and Schaake (5) to generate seasonal flows by disaggregating flows
of annual values into seasonal values. Extensions and improvements to this model were subsequently suggested by other authors.
Because of the hierarchical manner in which the flows are
generated, in general such disaggregation model preserves the year
to year and season to season covariances in addition to preserving
means and variances. Furthermore, it has the particular
characteristic that the generated seasonal flows add up to the
corresponding annual flow values. In this section the foregoing
statistical disaggregation model originally proposed for
hydrologic data generation is applied for two cases of reservoir
operation: the first in spatial sense, in order to determine the
operation of a multireservoir system based on the historical
operation of the same, and the second in temporal sense to find

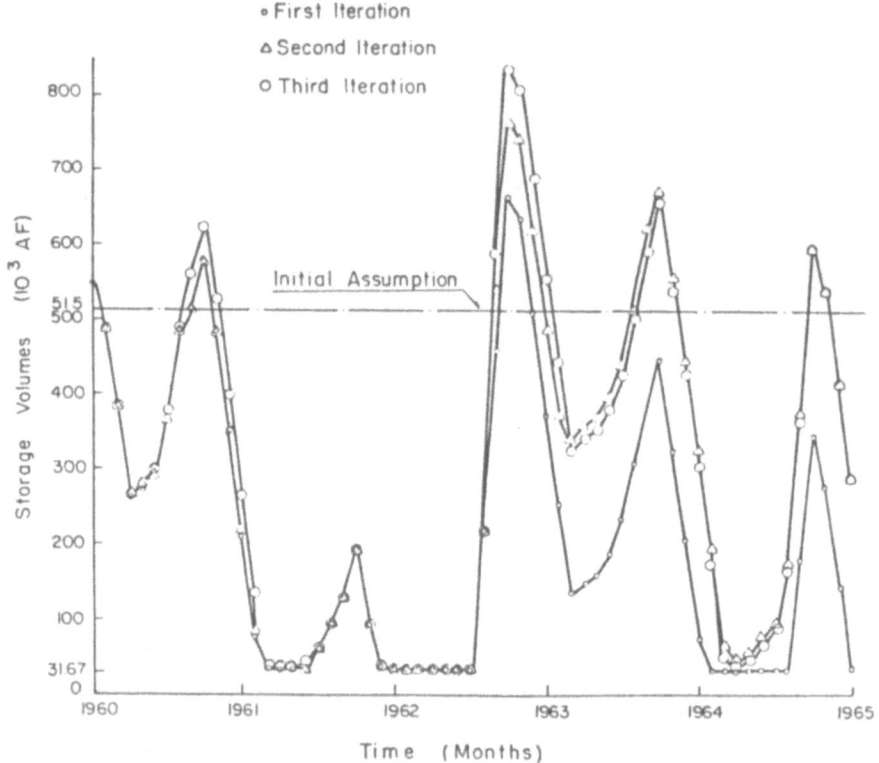

Figure 6. Optimal Operation of Seminoe Reservoir of the
North Platte Multireservoir System (4)

the operation of a reservoir system at a given time interval based
on the operation of the same reservoir made for a longer time
interval.

6.1 Spatial Disaggregation

Assume that $\underline{v}'_{j,t}$ is a vector of storage volumes for an
N-reservoir system for year j and season t and $V'_{j,t}$ is the
storage of the aggregated or equivalent reservoir. The
statistical disaggregation model may be written as

$$\underline{v}_{j,t} = A_t \, V_{j,t} + B_t \, \underline{a}_{j,t} \quad , \quad \begin{array}{l} t = 1,\ldots,M \\ j = 1,\ldots,T \end{array}$$

57

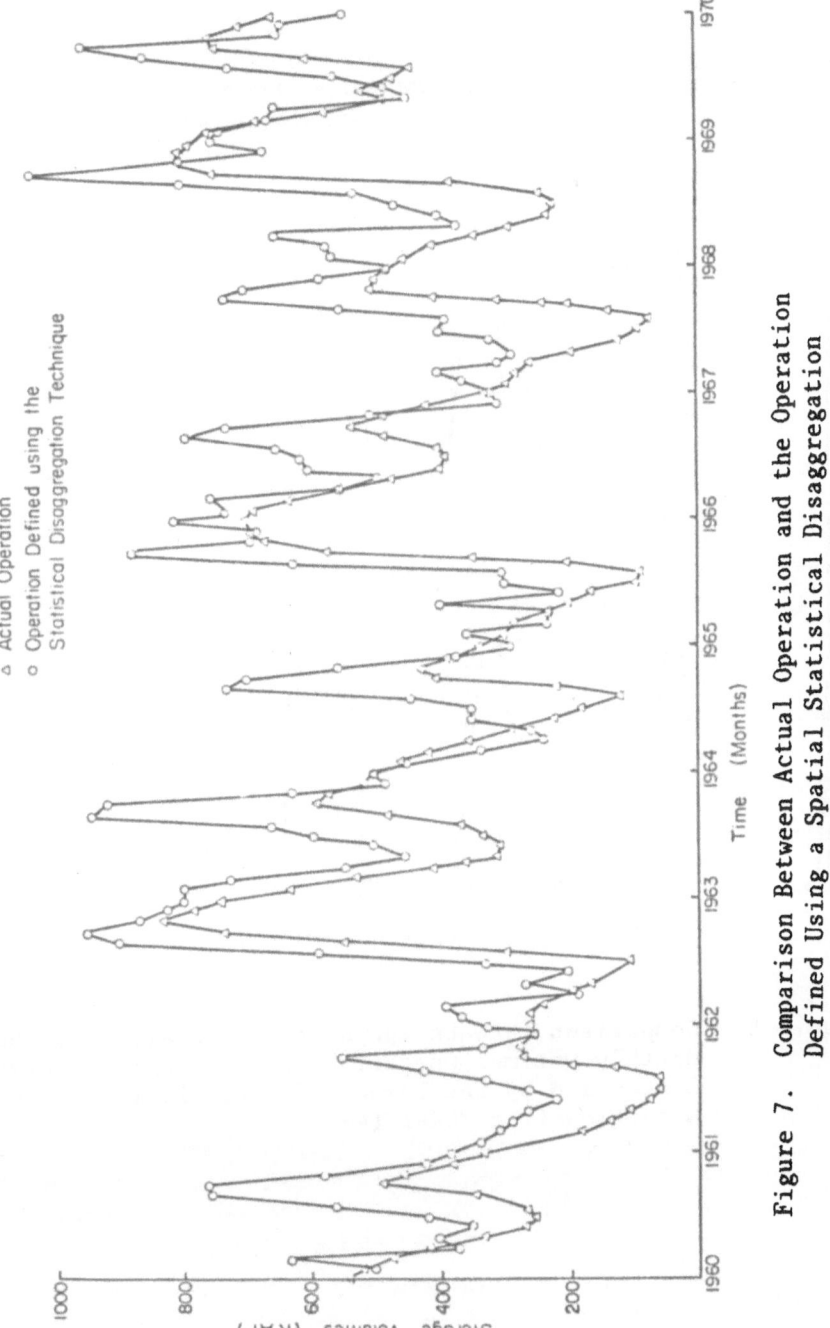

Figure 7. Comparison Between Actual Operation and the Operation Defined Using a Spatial Statistical Disaggregation Technique for Pathfinder Reservoir of the North Platte River System (4)

58

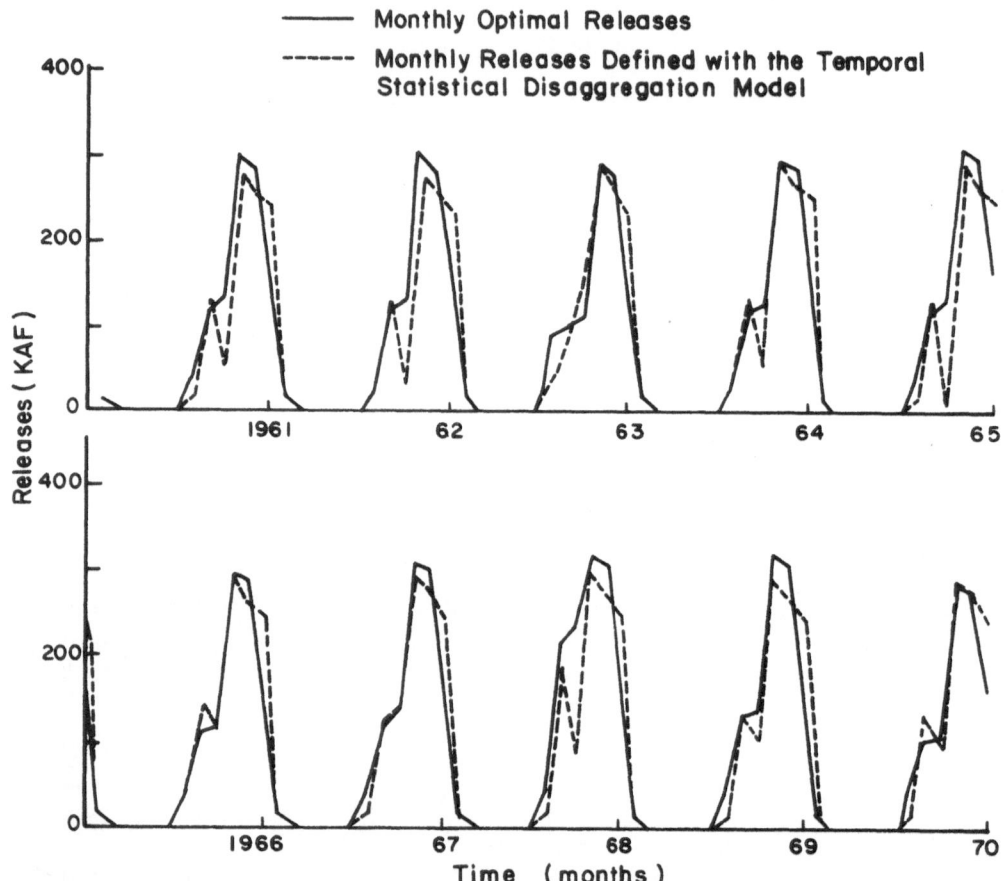

Figure 8. Comparison between the Releases Obtained by the
Monthly Optimal Operation of the Single Reservoir
System and by the Temporal Statistical
Disaggregation Model (4)

where $\underline{v}_{j,t}$ and $V_{j,t}$ have now zero means, A_t and B_t are the matrix parameters of the model and $\underline{a}_{j,t}$ is a random vector with zero mean and variance one. Assuming that historical storage volumes $\underline{v}'_{j,t}$ are available for a certain period of time and $V'_{j,t}$ can be determined by the optimal operation of the aggregated or equivalent reservoir, the parameters A_t and B_t can be simply estimated by the usual procedure of the method of moments. Figure 7 shows results of the application of the foregoing procedure for the North Platte River reservoir system.

6.2 Temporal Disaggregation

If the operation of a reservoir system is determined at two different time levels (say seasonal and monthly) then they can be related to each other by a statistical disaggregation model. The purpose of doing this would be to determine the monthly operation of a reservoir system based on the corresponding seasonal operation. To test the validity of the approach a single reservoir in the North Platte River system was used for which the optimal operation on a monthly and seasonal basis were obtained by dynamic programming. Then the two time level operations were related by the suggested statistical disaggregation model. Actual results of the model application are shown in Figure 8.

ACKNOWLEDGEMENT

The support of the projects "Conjunctive Water Uses of Complex Surface and Groundwater Systems" Bilateral U.S.-Spanish Project and "Investigation of Objective Functions and Operational Rules of Storage Reservoirs," OWRT Project B-195-COLO are gratefully acknowledged.

REFERENCES

1.   Arvantidis, N, and J. Rosing, "Optimal operation of multireservoir systems using a composite representation," IEEE Trans. Power App. and Syst., Vol. PAS-80(2), (1970). 327-335.

2.   Hall, W., "Optimal state dynamic programming for multireservoir hydroelectric systems," Class Notes, Colorado State University, Fort Collins, Colorado, (1971).

3.   Morlat, G., "Sur la consigne de'explotaton optimun des reservoirs saisonniers," La Houille Blanche, (1951), 497-509,

4.   Smith, R., "Aggregation-disaggregation techniques for the operation of multireservoir systems," Ph.D. Dissertation, Department of Civil Engineering, Colorado State University, Fort Collins, Colorado, (1981).

5.   Valencia, D. and J. Schaake, 1973, "Disaggregation process in stochastic hydrology," Water Res. Res. Vol. 9, (3), (1973), 580-585.

5. APPLICATION OF MATHEMATICAL SYSTEMS ANALYSIS

Lucien Duckstein
University of Arizona, Tucson

1. INTRODUCTION

The purpose of this chapter is to demonstrate how certain concepts of modern system theory can be used to study the structure of complex water resources systems and the operation of subsystems. The two main tools introduced here are Q-analysis, which is used to characterize the level of linkage between system elements, and bifurcation-catastrophe theory, which can describe system behavior at singularities. The Q-analysis approach is based on multidimensional graph theory and, just like ordinary two-dimensional graph theory, may be applied to problems of general system modelling, supply, demand, and decision making. So far, the main application of Q-analysis has been to investigate the structure of a system, such as an ecosystem or a network of roads.

The bifurcation-catastrophe technique essentailly analyzes instabilities of engineering, social and also ecological systems. The approach essentially reduces a singularity described by many parameters to a standard or canonical form characterized by only two or three parameters. Thus, if the system is to be controlled near the singularity, the number of control parameters has been reduced to a manageable size.

As an example of use of Q-analysis, consider a water distribution network which has resulted from growth without water master plan. As a case, the network in the city of Tucson, Arizona has resulted from a very small original system, to which sources and users were added as the city grew and as new areas were annexed or small water companies were bought out by the municipal water and sewers service. The same statement can be made for New York City and many other cities and towns. At one point in time, the problem arises to determine the structural characteristics of such systems, namely, the weak points, the locations where redundancy or looping exists, and those where it does not.

Another example where the structure of water network needs to be known is concerned with preventive measures against drought. In this case it is essential to know which reservoirs (vertices) or links (simplicial faces) "control" the system.

In the first part of this chapter it is shown how Q-analysis may provide a simple tool to perform such tasks. In the second part, Q-analysis is combined with fuzzy set theory to provide a decision-making aid to operate the subsystems. A simplified example of water treatment plant operation in the San Francisco Bay area is provided to illustrate the concepts.

In the third part of the study, another tool of modern system theory, namely, bifurcation analysis, is introduced. In particular, the catastrophe theory description of system stability is used to describe how a small disturbance may be sufficient for a dam to collapse or sudden eutrophication to occur in a water body.

2. STUDY OF SYSTEM STRUCTURE BY Q-ANALYSIS

The concept of Q-analysis, also called polyhedral dynamics (Atkin, 1974; Atkin and Casti, 1977; Shea, 1981) is based on a binary relationship between two sets A and B called, respectively, simplex set and vertex set. For example, A may be a set of water sources $(A(1), A(2)...,A(n))$ and B, a set of sinks or users $(B(1)...,B(n))$. This section outlines the mathematics of Q-analysis.

2.1 Incidence Matrix and Structure of Complex Systems

Let λ be the set of linkages between the elements of A and the set of B. An incidence matrix $\Lambda = [x(j,k)]$ can thus be defined as follows:

$$x(j,k) = \begin{cases} 1 & \text{if } (A(j), B(k)) \; \varepsilon \; \lambda \\ 0 & \text{otherwise} \end{cases} \tag{1}$$

This relationship defines a complex K denoted explicitly as $K_A(B,\lambda)$. If the roles of the two sets are interchanged, the conjugate complex $K_B(A,\lambda^{-1})$ is obtained.

Sometimes the matrix describing the relationship between sets A and B is initially composed of real numbers X(j,k) such as flows. In this case, one may define a threshold X* such that

$$x(j,k) = \begin{cases} 0 & \text{if } X(j,k) < X^* \\ 1 & \text{if } X(j,k) > X^* \end{cases}$$

To illustrate the correspondence between Q-analysis and polyhedra, the relationships between the sets A and B is represented geometrically as a polyhedron with vertices B and simplices (faces, edges, or points) A.

The complex K only shows the global relationship between the sets A and B. For a more detailed investigation of the relationship between simplices forming the complex, the notion of q-connectivity is introduced (Atkin, 1974).

Definition: Two simplices $\sigma(a)$ and $\sigma(b)$ are said to be q-connected in the complex K if and only if there exists a finite sequence of simplices $\{\sigma(\alpha(i)); \quad i = 1,2,\dots,p$ in K such that

(1) $\sigma(a)$ is a face of $\sigma(\alpha(1))$

(2) $\sigma(\alpha(p))$ is a face of $\sigma(b)$

(3) $\sigma(\alpha(i))$ and $\sigma(\alpha(i+1))$ share a face of dimension $\beta(i)$, where dimension of a face is its number of vertices minus one

(4) $q = \min \{a, \beta(1), \beta(2), \dots, \beta(p-1), b\}$.

The q-connectivity may be explained as follows: if two simplices have n vertices in common, they are said to form a single component at the (n-1) dimensional level.

2.2 Example 1

Consider the sets

A = (S1, S2, S3) - water sources

B = (AGR, IND, MUN, DOM) - water users (agricultural, industrial, municipal and domestic)

Let the matrix X showing the flow capacities be:

	AGR	IND	MUN	DOM
S1	405	--	15	--
S2	200	60	115	110
S3	220	--	48	17

Let us use the threshold function

$$x(j,k) = \begin{cases} 0 & \text{if } X(j,k) < 100 \\ 1 & \text{if } X(j,k) \geq 100 \end{cases}$$

The incidence matrix is

	AGR	IND	MUN	DOM
S1	1	0	0	0
S2	1	1	1	1
S3	1	0	1	1

and the complex is defined by

	S1	S2	S3
S1	0	0	0
S2		3	2
S3			2

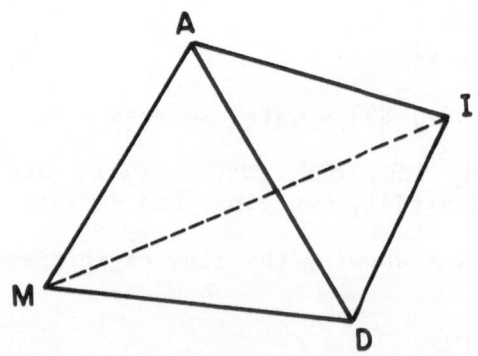

The 3 in element (2,2) of the complex matrix corresponds to the fact that source S2 supplies all n=4 users. Geometrically, source S2 is represented by the 3-dimensional polyhedron ADMI shown above. Similarly, source S3 corresponds to triangle ADM. Source S2 and S3 thus have the 2-dimensional figure ADM in common. By definition, it is said that:

>>S2 is 2-connected to S3 via <AGR, DOM, MUN>

Similarly

>>S1 is 0-connected to S3 via <AGR>

The results of Q-analysis are summarized as follows:

q = 3	Q3 = 1	<S2>
q = 2	Q2 = 1	<S2, S3>
q = 1	Q1 = 1	<S2, S3>
q = 0	Q0 = 1	all

The algorithm for reaching such a listing in the general case is given next.

2.3 Algorithm for Q-analysis

If $\sigma(a)$ and $\sigma(b)$ are q-connected, the they are also $(q-s)$-connected, $s = 1,...,q$. Atkin (1974) demonstrates that q-connectivity is an equivalence relationship on K and gives an algorithm to perform Q-analysis:

(1) Form Λ^T (an mxm matrix), where Λ = the incidence matrix.

(2) Evaluate $\Lambda^T -\Omega$, where Ω is an mxm matrix, with all entries equal to 1.

(3) Retain only the upper triangular part (including the main diagonal) of the symmetrical matrix $\Lambda^T -\Omega$.

(4) By reading from the main diagonal (up and to the left or to the right and down), the structure vector Q is obtained.

2.4 Eccentricity

The relationship between a given simplex σ and its complex K is now examined from two viewpoints (Atkin and Casti, 1977):

(a) to how many distinct elements of the complex is σ related?

(b) how well is σ integrated into K?

These two viewpoints are needed because, even though a particular simplex may have a high dimensionality (related to many vertices), it may not be related to the other simplices in K. An index synthesizing these two viewpoints is the so-called eccentricity of σ, denoted ecc(σ), and defined as follows (Atkin 1974):

$$ecc(\sigma) = \frac{\hat{q} - q^*}{q^* + 1}$$

$\hat{q}$ = the dimension of the simplex σ

q^* = the largest q-value for which σ appears in a
        component with another distinct simplex.

Upon inspection it can be seen that $\hat{q} - q^*$ tells how many vertices the simplex σ does not share with any other simplex in the complex. But, for example, $\hat{q} - q^* \rightarrow 7-5$, and $\hat{q} - q \rightarrow 6-4$ yield the same result, so that it is appropriate to introduce a normalization factor $q^* + 1 \geq 0$, yielding $\frac{7-5}{5+1} = \frac{1}{3}$ and $\frac{6-4}{4+1} = \frac{2}{5}$. The former simplex is thus better integrated into K than the latter. Note that ecc(σ)=∞ means that σ is totally disconnected from the complex.

2.5 Example 2

Let the two sets of example 1 be:

<u>Source or Simplex Set S</u> = {G1,G2,G3,S1,S2,S3,T} where,

    G1,G2,G3 are groundwater sources

    S1,S2,S3 are surface reservoirs

    T is tertiary treated effluent.

<u>User or Vertex Set F</u> = {AG1,AG2,IND,DOM,MUN} where,

    AG1 is agricultural user #1 (fruit and vegetables)
    AG2 is agricultural user #2 (general crops)

Let the system matrix X representing the flow capacities, be

	AG1	AG2	IND	DOM	MUN
G1	9	1	9	0	5
G2	2	3	2	0	3
G3	7	1	15	0	9
S1	6	7	11	0	12
S2	4	10	3	0	5
S3	8	5	7	5	11
T	30	18	27	0	25

To transform the system matrix m to a (0,1) incidence matrix, let the threshold function be

$$x(j,k) = \begin{cases} 0, & X(j,k) < 2 \\ 1, & X(j,k) \geq 2 \end{cases}$$

$$j = 1, \ldots, 5, \qquad k = 1, \ldots, 7$$

The incidence matrix is thus:

$$\Lambda = \begin{bmatrix} 1 & 0 & 1 & 0 & 1 \\ 1 & 1 & 1 & 0 & 1 \\ 1 & 0 & 1 & 0 & 1 \\ 1 & 1 & 1 & 0 & 1 \\ 1 & 1 & 1 & 0 & 1 \\ 1 & 1 & 1 & 1 & 1 \\ 1 & 1 & 1 & 0 & 1 \end{bmatrix}$$

Performing the q-analysis algorithm described earlier, one obtains the "shared-face" matrix:

	G1	G2	G3	S1	S2	S3	T
G1	2	2	2	2	2	2	2
G2		3	2	3	3	3	3
G3			2	2	2	2	2
S1				3	3	3	3
S2					3	3	3
S3						4	3
T							3

and the Q-analysis algorithm yields the summary table:

q = 4	Q4 = 1	<S3>
q = 3	Q3 = 1	<G2,S1,S2,S3,T>
q = 2	Q2 = 1	<All>
q = 1	Q1 = 1	<All>
q = 0	Q0 = 1	<All>

The highest q-connectivity between different sources is 3 (surface source S3 is 4-connected to itself). For example,

G2 is 3-connected to S1, via <AG1,AG2,IND,MUN>

What this means is that both sources G2 and S1 supply at least in part 4 out of 5 users. Note that Q3, which is the number of 3-connected components of K, equals 1. The reason for this is that if two or more simplices have n elements (vertices) in common, the two simplices are considered as a single component at the (n-1) dimensional level. At the level q = 2, all of the elements come into the complex, which means that 3 out of 5 users receive water from all seven sources.

To find out how the individual simplices--or--subsystems--are integrated into the complex, Eq. (3) defining eccentricity is used, yielding that all eccentricities are zero, except:

$$\text{ecc (S3)} = \frac{4-3}{3+1} = \frac{1}{4}$$

Thus the complex appears to be quite homogeneous. The only exception is S3 which is not quite as well integrated as the other sources. For this example, the Q-analysis has revealed that this water supply system has been well designed.

Regarding the choice of a threshold function, great care must be taken that relevant data are not discarded when computing the (0,1) incidence matrix. As an illustration, let us analyze the same example using a threshold $X^* = 9$ instead of $X^* = 2$.

The incidence matrix now becomes:

$$\begin{bmatrix} 1 & 0 & 1 & 0 & 0 \\ 0 & 0 & 0 & 0 & 0 \\ 0 & 0 & 1 & 0 & 1 \\ 0 & 0 & 1 & 0 & 1 \\ 0 & 1 & 0 & 0 & 0 \\ 0 & 0 & 0 & 0 & 1 \\ 1 & 1 & 1 & 0 & 1 \end{bmatrix}$$

and the application of Q-analysis yields the matrix

	G1	G2	G3	S1	S2	S3	T
G1	1	-1	0	0	-1	-1	1
G2		-1	-1	-1	-1	-1	-1
G3			1	1	-1	0	1
S1				1	-1	0	1
S2					0	-1	0
S3						0	0
T							3

Since a (-1) in position (j,k) denotes the j and k
are not connected, the results of Q-analysis are misleading; this
is caused by a poor choice of threshold function.

2.6 Patterns

Operation of a water resources system is dynamic, whereas the
elements of Q-analysis presented so far deal only with static
features. The concept of pattern has been introduced in Atkin and
Casti (1977) to model certain dynamic aspects of system structure.

Let $\Pi(i)$ be an integer-valued mapping defined on the subset
of i-dimensional simplices. The pattern Π is defined as
follows:

<u>Definition</u>: Let group Π contain the subgroups
$\Pi_0, \Pi_1, \ldots, \Pi_K$. Suppose that the only element common to all sub-
groups is zero and every element of Π is the sum of one element
of each subgroup $\Pi_0, \Pi_1, \ldots, \Pi_K$. Then Π is called the
<u>direct sum</u> of the subgroups and is written

$$\Pi = \Pi(0) + \Pi(1) + \ldots + \Pi(K).$$

In the previous example, the sets $\Pi(i)$ could be the
mappings from the i-dimensional sources to the flow from those
sources. Thus:

$$\Pi(2) \begin{cases} G1 \rightarrow \text{flow from source G1} \\ G2 \rightarrow \text{flow from source G2.} \end{cases}$$

The dynamics is introduced by means of pattern change $\delta\Pi$
which, in our case, simply changes the amounts flowing from
various sources. Thus, the original pattern Π is composed of
mappings defined on the following sources:

$\Pi(0)$: empty

$\Pi(1)$: empty

$\Pi(2)$: G1, G2

$\Pi(3)$: G3, S1, S2, T

$\Pi(4)$: S3

The new pattern $\Pi + \delta\Pi$ may be

$\Pi(0)$: empty

$\Pi(1)$: empty

$\Pi(2)$:   G1, G2, S1

$\Pi(3)$:   G3, S2

$\Pi(4)$:   S3, T

which can be interpreted physically by noticing that S1 become 2-connected instead of 3 and T, 4-connected instead of 3.

2.7 Obstruction Vector and Complexity

The obstruction vector is a measure of resistance to change of pattern (or obstruction to change) at level q. Let W be a vector of whose components are all 1's; the obstruction vector is defined as $\hat{Q} = Q - W$, in which Q is the vector of components $\hat{Q}(i)$ found from the Q-analysis. The obstruction vector may be used to identify the flexibility of a complex water recource system in case of emergency (floods, drought).

The concept of complexity describes the density of interconnections between simplices and is defined in Casti (1979) as having three properties:

(1) A system consisting of a single simplex has a complexity equal to 1.

(2) A subsystem (subcomplex) has complexity no greater than that of the entire complex.

(3) The combination of two complexes results in a level of complexity no greater than the sum of the complexities of the components.

If N is the dimension of the complex K, Q(i) then a measure of complexity satisfying the axioms is

$$\psi(k) = 2[\sum_{i=0}^{N} (i+1)\ Q(i)]\ /\ (N+1)\ (N+2) \tag{4}$$

Consider for example a subsystem of water distribution from sources S to users U

	U1	U2	U3	U4
S1	0	0	1	0
S2	1	1	1	0
S3	1	0	1	1
S4	1	0	1	1
S5	1	1	1	1
S6	1	1	0	1

The structure vector of the source-user problem is

$$Q(S,U) = (\overset{3}{1} \ 1 \ 1 \ \overset{0}{1} \ )$$

whereas the structure vector for the conjugate problem user-source is

$$Q(U,S) = (\overset{4}{2} \ 1 \ 1 \ 1 \ 1)$$

Thus, using Eq. (4) yields $\psi(S,U) = 1$ and $\psi(U,S) = 4/3$. Physically, it appears that the system user-source is more complex than the system source-user which, according to Casti (1979) means that it may be less stable.

3. DECISION-MAKING ON SUBSYSTEMS BY Q-ANALYSIS AND FUZZY SETS

In this section a multicriterion decision-making technique (MCDM) combining Q-analysis (Atkin, 1974a) and fuzzy sets (Kaufman, 1975), is developed after Kempf et al. (1978) and Armijo et al. (1978). The planning and operation of control schemes for lake and estuarine pollution is used to illustrate the approach. Although pollution problems are often of a multicriterion nature, the emphasis in the past has primarily been on modelling the aquatic ecosystem (Chen and Orlob, 1971) in order to quantify the effects of various pollution control plans on improvements in ecosystem response.

There exists a large body of multicriterion or multiobjective techniques which could be brought to bear on the problem (Cohon, 1978; Gershon, 1981; Zeleny, 1981). For example, the MCDM technique ELECTRE I (Roy, 1977a) induces a partial ordering on the set of alternative actions (or decisions) under consideration, in order to reduce the choice set. In the case of ELECTRE, pair-wise comparisons are used to develop a graph which represents, in some sense, the decision maker's preferences, while in concordance analysis, the systems are compared pair-wise and the criteria on which one system is preferred are grouped into a concordance set form which indices of concordance are calculated. The present method has been inspired in part by ELECTRE I, which is based on two-dimensional graph theory. The basis of Q-analysis may be seen as multi-dimensional graph theory, as illustrated, for example, in Shea (1981).

3.1 Incorporation of Fuzzy Sets into Q-analysis

So far, the sets used to define the relation between simplices and vertices represented by the incidence matrix Λ have been of the type which Kaufman (1975) calls "ordinary". For such a set, called A, a membership or indicator $\mu(A,x)$ can be used to indicate whether or not an element, x, is a member of the set, as follows:

$$\mu(A,x) = \begin{cases} 1 & \text{if } x \in A \\ 0 & \text{if } x \notin A \end{cases} \tag{5}$$

Note that this is analogous to Eq. 1 defining $x(j,k)$.

If, now, the membership function A consists of a set of ordered pairs:

$$(x, \mu(A,x)), \quad x \in A \tag{6}$$

where

$$\mu(A,x) \in \mu$$

then A is called a fuzzy set. The set μ is a totally ordered set, not necessarily $\{0,1\}$, which indicates the degree or level of membership of the element x in the set A. If $\mu = \{0,1\}$, then the "fuzzy" set is actually an ordinary set. In most cases, μ is defined to be the closed interval $[0,1]$.

The membership function can thus be used to define at what level a particular element is a member of the set A, which makes fuzzy sets useful for quantifying vagueness. Applications are numerous and growing in subjects such as decision theory (Bellman and Zadeh, 1970), and others where membership in some set of interest is ill defined. Fuzzy set concepts have been used by Roy (1977b, 1977c) as a basis for developing ELECTRE III, in Nijkamp (1978) to analyze conflicts among various interest groups and in Bezdek et al. (1978) to arrive at a model for group decision theory.

3.2 Methodology: Multi-Criterion Q-Analysis (MCQA)

Let $d(i,j)$ be the numerical rating of the ith action on the jth criterion; a decision impact matrix D can thus be defined as

$$D = \{d(i,j)\} \qquad i = 1,2,\ldots,m; \quad j = 1,2,\ldots,n$$

where

$$m = Card(P) = \text{number of projects under consideration}$$

$$n = Card(C) = \text{number of criteria}$$

In other words, the set of actions is taken as the simplex set and the set of criteria as the vertex set. As pointed out in Duckstein (1978), such a matrix D, also called a cost-effectiveness array, payoff matrix, etc., may be considered as the core of any multicriterion decision making scheme.

If the decision maker has some idea, however vague, of a "satisficing" outcome (Simon, 1957), linguistic variables (Zadeh, 1973) can be used to assign a membership grade to each of the entries in the project impact matrix. Let (-) represent any one of "slightly," "moderately," or some other adjective modifier which the decision maker can easily relate to his internal preference set (Kochen and Badre, 1974). Then membership grades will take the following form:

$$\mu(A,d(i,j)) = r - s, \text{ if outcome } d(i,j) \text{ is } (-)\text{to } (-) \text{ satisfactory}$$

$$r, s \in [0,1]$$

For example:

$$\mu(A,d(i,j)) = .3 - .5, \text{ if outcome } d(i,j) \text{ is}$$

(slightly) to (moderately) satisfactory

The membership grade indicates to what degree the outcome $d(i,j)$ is a member of the decision maker's set of satisficing or desired outcomes on each of the criteria.

The membership grades $\mu(A,d(i,j))$ define a matrix U(D) from which a series of multidimensional graphs or simplicial complexes can be constructed. Specifically, a threshold function α defined as in Eq. 2, yields an incidence matrix $\Lambda(\alpha)$ whose (i,j)th element is

$$\lambda(i,j) \begin{cases} 1, & \text{if } \mu(d(i,j)) \geq \alpha \\ 0, & \text{if } \mu(d(i,j)) < \alpha \end{cases}$$

$$\alpha \ \varepsilon \ [0,1]$$

In terms of the decision, the simplicial complex becomes a representation of U(D) for a particular overall decision maker satisfaction level α.

Q-analysis can then be performed on the complex. Those project simplices which are high dimensional (i.e., show up at a high q-level) are desirable on more criteria than those at lower q levels and therefore will be preferred. However, two projects which are q-connected at the ith level have desirable outcomes on exactly i + 1 criteria, and therefore are indistinguishable on those criteria. The objective, then, may be to find the highest decision-maker satisfaction level α at which a single, non-q-connected simplex occupies the highest Q vector position. The project represented by this simplex would be preferred on more criteria than the others.

In the event that no distinct simplex occupies the highest structure vector position over the entire range of satisfaction levels, either one of the following two possibilities could occur:

(1) Two or more non-q-connected simplices occur at the higest vector position, i.e., the highest component in the Q vector is greater than 1, which is the case, for example, in Table 3 for α = .6. This indicates that the projects represented by the simplices are satisficing on an equal number of criteria, but do not share any common criterion.

(2) A q-chain of two or more simplices occurs at the high vector position; i.e., the highest component of the structure vector is 1, which is the case in Table 3 for α = .2. This indicates that the projects in the q-chain share an equal number of satisficing criteria.

The results of the Q-analysis can be used to reduce the choice set by eliminating those projects which do not occupy the high structure vector position and continuing the process for the high q projects, perhaps with more detailed information.

Alternatively, the α level at which a project first appears alone at the highest structure vector position could be low, indicating the decision-maker's dissatisfaction with the outcomes. In this case, either another project with more satisfactory outcomes could be sought or the decision-maker might modify his preferences and undertake another iteration to arrive at a choice.

Q-analysis of the conjugate complex, formed by using the criteria as simplices and the projects as vertices, can provide information on how worthwhile a particular criterion is for differentiating between the projects; this is similar to duality in linear programming. Those criterion simplices which first appear in the structure vector at high dimensional levels were found satisficing for more projects than those which first appear at lower dimensional levels. Thus, higher dimensional criteria are of less use for differentiating between the projects, since many projects share satisfaction of the same criterion.

The dimensional level at which the criterion simplices first become q-connected provides information on how well the criteria are being used to judge the projects. If a criterion simplex first becomes connected at any dimensional level above zero, as mentioned above, then more than one project was satisfactory for that criterion and that criterion cannot be used for a good discrimination between the projects. If, on the other hand, a criterion simplex first becomes connected at the (-1)-dimensional level, no project was found satisfactory for that criterion. Ideally, all the criterion simplices should be connected at the zero-dimensional level, if a single project is to be selected. In such a case, the projects are being compared and the most preferred project will only share one desirable criterion with each of the other projects. Q-analysis of the criterion complex could also be used to determine which criteria should be dropped, if another iteration were required. Finally, the role of the discordance index of ELECTRE (Roy, 1977a; Gershon et al., 1980) can be played by a Q-analysis performed on the matrix $\Omega - U(D)$, where elements of Ω are all equal to 1.

3.3 Example

This example deals with the design of a water quality control scheme, but the methodology can be used as is for operation, provided the decision variables are discretized. Thus, consider a simplified problem involving control of waste water discharge into San Francisco Bay from the city of San José, based on simulation results in Chen and Orlob (1971) and general information on the costs and effectiveness of various sewage treatment systems in Whipple (1977). The decision maker was presented with a written description of the situation as summarized below, the project impact matrix of Table 1, and a rough map of San Francisco Bay showing the location of each treatment plan (Figure 1.); he was then asked to assign a grade of membership to each outcome.

Table 1. Project Impact Matrix

	C1	C2	C3	C4
P1	0	0.5	34	4664
P2	4	0.28	40	10070
P3	8	0.18	15	25330
P4	6	0.14	146	6000

Projects:

P1-present situation, primary treatment only
P2-secondary treatment
P3-tertiary treatment
P4-primary treatment with diversion to North Bay

Criteria:

C1-recreational Potential
C2-Nh_3-N, mg/l
C3-Land Use, hectares
C4-Treatment Cost, $/day

The problem is described as follows. The city of San José is planning a waste treatment plant in response to citizen complaints about the deterioration of water quality in the South Bay. The capital costs of constructing the plant will be covered through a bond issue and outside funding; however, the city must pay operating costs out of tax revenue. Presently, wastes are given primary treatment only, i.e. 30 percent of the BOD removal. Three plans for pollution control are under consideration: secondary treatment (removing 85 percent of the BOD), tertiary treatment (removing 100 percent of the BOD, 95 percent of the phosphorus, and 90 percent of the nitrogen), and primary treatment with export to North Bay, where flushing action from the ocean could more easily maintain water quality.

The present situation and the plans involving primary and secondary treatment would also require a 20 hectare sludge dump in the county, while an incinerator, which would pay for itself in generated electricity, would be included in the tertiary treatment plan.

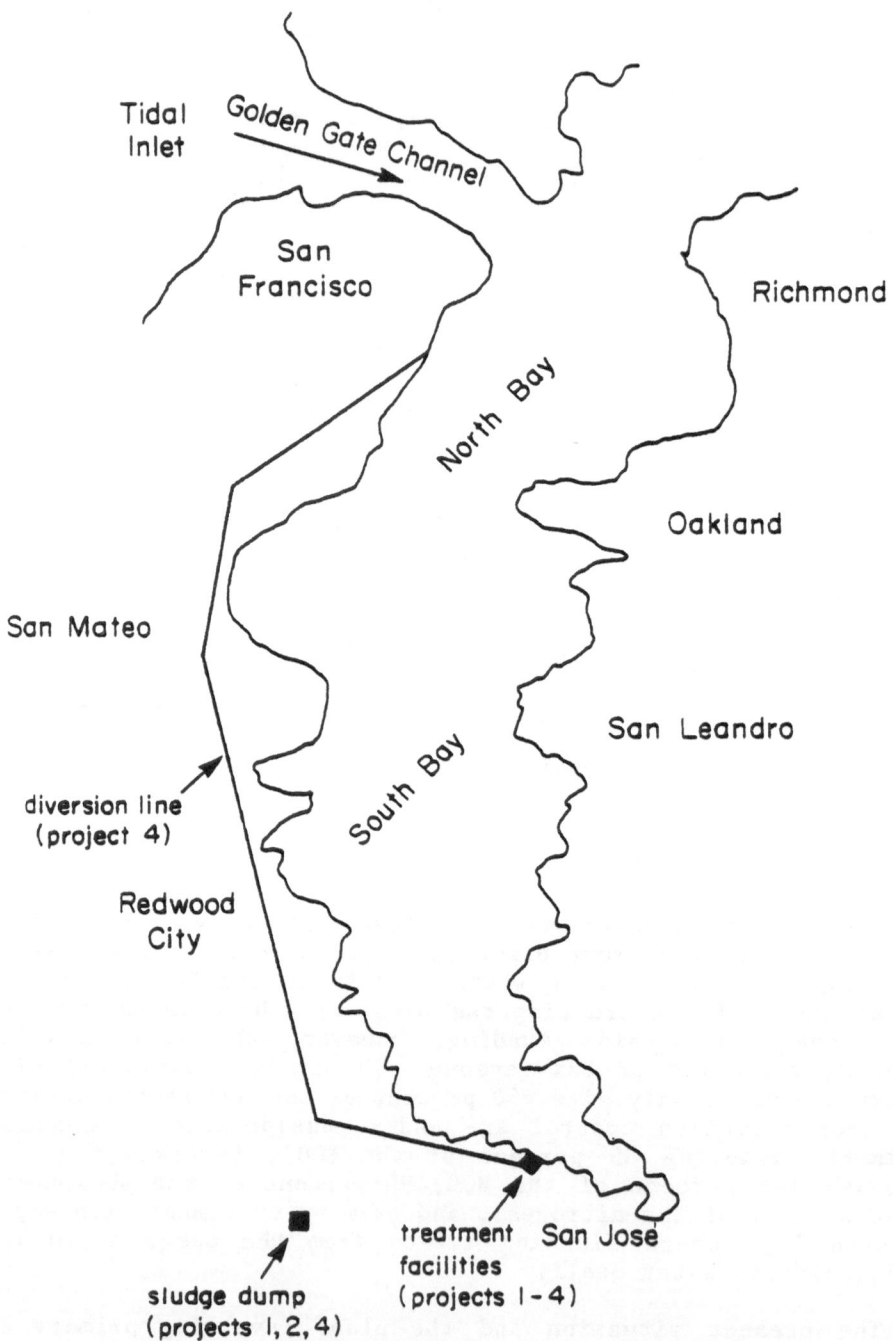

Figure 1.   Schematic Map of San Francisco Bay

Four criteria have been selected to rate project performance:

(1) <u>Recreational potential</u>--rated on a scale from 0 - 10. The scale was composed by considering improvement in the fish population resulting from increased dissolved oxygen concentration and also from decreased fecal coliform count, which would make the water safe for contact.

(2) <u>NH$_3$-N Concentration</u>--in mg/l, used as an indicator of trophic state.

(3) <u>Land use</u>--in hectares disturbed or used. Impact was estimated on the basis of the sum of a 20 hectare sludge dump, a 25 yard by 40 mile long right of way for the diversion line in project 4, and the area occupied by the plants themselves.

(4) <u>Treatment costs</u>--in dollars/day, estimated from incremental costs in Whipple (1977).

The decision-maker, a water resources engineer familiar with the case area, was shown scales which were marked with the ranges for criteria (1) and criteria (2). The consequences in terms of those criteria (e.g., a 0 recreational potential rating corresponds to no fishing, no swimming, etc.) were also given. The ratings on the last two criteria were explained directly.

The decision-maker was then asked to assign a membership grade according to the following scale:

$$\mu(d_{ij}) = \begin{cases} 0 - .25 & \text{if the rating is undesirable to slightly desirable} \\ .25 - .50 & \text{if the rating is slightly to moderately desirable} \\ .50 - .75 & \text{if the outcome is moderately to highly desirable} \\ .75 - 1 & \text{if the outcome is highly desirable to essential} \end{cases}$$

The resulting preference matrix is shown in Table 2. This preference matrix was sliced using intervals of .1 whenever a change in the simplicial complex could be expected. The

Table 2. Preference Matrix

	C1	C2	C3	C4
P1	0	0	.8	.9
P2	.2	.6	.75	.6
P3	.75	.8	.9	.3
P4	.6	.85	.2	.85

Q-analysis results are shown in Table 3.   Notice that at high α levels ($\sim.9$), very few of the criteria are satisfied in the Q-analysis structure vectors, while at low α levels ($\sim.1$), all of the criteria are satisfied, as confirmed by Q-analysis on the conjugate complex, Table 4.

At $\alpha = .7$ (moderately to highly desirable), Q-analysis provides a good discrimination between the projects.   Project P3 occupies the high position (q=2), having desirable outcomes on criteria 1, 2 and 3.   P1 and P4 enter at the q=1 level, desirable on C3 and C4, for P1, and C2 and C4, for P4.   At the 0 level, all projects are members of the same q-chain.   When presented with the results, the decision-maker indicated his satisfaction with project 3.

From the results of Q-analysis on the conjugate complex, displayed in Table 4, the following observations on the criteria are apparent: only for $\alpha = .7$ are the criterion simplices first q-connected at the zero-dimensional level.   As discussed previously, this indicates that the criteria provide very good differentiation between the projects and that every criterion is being used to make the decision.   In addition, C1 was most important in making the decision, since the decision maker found P3 satisfactory for C1 only, while C3 was of least importance, since P1, P2, and P3 were all satisficing C3.

Table 3.   Results of Q-analysis for Case Example

At $\alpha = .9$	At $\alpha = .8$
$q = 0$, $Q_0 = 2$ (P1), P3)	$q = 1$, $Q_1 = 2$ (P1), (P3)
$q = -1$, $Q_{-1} = 1$ all	$q = 0$, $Q_0 = 1$ (P1, P3, P4)
	$q = -1$, $Q_{-1} = 1$ all

At $\alpha = .7$	At $\alpha = .6$
$q = 2$, $Q_2 = 1$ (P3)	$q=2$, $Q_2 = 3$ (P4),(P2),(P3)
$q = 1$, $Q_1 = 3$ (P3),(P1),(P4)	$q=1$, $Q_1 = 1$ all
$q = 0,-1$ $Q_0 = 1$ all	$q=0,-1$ Q_0 and $Q_{-1} = 1$ all

no change between .6 and .3

At $\alpha = .3$	At $\alpha = .2$
$q = 3$, $Q_3 = 1$, (P3)	$q = 3$, $Q_3 = 1$ (P2,P3,P4)
$q = 2$, $Q_2 = 1$, (P2,P3,P4)	$q = 2$, $Q_2 = 1$ (P2,P3,P4)
$q = 1$, $Q_1 = 1$, (P1,P2,P3,P4)	$q = 1$, $Q_1 = 1$ (P1,P2,P3,P4)
$q = 0, -1$, Q_0 and $Q_{-1} = 1$ all	$q = 0, -1 Q_0$ and $Q_{-1} = 1$ all

Table 4.   Q-analysis on Conjugate Complex for Case Example

At $\alpha = .9$	At $= .8$
$q = 0$, $Q_0 = 2$ $(C_3),(C_4)$	$q = 1$, $Q_1 = 3$ $(C_2),(C_3),(C_4)$
$q = -1$, $Q_{-1} = 1$ (all)	$q = 0$, $Q_0 = 1$ (C_2,C_3,C_4)
	$q = 1$, $Q_{-1} = 1$ (all)

At $\alpha = .7$	At $\alpha = .6$
$q = 2$, $Q_2 = 1$ (C_3)	$q = 2$, $Q_2 = 2$ (C_2), (C_3)
$q = 1$, $Q_1 = 3$ $(C_2),(C_3),(C_4)$	$q = 1$, $Q_1 = 2$ $(C_1,C_2,C_3),(C_1)$
$q = 0,-1$ Q_0 and $Q_{-1} = 1$ (all)	$q = 0,-1$ Q_0 and $Q_{-1} = 1$(all)

no change between .6 and .3

At $\alpha = .3$	At $\alpha = .2$
$q = 2$, $Q_2 = 3$ $(C_2),C_3),C_4)$	$q = 3$, $Q_3 = 1$ (C_4)
$q = 1$, $Q_1 = 1$ (all)	$q = 2$, $Q_2 = 1$ (all)
$q = 0,-1$ Q_0, $Q_{-1} = 1$ (all)	$q = 1,0,-1$ Q_1, $Q_0 = 1$ and Q_{-1}(all)

3.4 Discussion

The results of the case example point out number of areas which refinement and testing are necessary. They include the points which follow.

(1) Before applying the procedure to any large scale design or operation problem, testing on some problem with which people are more familiar, such as buying a car, might be advisable. People in general would relate more easily to an everyday problem, and therefore the use of fuzzy sets or grades of membership to quantify people's feelings about the decision outcome could be tested. There is empirical evidence that grades of membership are useful in this regard (Kochen and Badre, 1974).

(2) Extension to group decision-making is possible in a number of ways, described as follows:

a. The group could be presented with the project impact matrix and asked to work out a joint preference matrix. Analysis would then proceed as in the single decision-maker case. This is the procedure used and recommended in Krzysztofowicz (1978) to assess group utility.

b. Each member of the group could be assessed individually and after construction of the 0-1 incidence matrices, the resulting group matrix could be constructed using any of a number of decision rules. For example:

(i) Single person veto - The group matrix would consist of the intersection of the individual matrices. Thus, an entry in the group matrix would be 1 only if all individual entries were 1.

(ii) Majority rule - The group matrix would be the union of all the individual matrices, cut at the level of the majority. All entries in the union matrix greater than k, where k = number of prople in the majority, would be 1 and all those less than k would be zero.

(3) While uncertainities have not been considered explicitly in the technique as outlined, the effects of informational uncertainty will show up in the Q-analysis as a failure to resolve the projects into a single, high-dimensional project at a high desirability level. How to quantify this and other uncertainties requires further research.

Considering the large spectrum of multiobjective and MCDM techniques which already exist (Cohon, 1978; Goicoechea et al., 1981), the question may arise as to whether or not another technique is really useful. However, MCQA seems to have two advantages over previous techniques. First, several techniques use weights associated with the criteria to generate desirability and undesirability indices for the projects or use simple dominance criteria (e.g., project j is preferred to project i if outcome j is greater than outcome i for criterion k). The grades of membership used in MCQA allow the decision-maker considerable flexibility in expressing his preferences, even if certain intransitivites should exist, and the final decision is made on the basis of how many criteria the decision-maker finds satisfactory. As noticed earlier, one may wish to incorporate the notion of discordance in MCQA so as to account for "discomfort" created by the criteria that do not agree with the "majority." In any case, no assumptions have been made about the decision-maker's preference structure except that it exists; the decision-maker is allowed to judge the project outcomes directly.

Primary computational efforts in other multicriterion decision-making techniques are focused on generation of indices for the alternatives from weight vectors and pairwise comparison of these ratings, while the computational effort in MCQA is directed toward generating the 0-1 incidence matrices from the project impact matrix and finding structure vectors. Essentially, MCQA accomplishes a global comparison between all the projects simultaneously, instead of requiring pairwise comparisons. The resulting savings in computational effort allows examination of the problem at a number of decision-maker satisfaction levels and provides an indicator of which criteria were most important even for complex water resource operation problems. Thus, unimportant criteria can be eliminated in further iterations.

3.5 Summary

An example applying multicriterion Q-analysis to the planning of a sewage disposal system has been presented. Transferring the approach to an operation problem appears to be straightforward. As noticed above and discussed further in Kempf et al., (1978), MCQA seems to have two primary advantages over other multicriterion techniques. First, in terms of the relation between the decision-maker and the problem, the decision-maker is asked to judge the project outcomes directly, rather than assign weights to criteria. The result allows the decision-maker more flexibility in expressing his preferences. Second, Q-analysis establishes an ordering over a set in an efficient manner: global comparison of all the projects allows the problem to be examined at a number of decision-maker satisfaction levels and provides an indication of how useful particular criteria are in the decision.

4. SUBSYSTEM STABILITY USING BIFURCATION-CATASTROPHE THEORY

The second theoretical system tool presented in this chapter is catastrophe theory. This tool has been used to model a growing number of processes in the physical, biological, and social sciences, in which large extremes of some "state" variable occur. The theory was originally developed by Thom (1975) as a means of classifying the singularities of smooth maps. The classification theorem itself is of little help in formulating models, as Sussman and Zahler (1978) point out; of greater usefulness are the implications of the theorem and the linkage between catastrophe theory and the classical theory of dynamic systems (Zeeman, 1973; Poston and Stewart, 1977, 1978). The basic idea of the theory is to shift back and forth between a physically meaningful coordinate system and a transformed one that can be handled mathematically. Such a shift is possible because one deals with smooth diffeomorphisms, that is, types of transformations which are one-to-one and onto.

The first example is presented in section 4.1; it deals with a physical model as a simplified representation of a gravity dam. The "catastrophe" corresponds to collapse of the dam. The second example, in Section 4.2, demonstrates how the stability of water quality in a small eutrophic pond can be modeled by means of catastrophe theory.

4.1 Collapse of a Gravity Dam

The Euler arch shown below (Poston and Stewart, 1978) may be considered as a schematic representation of a gravity dam (Casti, 1981). The weight of the dam is α, and the anchors hold the slope at an angle θ by means of resistance forces β, and stress in the dam originating from both internal and external forces is represented by a compressed spring of constant k. The critical combinations of parameters α, β, μ which lead to instability--hence possible collapse--of the dam, are sought.

The potential function is composed of the sum of three terms representing, respectively, the energy stored in the spring, the energy due to the load α and the work done by the compression forces β:

$$V(\theta;\alpha,\beta,\mu) = 2\mu\theta^2 + \alpha \sin\theta - 2\beta(1-\cos\theta) \qquad (8)$$

The first derivative is

$$V'(\theta) = 4\mu\theta + \alpha\cos\theta - 2\beta \sin\theta \qquad (9)$$

It vanishes for $\theta = 0$, $\alpha = 0$, which is a singular point. To investigate further, equate the second derivative to zero, to obtain:

$$V''(\theta) = 4\mu - \alpha\sin\theta - 2\beta\cos\theta = 0 \qquad (10)$$

If $\alpha = 0$, $V''(0)$ vanishes for $2\mu = \beta$. Whenever $V''(0)$ vanishes, the singularity is said to be non-Morse and no coordinate transformation can be found to isolate the singularity. Note that the point $\alpha = 0$, $2\mu = \beta$ would correspond to a dam with zero weight, and collapsing forces equal to resistance.

Consider the more realistic case when

$$\alpha \neq 0, \ \alpha \ll 1, \ \beta \sim 2\mu.$$

Then Equation (8) may be approximated by the Taylor series

$$V(\theta) \sim \alpha\theta - (\beta-2\mu)\theta^2 - \frac{1}{6} \alpha\theta^3 + \frac{1}{12} \beta\theta^4 + \ldots \qquad (11)$$

This is the potential function for the cusp catastrophe. To transform it into the canonical form, let $x = \theta - \frac{\alpha}{2\beta}$. Then,

algebraic maniuplations yield:

$$V(x) = (1/6) \mu x^4 - (\beta-2\mu) x^2 + \alpha x$$
$$\sim x^4 + ax^2 + bx \qquad (12)$$

with

$$a = -(6/\mu) \ (\beta-2\mu) \ \Big\}$$
$$b = (6\alpha/\mu) \qquad (13)$$

In terms of standard geometry, a cusp catastrophe surface is projected in the (a,b) plane as the cubic of equation

$$27a^3 + 4b^2 = 0 \qquad (14)$$

The "catastrophe" or dam collapse occurs near $x = 0$ when the parameter values are in the particular combination of Eq. (14).

Note that the original problem had three parameters, whereas only two are present in the cusp catastrophe model. It is thus seen how this formualtion may simplify the investigation of the stability of a subsystem.

4.2 Eutrophication Example

The purpose of this example, which summarizes the study in Duckstein et al., (1979) is to demonstrate how a differential equation model of phytoplankton dynamics, formulated on the basis of physical and biological considerations, can be reduced analy-tically to the canonical equation of the cusp catastrophe. The model is shown in turn to be usable for fishpond management.

The type of situation considered is the one often encountered in small, highly eutrophic ponds or in embayments of larger water bodies with restricted flow and high nutrient levels. In these lakes and ponds, extreme algal blooms and dieoffs tend to occur over a period of several weeks, and the dominant algae tend to be of the Cyanophyceae type. Very few zooplankton which would tend to control algae population, are present. Because water bodies of the latter type are often used for raising fish (Barica, 1973; Boyd et al., 1975) or are part of a larger lake experiencing severe eutrophication problems (Thomas, 1960), some type of pre-dictive model would be highly desirable (Reckhow, 1978).

For a given pond, notation is defined as follows:

x = total algae concentration

b = <u>Anabaena</u> (a key algae species) concentration

a = soluble phosphate concentration

a_0 = equilibrium value of a (.01 in the example)

C_j = constants, $j = 1,\ldots,6$.

The model for phytoplankton dynamics derived in Duckstein et al., (1979), based on the process shown in Figure 2, consists of the following three equations:

$$dx/dt \quad -(C_1 x^3 - C_2 ax + C_3 b) \qquad (15)$$

$$da/dt = C_4 x(a - a_0) \qquad (16)$$

$$db/dt = C_5 ab - C_6 bx \qquad (17)$$

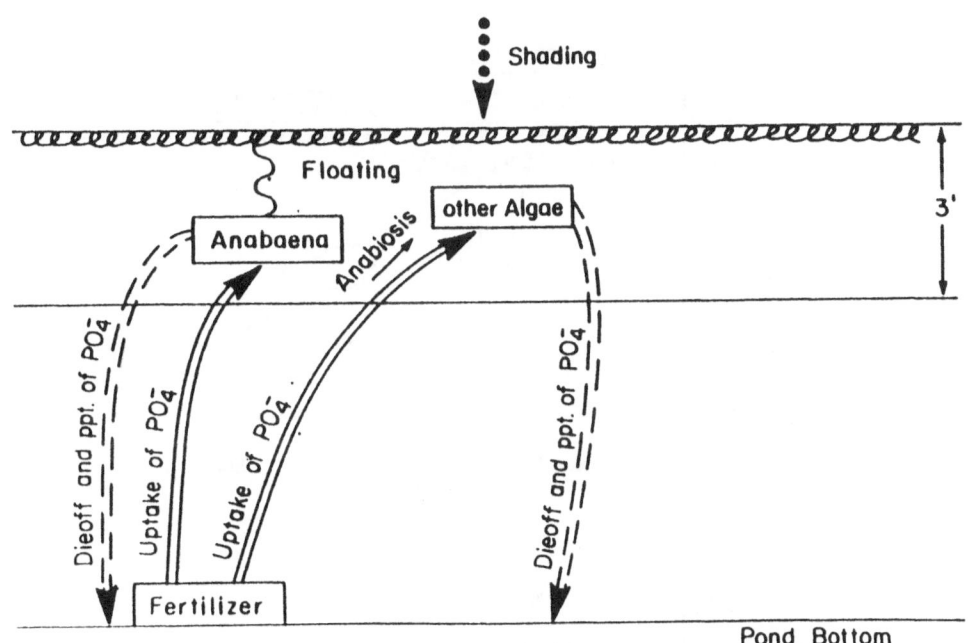

Figure 2. Process Diagram for Algae Cycles

Computer integration of this seven-parameter model has been calibrated against data taken during a phytoplankton bloom (Parks et al., 1975) to demonstrate the physical validity of the catastrophe theory model. The match between the data and the integrated model is fairly good for Anabaena and total algae concentrations (Figure 3a and 3b) except for a large dip in the date which does not occur in the simulation, during the fifth time unit. This dip corresponds with an increase in average wind velocity, according to the data Parks et al., (1975) and could be the result of increased vertical or horizontal mixing removing algal cells from the top layer of the pond at the measuring station. The simulation thus provides an average trend of the algal bloom-dieoff sequence which may be disturbed by stochastic processes such as weather conditions.

However, the plot for soluble orthophosphate concentration is less exact, as shown in Figure 3c. The simulated soluble orthophosphate concentration decreases more slowly from the initial value than the data. As a comparison, a graph of soluble phosphate vs. time from DiToro et al., (1977) for a laboratory experiment involving fertilization has been included. Note that the shape of the curve for the laboratory data is closer to the simulation than that of the pond. Several explanations for this discrepancy can be given (Duckstein et al., 1979) including the possibility that model represented by Eq. (15) may be erroneous.

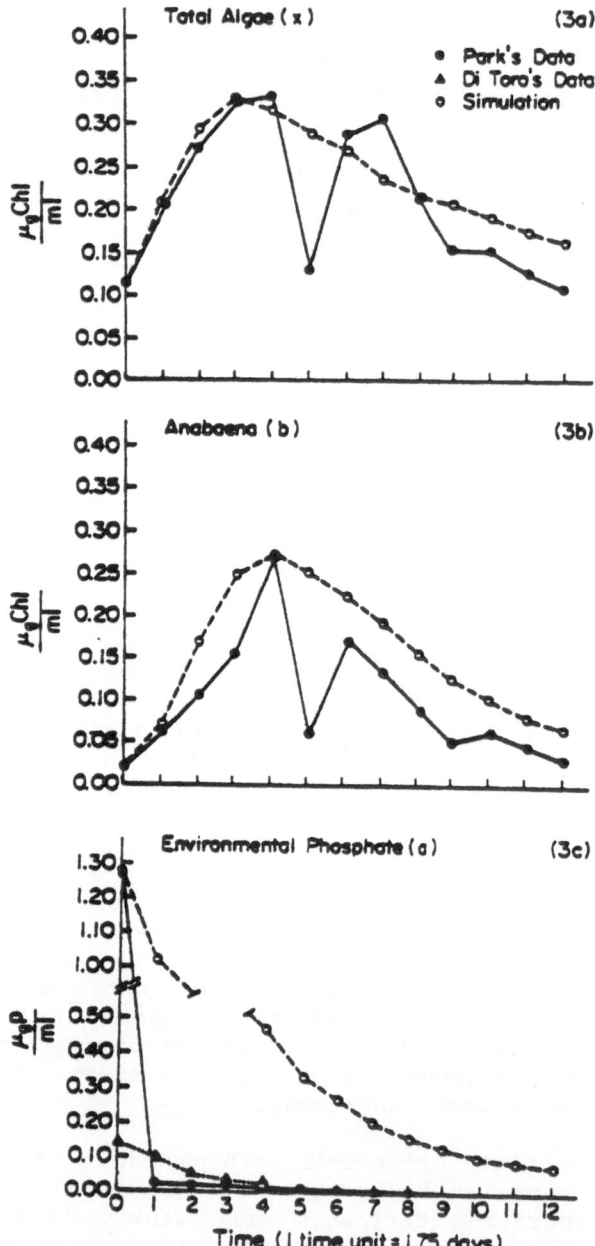

Figure 3.   Comparison of Data in Parks et al., (1975)   with
            Catastrophe Theory Based Simulation
            (after Duckstein et al., 1979).

The above theory is now used to provide insights into the process of algal production with implications for subsystem water quality management; in this case, a fishpond.

In the dam collapse example (section 4.2), a potential function based on energy considerations was explicitly determined. This cannot be done here; however, the differential equation for the total algal concentration, i.e., Eq. (15), can be thought of as the differential equation describing the "state" variable in the system. The state of the pond is being controlled by two parameters--the concentration of Anabaena and the concentration of orthophosphate. Thus, if a potential function exists, it will belong to a two-parameter family (Poston and Stewart, 1977) and the system whould tend toward those states which minimize the potential function.

From dynamic system theory (Poston and Stewart, 1978), Eq. (15) can be viewed as the gradient of the potential function

$$V(x,a,b) = (\frac{C_1 x^4}{4} - \frac{C2^{ax^2}}{2} + C_3 bx) + K$$

where the constant of integration K can be ignored because it only serves to locate the origin.

To transform Eq. (18) into canonical form, a linear co-ordinate transform with non-zero Jacobian matrix is used, to yield the two-parameter function:

$$V(z,P_1,P_2) = (\frac{z^4}{4} - \frac{P_1 z^2}{2} + P_2 z) \tag{19}$$

If Eq. (19) as well as its derivative are set equal to zero,

$$V(z,P_1,P_2) = 0$$

$$\frac{\partial V}{\partial x} = (z^3 - P_1 z + P_2) = 0 \tag{20}$$

then one obtains the canonical equations for the potential function and behavior surface, respectively, of the simple cusp catastrophe. Plotting of all solutions of Eq. (19) in x-a-b space for real a, b will result in the well-known "cusp" manifold (see Woodock and Poston, 1974, for a geometrical study of the cusp). The cusp manifold is the set of critical points for Eq. (19).

System constraints are non-negativity of Z, $(Z-P_2)$, P_1, P_2.

The geometrical behavior of system is depicted qualitatively in Figure 4 without attempting to calibrate the manifold.

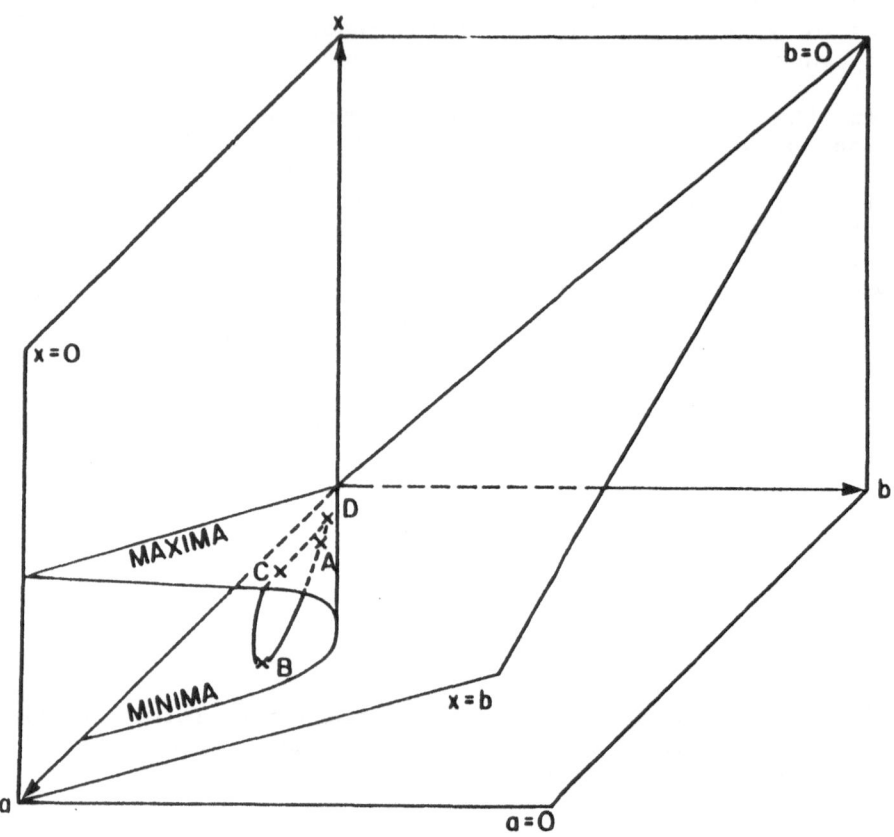

Figure 4.   Behavior of System on Cusp Manifold

Notice that the constraints have removed all but a pocket of
the manifold from consideration as possible critical points. The
system starts at point A on the lower sheet of the cusp, where
phosphate, blue-green algae and total algae concentrations are
low.   Fertilization moves the system from A to B, out into the
pocket and off the sheet of potential function minima.   The
dynamics of the system follow the trajectory traced out by the
curve from C, where blue-green and total algae have maxima on the
center sheet, back to D, a point close to A on the equilibrium
sheet.

To show how the catastrophe model can be used practically,
consider its implications for fishpond management.   The analysis

is very similar to that of a physical system which has been extensively optimized and in which catastrophe dynamics play a role (Poston and Stewart, 1978). Addition of phosphate to the pond while blue-green algae are present serves to throw the system off the equilibrium sheet of the cusp manifold in x-a-b behavior space, resulting in a bloom-dieoff sequence which can be accentuated by weather and other stochastic factors. Without extensive overfertilization, the fishpond represents an ecosystem which has been optimized to increase fish yield. Fertilization is an attempt to "overoptimize" the system and, due to the underlying dynamics, such an action results instead in a decrease of fish yield from diebacks.

5. CONCLUSIONS

Two mathematical tools stemming from modern system theory--Q-analysis and catastrophe theory--have been briefly introduced and then applied to the central theme of this chapter, namely, the identification of subsystems in the operation of complex water resources systems.

The first tool, Q-analysis, provides a simple approach to analyzing the structure or level of connectivity between system elements classified into two subsets, simplices and vertices. In a first example, the simplex set is a set of water supply sources, and the vertex set is the set of sinks, including connected reservoirs and users. In a second example, discretized operational schemes for water quality control (simplices) are evaluated from the viewpoint of multiple criteria (vertices) using fuzzy sets to create an incidence matrix.

The second tool, catastrophe theory, can be used to study singularity points of both engineering and biological subsytems. An example of the former is a gravity dam represented schematically as a Euler Arch; the dam collapse phenomenon can be modelled as a cusp catastrophe. An example of stability of a biological subsystem is phytoplankton dieoff in a eutrophic pond, which can also be modelled as a cusp catastrophe having two parameters instead of the original seven-parameter model. Note that other types of catastrophe surfaces, such as the butterfly or the swallowtail, may provide valid representations of water ecosystems (Kempf and Duckstein, 1980).

The purpose of the chapter was to demonstrate the usefulness of modern theoretical systems analysis tools applied to problems of complex water systems. The mathematical framework provides a rationale for more systematic decision making. The engineering use of Q-analysis includes both descriptive and management aspects. The former aspect makes it possible to investigate the structure, linkages, or connectivity of a water supply network, using a set of easily obtainable indices. The management aspect is illustrated by the search for a satisfactory water quality control plan in the presence of several alternatives evaluated on the basis of multiple criteria.

The bifurcation-castastrophe analysis may be used to predict when sudden changes in the trophic state of a lake are likely to occur. Also, the method is useful to predict and hopefully prevent the collapse of complex structures, such as dams.

ACKNOWLEDGEMENTS

The substantial contributions of John Casti, Jim Kempf and Richard Shea to the material presented herein are gratefully acknowledged.

REFERENCES

Armijo, R., J. Casti and L. Duckstein (1979). "Multicriterion water resources system design by Q-analysis," presented, Joint National TIMS/ORSA Meeting, April 30-May 2, New Orleans.

Atkin, R. (1974). Mathematical Structure in Human Affairs, Heinemann Publishing Co., London.

Atkin, R. and J. Casti (1977). "Polyhedral dynamics and the geometry of systems," International Institute for Applied Systems Analysis, Research Report RR-77-6.

Barica, J. (1973). "Changes in water chemistry accompanying summer fish kills in shallow, eutrophic lakes of Southwest Manitoba," Proceedings, Symp. on Lakes of Western Canada, University of Alberta Water Resources Centre, Publication No. 2, pp. 227-240.

Bellman, R. and L. Zadeh (1970). "Decision making in a fuzzy environment," Management Science, Vol. 17, No. 4, pp. 141 -164.

Bezdek, J., B. Spillman and R. Spillman (1978). "A fuzzy relation space for group decision theory," Fuzzy Sets and Systems, Vol. 1, pp. 255-268.

Boyd, C., E. Prather and R. Parks (1975). "Sudden mortality of a massive phytoplankton bloom," Weed Science, 23, pp. 61-66.

Casti, J. (1979). Connectivity, Complexity and Catastrophe in Large-Scale Systems, International Series on Applied Systems Analysis, Vol. 7, John Wiley & Sons, New York, 203 p.

Casti, J. (1981). "Topological methods for social and behavioral systems," presented, NSF-DOT Conf. on Nonlinear Dynamics General Urban/Rgeional Systems Theory, Washington, D.C., May 27, 28, 1981. Available as paper #81-13, Department of Systems & Industrial Eng., Univ. of Arizona, Tucson.

Chen, C. and G. Orlob (1971). "Ecological simulation for aquatic environments," Water Resources Engineers, Inc., Walnut Creek, California.

Cohon, J. (1978). Mulitobjective Programming and Planning, Academic Press, New York, 333 p.

de Neufville, R. and J. Stafford (1971). Systems Analysis for Engineers and Managers," McGraw-Hill Book Company, 353 p.

DiToro, D., R. Thomann, D. O'Connor and J. Mancini (1977). "Estuarine phytoplankton biomass models--verification and preliminary applications," The Sea, Vol. 6, John Wiley and Sons, New York, pp. 969-1020.

Duckstein, L. (1978). "Imbedding uncertainties into multi-objective decision models in water resources," Keynote paper, Session T1, Int'l. Symp. on Risk & Reliability in Water Resources, June 26-28, Waterloo, Ontario.

Duckstein, L., J. Casti and J. Kempf (1979). "Modelling phytoplankton dynamics using catastrophe theory," Water Resources Research, Vol. 15, No. 5, pp. 1189-1194.

Gershon, M. (1981). "Multiobjective model choice in natural resource systems," Unpublished Ph.D. Dissertation, Department of Systems & Industrial Engineering, University of Arizona, Tucson, Arizona 85721.

Gershon, M., R. McAniff and L. Duckstein (1980). "Multiobjective river basin planning with qualitative criteria," Working paper #80-18, Department of Systems & Industrial Engineering, University of Arizona, Tucson, Arizona 85721.

Goicoechea, A., D. Hansen and L. Duckstein (1981). Introduction to Multiobjective Analysis with Engineering and Business Applications, to appear, John Wiley, New York.

Kaufman, A. (1975). Introduction to the Theory of Fuzzy Subsets, Vol. I, Academic Press, New York.

Kempf, J., J. Casti and L. Duckstein (1978). "Monitoring lake ecosystem response to pollution inputs using Q-analysis," Paper, Joint TIMS/ORSA Meeting, May 1-3, New York.

Kempf, J. and L. Duckstein (1980). "Reduction of a phytoplankton model to swallowtail catastrophe form," Working paper #80-12, Department of Systems & Industrial Engineering, University of Arizona, Tucson, Arizona 85721.

Kochen, M. and A. Badre (1974). "On the precision of adjectives which denote fuzzy sets," J. of Cybernetics, Vol. 4, No. 1, pp. 49-59.

Krzysztofowicz, R. (1978). "Preference criterion and group utility model for reservoir control under uncertainty," Natural Resource Systems Technical Report Series #30,

Department of Hydrology & Water Resources, University of Arizona, Tucson, Arizona 85721.

Nijkamp, P. (1978). "Conflict patterns and compromise solutions in fuzzy choice theory: An analysis and application," Research Memorandum No. 1978-7, Department of Economics, Free University of Amsterdam, Amsterdam, The Netherlands.

Nijkamp, P. and J. Vos (1977). "A multicriteria analysis for water resource and land use development," Water Resources Research, Vol. 13, No. 3, pp. 513-518.

Parks, R., E. Scarsbrook and C. Boyd (1975). "Phytoplankton and water quality in a fertilized fish pond," Circular 224, Agricultural Exper. Station, Auburn University, Auburn, Alabama, 16 p.

Poston, T. and I. Stewart (1977). Taylor Expansions and Catastrophes, Research Notes in Mathematics, Vol. 7, Pitman, London.

Poston, T. and I. Stewart (1978). Catastrophe Theory and its Applications, Pitman, London, 408 p.

Reckhow, K. (1978). "Quantitative techniques for the assessment of lake quality," Prepared for Land Resource Programs Div., Inland Lake Management Unit and the Water Quality Div., Michigan Department of Natural Resources, by K. H. Reckhow, Department of Resource Development, Michigan State University, E. Lansing, Michigan.

Roy, B. (1977a). "A conceptual framework for a prescriptive theory of 'Decision Aid'," in Multiple Criteria Decision Making, M. K. Starr and M. Zeleny (eds), North-Holland Publishing Co., Amsterday-New York, pp. 179-210.

Roy, B. (1977b). "ELECTRE III: Un algorithme de classements fonde sur une representation floue des preferences en presence de criteres multiples," Paper, Lamsade, Universite Paris-IX, France.

Roy, B. (1977c). "Partial preference analysis and decision-aid: The fuzzy outranking relation concept," in Conflicting Objectives in Decisions, D. Bell, R. Keeney and H. Raiffa (eds.), Chap. 2, pp. 40-75, John Wiley & Sons, New York.

Shea, R. (1981). "A handbook of polyhedral dynamics," Master's report, Department of Systems & Industrial Engineering, University of Arizona, Tucson, Arizona 85721.

Simon, H. (1957). Models of Man: Social and Rational, John Wiley, New York.

Sussman, H. and R. Zahler (1978). "Catastrophe theory as applied to social and biological sciences: A critique," Synthese, 37(2), pp. 117-216.

Thom, R. (1975). Structural Stability and Morphogenesis, W. A. Benjamin, Inc., Reading Mass., 296 p.

Thomas, E. (1960). "Sauerstoffminima und Stoffkreislaufe in ufernahen Oberflachen Wasser des Zurchersees (Cladophora und Phragmites-Gurtel)." Montasbull. Schweiz Verein Gas -Wasserfachmannern, 6, pp. 1-8.

Whipple, W. (1977). Planning Water Quality Systems, D. C. Heath and Co., Toronto.

Zadeh, L. (1973). "Outline of a new approach to the analysis of complex systems and decision processes," IEEE Transactions on Systems, Man and Cybernetics, SMC-3, 28.

Zeeman, E. (1973). "Differential equations for the heartbeat and nerve impulses," in Dynamical Systems, M. M. Peixoto (ed)., Academic Press, New York, pp. 683-741.

Zeeman, E. (1976). "Catastrophe theory," Scientific American, 243, pp. 65-83.

Zeleny, M. (1981). Multiple Criteria Decision Making, McGraww-Hill Book Company, New York, 512 p.

6. OPTIMAL OPERATION OF RESERVOIRS BY DYNAMIC PROGRAMMING

Ricardo Harboe
Ruhr-University, Bochum, Federal Republic of Germany

1. INTRODUCTION

In most optimizing models for the operation of reservoirs it is assumed that the objective function is a sum of returns in different time periods. Tihs is true for most economic objective functions. In several cases, however, the objectve is formulated in other terms, but are and can be solved by deterministic dynamic programming. Consider the following examples:

(a) A decision-maker wants to implement a more conservative criteria than minimizing the sum of losses from a given flood control project. He could try to minimize the maximum loss that could occur over a given time horizon. In this case we would have a min-max type of objective function.

(b) The production and sale of on-peak firm energy from a hydroelectric power plant operated in conjunction with a reservoir brings high returns because firm energy means we can save not only operation costs but also capital costs elsewhere. But, firm energy is determined by a critical period in the operation of the system, which is the basis for the "firm energy contract." Our objective function should try to maximize this firm energy contract; i.e., maximize the minimum amount of peak energy over the time periods. In this case, we have a max-min type of objective function.

(c) A decision-maker is in charge of a reservoir for low-flow augmentation. If during the operation of the reservoir, the flow is above a certain rate, nobody will care; but, if in one year (maybe one out of 50 years) the flow is too low, everybody (including the newspapers) will talk about a failure in the operation of the reservoir and possibly criticize the decision-maker. Therefore, it is reasonable for him to find the optimal operation rule for the reservoir using an objective function that maximizes the lowest flow; i.e., maximize the minimum flow over the planning horizon. Again, we have a max-min type of objective function.

(d)   The optimal operation of an irrigation project presents yet another type of objective function, which has a product form. If we consider the distribution of water deficits during an irrigation season, the total crop yield is in some cases the product of some functions of the monthly deficits. The objective is thus to maximize the product of monthly functions.

(e)   There are several problems which can be formulated in probabilistic terms and therefore lead to a product form of the objective function. When the global probability of success in a given operation process is to be maximized, the product of individual success probabilities in each month is equivalent to the global probability. In this case, we assume the events in each month are independent so that the joint probability is the product of individual probabilities. So, our objective function is to maximize the product of some functions of probability of success. The same argument would apply if we wish to minimize the probability of failure.

2. FORMULATION OF THE PROBLEMS

We have seen the types of problems that gives rise to max-min and product forms of the objective functions. Let us now formulate them as mathematical optimization problems, but in general terms.

We define the following variables in a dynamic programming model:

x_i = state variable at the beginning of stage i (input): brings all necessary information from previous stages.

u_i = decision variable in stage i (input): can be controlled by decision-maker.

$f_i(x_i,u_i)$ = return for stage i as a function of state
and

decision variables (output): in economic
(benefits and costs) or technical terms.

x_{i+1} = state variable at the end of stage i (output).

These inputs and outputs in a given stage are represented as in Figure 1. The state transformation is given as:

$$x_{i+1} = g_i(x_i, u_i) \tag{1}$$

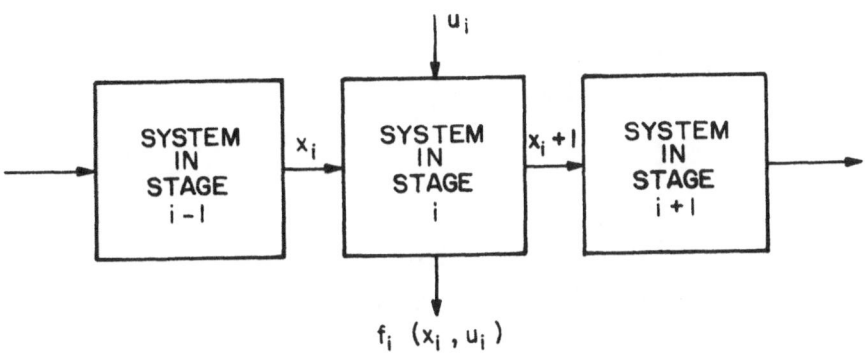

Figure 1. One Stage System Variables

The two forms of objective functions for such a system i.e., "max-min" and "product" are formulated in mathematical terms in the following sections.

2.1 Objective Function for Max-Min Problem

$$\max_{u_i} z = \min[f_1(x_1,u_1),f_2(x_2,u_2),\ldots,f_N(x_N,u_N)] \tag{2}$$

$$\max_{u_i} z = \min_i[f_i(x_i,u_i)] \tag{3}$$

As an example of a max-min type of problem, consider the low-flow augmentation problem (4). If we have a single reservoir, as in Figure 2, with no lateral inflow and no delay between the

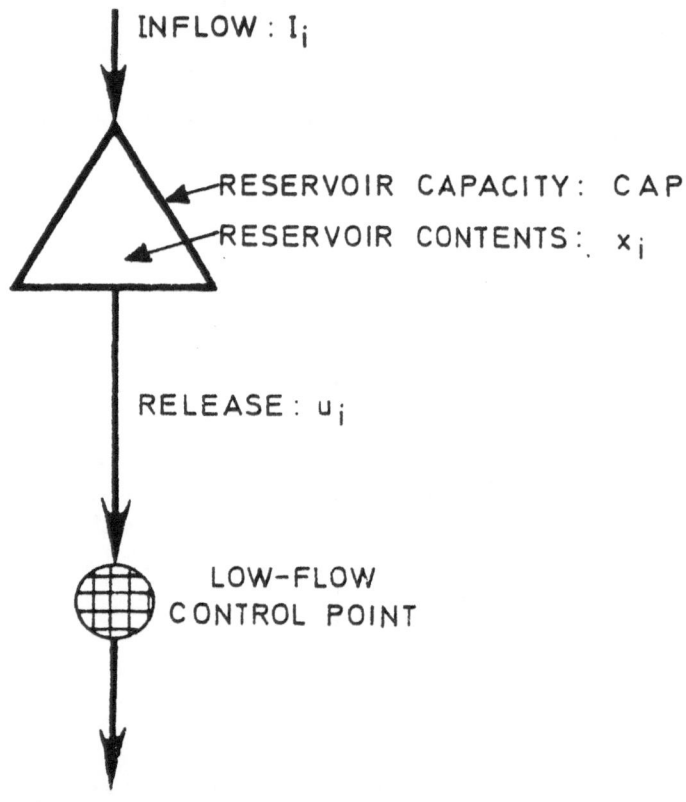

Figure 2. Reservoir for Low-Flow Augmentation

reservoir and the low flow control gage, the return function for a given month i would be:

$$f_i(x_i, u_i) = u_i \qquad (4)$$

The return in each month is equal to the release from the reservoir. The objective function would be to maximize the minimum flow at the control gage, i.e.:

$$\max_{u_i} z = \min[u_1, u_2, \ldots, u_N] \tag{5}$$

$$= \min_i [u_i]. \tag{6}$$

The minimum release over the time horizon (N months) has to be maximized.

The state transformation function would be:

$$x_{i+1} = x_i + I_i - u_i, \tag{7}$$

A typical constraint is:

$$x_i \leq CAP, \text{ for all } i \tag{8}$$

and $u_i \leq x_i + I_i$, for all i. $\tag{9}$

2.2 Objective Function for Problem of Product

$$\max_{u_i} z = f_1(x_1, u_1) \cdot f_2(x_2, u_2) \cdots f_N(x_N, u_N) \tag{10}$$

$$= \prod_{i=1}^{N} f_i(x_i, u_i) \tag{11}$$

As an example, consider the irrigation timing problem (2), where a given deficit has to be distributed over several months of the irrigation season. Define:

x_i = state variable representing the amount of water left over for stages i, i+1,...,N.

u_i = amount of water applied to irrigation area during stage i.

d_i = water demand for irrigation during stage i (if full crop is to be attained).

$f_i(x_i, u_i) = a_i(d_i - u_i)$ = proportion of final crop that can be obtained if water deficit in stage i is $(d_i - u_i)$.

The objective function that maximized crop production would be:

$$\max_{u_i} \quad \overset{N}{\underset{i=1}{\pi}} \quad a_i(d_i - u_i). \tag{12}$$

Subject to $\quad x_{i+1} = x_i - u_i; \quad u_i \le x_i;$ and x_1 given. $\tag{13}$

3. DERIVATION OF THE RECURSIVE EQUATIONS

Let us consider the last stage, N, of the multistage sequential decision process, as shown in Figure 3, with objective function:

$$\max \quad \overset{N}{\underset{i=1}{\Sigma}} \quad f_i(x_i, u_i), \tag{14}$$

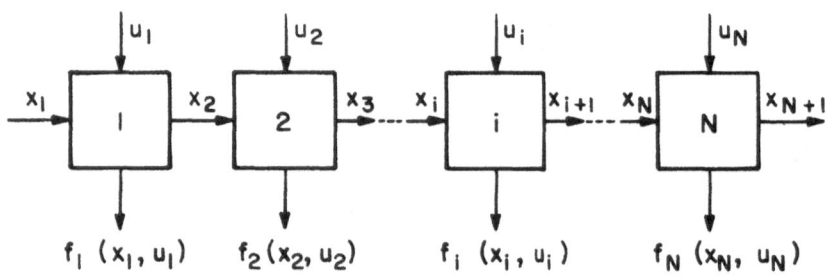

Figure 3.   Sequential Multistage Decision Process

We know that the optimal return (F_N) from a single stage is a function of the state variable (x_N); i.e.:

$$F_N(x_N) = \max_{u_N} \{f_N(x_N, u_N)\} \tag{15}$$

This equation is still valid for both new types of objective functions that were introduced i.e., max-min $(\max\{\min[f_i]\})$, or maximum of product form $(\max \underset{i}{\pi} f_i)$, because we still want to maximize f. Recall that while solving this maximization, we obtained $u_N^*(x_N)$, the optimal policy.

However, the most important step in the derivation of the recursive equations of dynamic programming, when considering the optimizaton of the last two stages (N-1 and N in Figure 3), was to transform (decompose) the objective funtion. Thus, starting with Eq. (16),

$$F_{N-1}(x_{N-1}) = \max_{u_{N-1}, u_N} \{f_{N-1}(x_{N-1}, u_{N-1}) + f_N(x_N, u_N)\} \quad (16)$$

where: $F_{N-1}(x_{N-1})$ = optimal return that can be obtained (with optimal decisions) from stages N-1 and N as a function of x_{N-1}. We transform into the equivalent problem:

$$F_{N-1}(x_{N-1}) = \max_{u_{N-1}} \{f_{N-1}(x_{N-1}, u_{N-1}) + \max_{u_N} f_N(x_N, u_N)\}. \quad (17)$$

And, by definition of $F_N(x_N)$, replacing we obtain:

$$F_{N-1}(x_{N-1}) = \max_{u_{N-1}} \{f_{N-1}(x_{N-1}, u_{N-1}) + F_N(x_N)\} \quad (18)$$

subject to: $x_N = g_{N-1}(x_{N-1}, u_{N-1})$ (19)

This step, namely taking the maximization over u_N into the brackets, was possible only because the function:

$$f_{N-1}(x_{N-1}, u_{N-1} + f_N(x_N, u_N)$$

was an increasing function of $f_N(x_N, u_N)$ for any given value of the function $f_{N-1}(x_{N-1}, u_{N-1})$; which is a "sufficient condition" (1). As a matter of fact, this is true for the addition of any two functions. Therefore, the addition of returns leads always to the same form of recursive equations (the maximization over u_N does not affect f_{N-1}).

In the case of max-min type objective function ($\max\{\min_i f_i\}$) we would have for the last two stages of the optimization:

$$F_{N-1}(x_{N-1}) = \max_{u_{N-1}, u_N} \{\min[f_{N-1}(x_{N-1}, u_{N-1}), f_N(x_N, u_N)]\}. \quad (20)$$

Since the function:

$$\min[f_{N-1}(x_{N-1}, u_{N-1}), f_N(x_N, u_N)]$$

is always an increasing function of $f_N(x_N, u_N)$ for any given value of the function $f_{N-1}(x_{N-1}, u_{N-1})$, decomposition is also possible and we obtain similarly:

$$F_{N-1}(x_{N-1}) = \max_{u_{N-1}} \{\min[f_{N-1}(x_{N-1}, u_{N-1}), \max_{u_N} f_N(x_N, u_N)]\} \quad (21)$$

because again, the max over u_N does not affect f_{N-1}.

Since by definition,

$$\max_{u_N} f_N(x_N, u_N) = F_N(x_N), \quad (22)$$

the following recursive equation is obtained:

$$F_{N-1}(x_{N-1}) = \max_{u_{N-1}} \{\min[f_{N-1}(x_{N-1}, u_{N-1}), F_N(x_N)]\} \quad (23)$$

subject to the state transformation equation:

$$x_N = g_{N-1}(x_{N-1}, u_{N-1}). \quad (24)$$

Similar recursive equations can be derived for any stage i ($1 \le i \le N$) as in the case of sum of returns, replacing i for N-1 and i+1 for N. The second condition, namely separability, is also satisfied; i.e., the objective function to be maximized can be written as:

$$h(f_1, f_2, \ldots, f_N) = h_1(f_1, h_2(f_2, f_3, \ldots, f_N)) \quad (25)$$

and

$$h_2(f_2, f_3, \ldots f_N) = h_3(f_2, h_4(f_3, f_4, \ldots, f_N)) \quad (26)$$

etc.

in the following form (h is now the min operator):

$$\min[f_1, f_2, \ldots f_N] = \min[f_1, \min[f_2, f_3, \ldots, f_N]] \quad (27)$$

and

$$\min[f_2, f_3, \ldots, f_N] = \min[f_2, \min[f_3, f_4, \ldots, f_N]] \quad (28)$$

etc.

In each step of the solution it is important to store the values of the optimal decision function $u_i^*(x_i)$.

For the product type of objective function, when considering the last two stages of the optimization, we have:

$$F_{N-1}(x_{N-1}) = \max_{u_{N-1}, u_N} \{[f_{N-1}, (x_{n-1}, u_{N-1}) \cdot f_N(x_N, u_N)\} \tag{29}$$

Since the function:

$$f_{N-1}(x_{N-1}, u_{N-1}) \cdot f_N(x_N, u_N)$$

is an increasing function of $f_N(x_N, u_N)$ only for positive values $f_{N-1}(x_{N-1}, u_{N-1})$, decomposition in the traditional form is only possible in this case:

$$f_{N-1}(x_{N-1}, u_{N-1}) \geq 0. \tag{30}$$

The recursive equation for stage $N-1$ would be:

$$F_{N-1}(x_{N-1}) = \max_{u_{N-1}} \{f_{N-1}(x_{N-1}) \cdot F_N(x_N)\}, \tag{31}$$

subject to:

$$x_N = g_{N-1}(x_{N-1}, u_{N-1}). \tag{32}$$

Similar equations can be derived for any stage i (replacing i for N-1) since the separability conditions are also satisfied. Special algorithms can be devised to solve the problems with product type objective functions when the return functions are unrestricted in sign. These require the parallel solution of two recursive equations in each stage and will not be discussed here.

The solution procedure for these new forms of recursive equations is the same as for standard dynamic programming algorithm, using discretized values of state and decision variables, in which the sum of:

$$f_i(x_i, u_i) + F_{i+1}(x_{i+1})$$

is replaced by:

$$\min[f_i(x_i, u_i), F_{i+1}(x_{i+1})]$$

or:

$$f_i(x_i, u_i) \cdot F_{i+1}(x_{i+1})$$

respectively.

4. EXAMPLE APPLICATION OF MAX-MIN OBJECTIVE FUNCTION

As an example of how the max-min objective function is applied, consider the problem of optimum operation of a multipurpose reservoir with a hydroelectric power plant (3). The objective is to maximize the return from energy production while satisfying other purposes (firm water supply, flood control, low-flow augmentation) as constraints. The capacities of the reservoir and power plant are given (existing project) and a long-term operation policy (monthly model) for planning purposes is sought. Since the returns from sale of on-peak firm energy are much higher than the returns from nonfirm and off-peak energy (dump energy), the economic objective function will consider only the maximization of an annual firm on-peak energy contract which can be sustained over a long planning horizon. Since the model is deterministic, a historical record or any synthetic hydrologic series can be used as reservoir inflow.

4.1 Mathematics

The mathematical model developed for this system has the following constraints:

a) $\quad XMIN \le x_i \le XMAX_i$; $\qquad i = 1,\ldots, N$ $\qquad$ (33)

b) $\qquad x_{i+1} \le XMAX_i$; $\qquad i = 1,\ldots, N$ $\qquad$ (34)

where:

$\quad x_i$ = amount of water in storage at beginning of month i (state variable).

$XMIN$ = dead storage of reservoir

$XMAX_i$ = maximum allowable storage level during month i due to flood control reservation

N = planning horizon

c) $u_i \ge \max[MR_i, FW_i]$ $\qquad\qquad\qquad\qquad$ (35)

where:

$\quad u_i$ = release from reservoir during month i (decision variable)

MR_i = mandatory release due to low-flow augmentation purpose during month i

FW_i = firm water supply during month i

($FW_i = \alpha_i \cdot AFW, \alpha_i$ = coefficient such that $\displaystyle\sum_{i=1}^{12} \alpha_i = 1$,

AFW = given annual firm water contract).

d) $\quad x_{i+1} = x_i + I_i - u_i - EVA_i$ \hfill (36)

This is the state transformation equation which is equivalent to the mass balance equation for the reservoir. The terms are:

I_i = inflow to reservoir during month i (historic or synthetic).

EVA_i = evaporation from lake surface during month i.

e) $\quad EVA_i = \bar{A}_i (x_i, u_i) \cdot ER_i$ \hfill (37)

in which: $\bar{A}_i(x_i, u_i)$ = average lake surface in month i (since it is a function of storage, release and also evaporation, it must be calculated by successive approximations),

ER_i = mean evaporation rate [mm/month] during month i

f) $\quad PE_i \doteq u_i \cdot EPR(x_i)$ \hfill (38)

where:

PE_i = potential energy production during month i

$EPR(x_i)$ = energy production rate as a function of reservoir contents (head of the power plant).

g) $\quad OEMAX_i = \eta \cdot PCAP \cdot OPH_i$ \hfill (39)

$OEMAX_i$ = maximum potential on-peak energy that can be produced during month i

PCAP = power plant capacity (given)

OPH_i = on-peak hours available during month i

108

$$\eta = \text{power plant overall efficiency}$$

h) $\qquad OE_i = \min[PE_i, OEMAX_i]$ (40)

$\qquad OE_i = $ actual on-peak energy production during month i

The objective function can be derived as follows. By definition, the annual on-peak firm energy contract (AFE) and its corresponding seasonal distribution ($\beta_i \cdot$ AFE) have to be satisfied all the time during the planning horizon, that is:

$$OE_i \geq \beta_i \cdot \text{AFE for all months i,} \tag{41}$$

where:

$\beta_i = $ proportion of annual on-peak firm energy contract that has to be delivered during month i

$$(\sum_{i=1}^{12} \beta = 1)$$

Therefore, we would have:

$$\text{AFE} = \min_i \left[\frac{OE_i}{\beta_i}\right], \text{ over } i = 1,\ldots, 12. \tag{42}$$

Since the objective is to maximize the annual on-peak firm energy contract, the objective function is:

$$\max \text{ AFE} = \max \left\{\min_i \left[\frac{OE_i}{\beta_i}\right]\right\} \tag{43}$$

$$= \max \left\{\min \left[\frac{OE_1}{\beta_1}, \frac{OE_2}{\beta_2}, \ldots \frac{OE_{12}}{\beta_{12}}\right]\right\}. \tag{44}$$

This is the max-min type of objective function that was previously described. The recursive equations for solving this particular maximization problem through dynamic programming are:

$$F_i(x_i) = \max_{u_i}\left\{\min\left[\frac{OE_i}{\beta_i}, F_{i+1}(x_{i+1})\right]\right\} \tag{45}$$

subject to the constraints mentioned above.

The optimal value of the annual firm energy contract is then:

$$AFE^* = F_1(x_1')$$ (46)

where: x_1' = given initial reservoir contents or the contents that maximizes F_1.

4.2 Operating Rule

It is interesting to see that with this deterministic dynamic programming model, it is of little interest to trace forward the optimal policy, because it is only valid for a given sequence of deterministic inflows. More important is the fact that an operating rule for actual operation of the reservoir, with any inflow record, can be stated as follows:

In any given month, the actual release is set equal to the largest of the following four calculated releases.

(1) Release necessary to produce monthly on-peak energy, i.e., $\beta_i \cdot AFE^*$, required by the optimal annual firm energy contract, AFE^*.

(2) Release necessary to satisfy the corresponding monthly firm water supply, i.e., $\alpha_i \cdot AFW$.

(3) Release necessary to satisfy the monthly mandatory release for low-flow, i.e., MR_i.

(4) Release necessary to satisfy the flood control constraint, i.e., to keep reservoir contents below $XMAX_i$.

A simulation of the operation of the reservoir with any given inflow record can be devised based upon this rule. If the same inflow record as used in the optimization (for example historical record) is used in the simulation, no failures in satisfying the four required releases will occur. If, however, another inflow series (for example an synthetic hydrological record) is used in the simulation, some failures will appear. This suggests an interesting way to evaluate this operating policy using several synthetic series and calculating some performance indexes for it (probability of failure). The optimal annual firm energy contract could also be obtained as a mean value or a percentile of the distribution of values of AFE obtained from several optimizations, each with a different synthetic series.

4.3 Example

Another useful application of the optimal operating policy is to obtain the dump energy production during the planning horizon. In the simulation run, whenever the first of the four releases listed above is not the controlling release, the actual release is larger than required by firm energy production, and there will be some dump energy production which is easy to evaluate.

This model was applied to find an optimum operating policy for Folsom Reservoir and its power plant on the American River in Northern California. The critical period of the historical inflow record (10 years) was used for optimization and the whole historical inflow record (50 years) was used for the simulation.

Some results for different values of the annual firm water (AFW) supply constraint are shown in Table 1. From these results,

Table 1. Results from Optimization and Simulation of Folsom Reservoir and Power Plant

Annual Firm Water Supply ($10^3 \cdot$ KAF)	Annual Firm Energy Contract ($10^3 \cdot$ MWh)	Average Dump Energy per year ($10^3 \cdot$ MWh)	Maximum Annual Dump Energy ($10^3 \cdot$ MWh)	Minimum Annual Dump Energy ($10^3 \cdot$ MWh)
400	208	423	822	34
500	208	423	822	34
600	204	427	828	40
700	196	436	843	50
800	183	449	858	62
900	161	470	878	84
1000	103	528	922	144

KAF - thousand acre-feet
MWh - megawat-hour

using appropriate prices for annual on-peak firm energy, dump energy and firm water supply, the optimal combination of these outputs can be obtained.

5. REFERENCES

1. Nemhauser, George L., Introduction to Dynamic Programming, John Wiley, London, 1966.

2. Hall, W. A. and Dracup, J. A. Water Resources Systems Engineering, McGraw-Hill, 1970.

3. Harboe, R. C., Mobasheri, F. and W. W. -G. Yeh. "Optimal policy for reservoir operation." Journal of the Hydraulics Division, A.S.C.E., HY 11 2297-2308, 1970.

4. Harboe, R. C. "Introduction to dynamic programming in water resources planning and operation." CNR, Istituto di Ricerca per la Protezione Idrogeologica Nell'Italia Centrale, Perugia, Italy, 1980.

7. SIMULATION MODELING OF A REGIONAL WATER SYSTEM

C. Bartolomei, P. Celico, A. Pecoraro, A. del Treste,
Cassa per il Mezzogiorno, Roma
and
Y. Emsellem, F. Mangano, D. Verney
ARLAB, Valbonne, France

1. INTRODUCTION

1.1 Traditional Hydrology

The relationship between rainfall and river discharge has been long established as an area of study in the field of hydrology. Regression models are commonly accepted and used in practice. Improvements in this basic relationship are possible by introducing co-variables such as soil moisture storage, ground water storage, etc. But universal application of the rainfall-runoff relationship is not possible as there are important exceptions. These include Karst regions and alluvial areas. The unit hydrograph is another approach, but it is also limited in scope for the same reason. The problem involves many interrelated variables of the hydrologic cycle which are not considered in the traditional models.

1.2 Development of Comprehensive Computer Models

The development of comprehensive hydrologic models has come with the advent of high speed computers. One such model, the CEQUEAU, developed in Quebec, divides a region into a grid. The various hydrologic parameters associated with each grid element are described by means of a field survey or literature search. The hydrologic linkages between each grid element are controls. Another such model, the ABRUZZI, described by Verney, del Treste, Celico, Mangano, and Emsellem at the 1980 Taormina Congress, incorporates a variety of hydrologic variables, including non-saturated flow through porous media, application of Darcy's law in describing groundwater flows, river discharge, evapotranspiration, storage in reservoirs, etc.

2. APPLICATION

The ABRUZZI hydrologic model was created for application by Cassa per il Mezzagiorno, an agency for the development of Southern Italy. The ABRUZZI model is a deterministic model, which calculates the hydrologic outputs of a region such as streamflow, groundwater yield, etc. as a function of the hydrologic inputs, such as rainfall history, temperature history, etc. The model itself represents the physical processes such as evapotranspiration, snowfall accumulation and melt, underground percolation, etc. Constraints such as laws and management rules are taken into account also.

To construct the model, the region is broken down into a grid. All items of hydrologic information, e.g., meteorological data, physical parameters, calculated measurements, etc., are related to each cell in the grid. At the same time, such information can be aggregated in terms of overall regional information. The time interval can be whatever is appropriate for the problem. Also, any convenient cell size can be chosen.

The ABRUZZI model, as developed for Cassa, represents a region about 6000 square kilometers, but with the band surrounding it, the area modeled is about 10,000 km^2. This was broken down into a grid of 2500 cells, each 2 km by 2 km in size. The time interval was the month, and the period of simulation was 50 years.

3. STRUCTURE OF ABRUZZI MODEL

The general structure of the ABRUZZ1 model is shown in Figure 1. All of the hydrologic processes shown are represented by each cell of the grid. The whole array of calculations indicated are performed for each cell in the grid, using the time increment selected. This is repeated time increment by time increment for the whole period of time to be simulated.

3.1 Components

The principal variables involved in the model are described in the following. Their description gives an idea of the comprehensiveness of the ABRUZZI model.

114

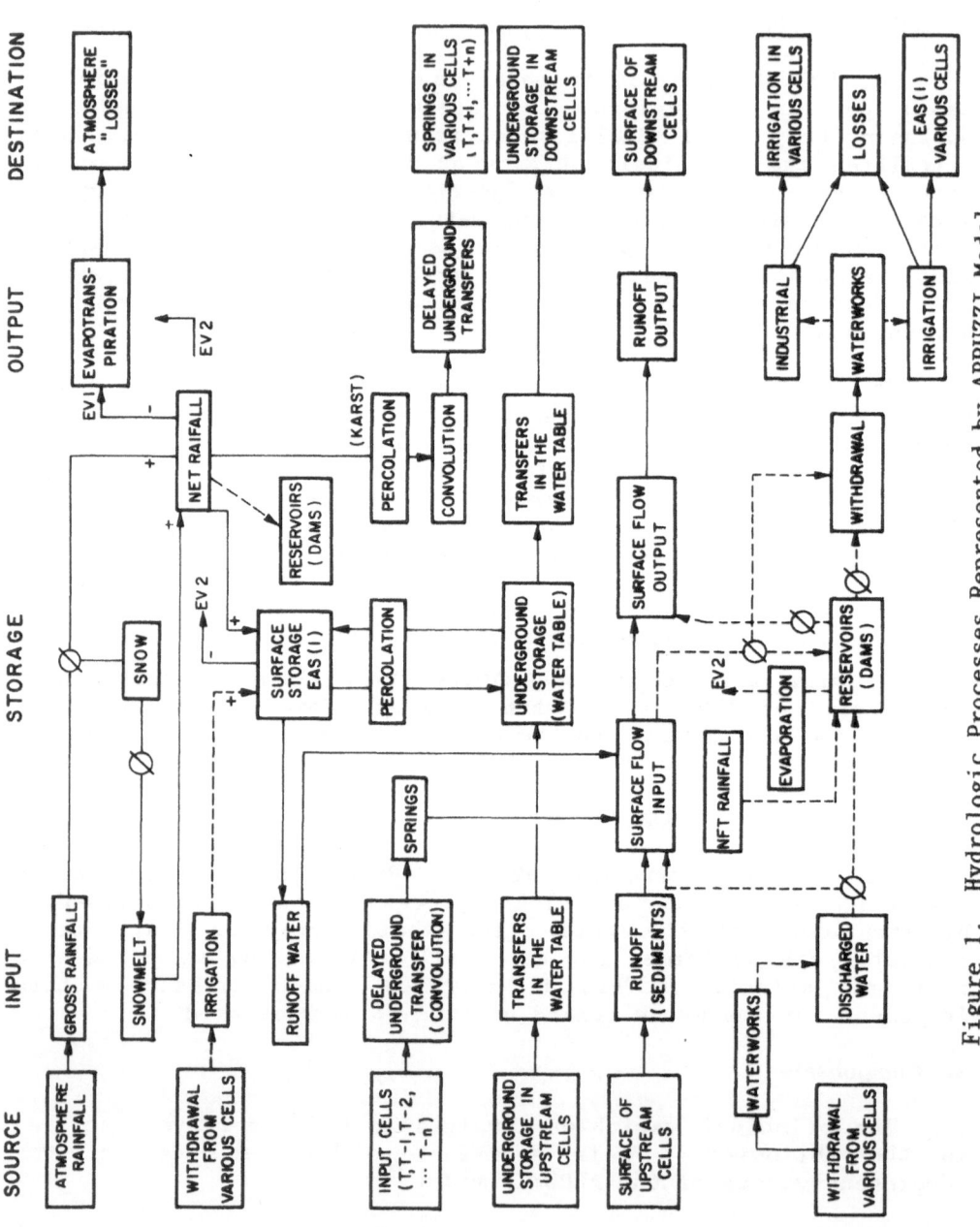

Figure 1. Hydrologic Processes Represented by ABRUZZI Model

1. Gross rainfall and mean temperature. Figure 2 is an isohyetal map for the region showing contours of mean rainfall for a winter month. The mapping provides input data for each cell for a given month. Computations are made for each month by "Krigeing."

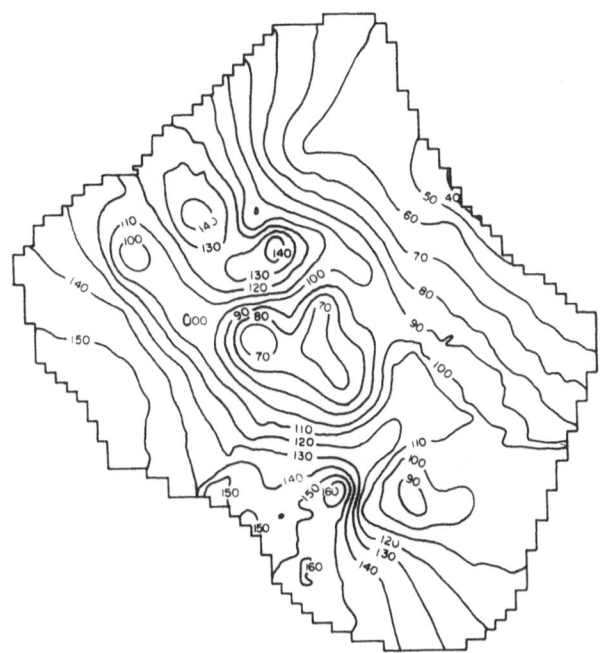

Figure 2.   Isohyetal Map of Mean Rainfall in Millimeters for a Winter Month for the Region Represented by the ABRUZZI Model

2. Snowfall. The quantity of snowfall for any given month is a function of altitude and temperature. For a monthly time increment temperature and altitude is sufficient. These variables were correlated with snow gage readings.

3. Evapotranspiration. This is dependent upon climatic variables such as temperature, humidity, sunshine, etc., and is related to them by established models incorporated into the ABRUZZI model.

4. Soil moisture. This is determined by a water balance for the root zone performed for each time increment.

5. Deep percolation. This is the excess above field capacity in the root zone which percolates to the water table.

6. Groundwater flows. A groundwater model was incorporated, which included permeability of aquifers, coefficients of storage, and elevations of groundwater contours. From this flows of springs and contribution of groundwater to streamflow was calculated.

7. Stream runoff. The main outputs of the model is the set of computed flows at the grid node points of special interest for water resource management.

8. Water diversions. The model incorporates existing management laws to permit calculation of withdrawals from surface or groundwaters within any grid element, commensurate with water availability, as determined by the simulation of the natural system, and coordinated with other withdrawals.

3.2 Computational Methods

There are two computational phases in each time step. In the first phase, a convolution method calculates for each Karst basin the total quantity of water undergoing long range transfer and allocates it among the principal transfers. This is done for the present time step and for the later time step corresponding to the transit time of the groundwater. This may be from a few months to a year or two, depending upon the size and type of underground network. Figure 3 shows a simulation of flow from a spring by convolution calculus. The convolution calculus method allows calculation of delayed phenomena provided their structural characteristics are known, as is the case with springs found in Karst systems, and for which a chronological sequence of flow measurements exist.

In the second computational phase, the water remaining in each cell is subject to percolation, behaving according to Darcy's law. In cases where the hydrogeologic characteristics are known, such as aquifer storage coefficients, aquifer depth, and transmissibility, then a second phase alone gives a better overall picture of underground transference. It requires flow measurements at frequent points of the surface network, and groundwater piezometric measurements. For the ABRUZZI model both methods were needed. The model permits different ways of combining them.

The ABRUZZI model also includes subsystems and subroutines reproducing different configurations according to the type of region studied, the results sought, and the kinds of presentation required. Several forms are possible, such as tables, graphs, maps, etc.

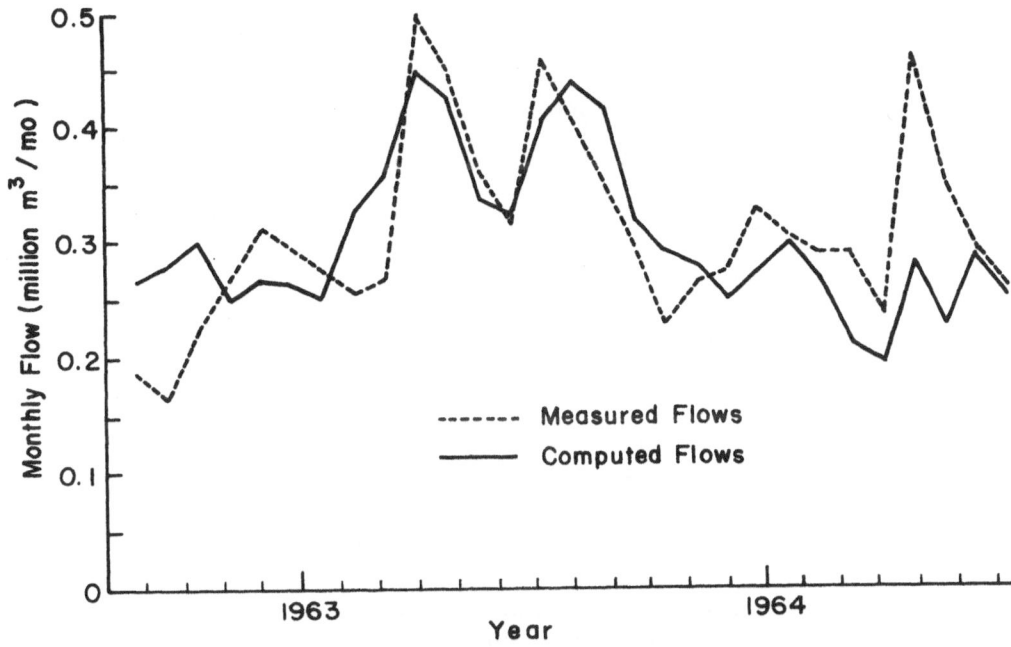

Figure 3.   Simulation of Monthly Flow from a Spring by Convolution
Calculus

4. MODEL CALIBRATION

To calibrate the model it is necessary to know the flow
histories at several points in the hydrologic network. The model
parameters are then adjusted until there is adequate fit between
the computed and measured flows. Figure 4 is an example of the
type of comparison made to achieve calibration. Measured flows
are compared with computed flows as seen by the two traces.

In setting up the model, there must be a field survey to gain
information about the hydrologic parameters that constitute it.
If this is done carefully for each cell then there will be
good agreement between the measured and computed results.

The model can be useful for examining either local cells or
large areas. The model can simulate global amounts of water over
large areas and for several months duration. The flow at the
mouths of large rivers are well simulated, e.g., within ten
percent for low water flows.

118

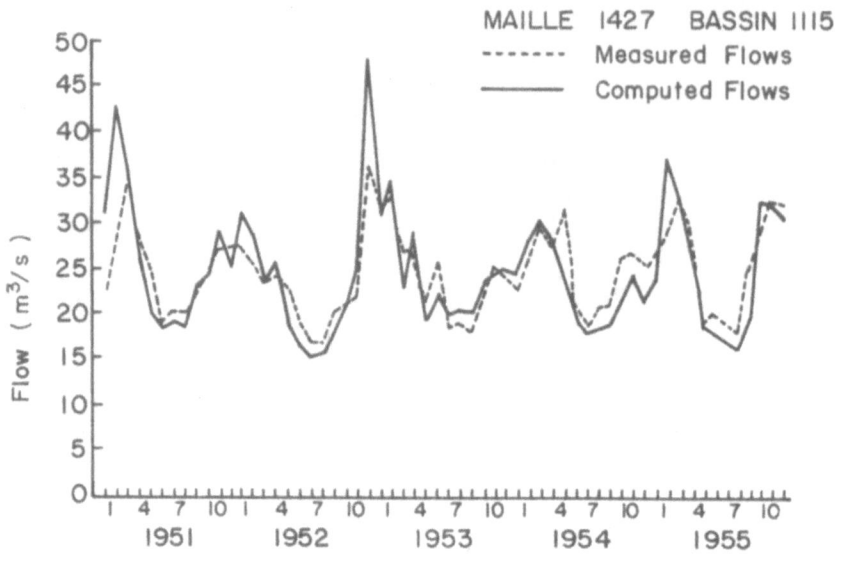

Figure 4.   Comparison of Measured and Computed Flows for
             Calibration of ABRUZZI Model

The difficulties are caused when it is necessary to
over-simplify the hydrologic processes, or if there is too much
uncertainty in estimation of the associated parameters, or if
there are too many gaps in the data. Most often the latter are
due to lack of hydrogeological data, questionable measurements of
flow, insufficient delineation of sources of water, insufficient
precipitation data, and uncertainties in rainfall measurements.

5. DEVELOPMENTS

The ABRUZZI model has been operational since 1980. Its use
has revealed areas where further improvements can be made. For
example, instead of representing the area by square grid cells, it
would be better to use drainage basins as the cellular unit, or
perhaps even administrative units. An example of this further
improvement was developed, by ARLAB, for the ground water supplies
in the whole of France. It considers administrative units, water
authorities, hydrographic basins, and hydrogeological units.
Digitized maps cover the entire country by this scheme. This work
was executed for the European Communities Commission.

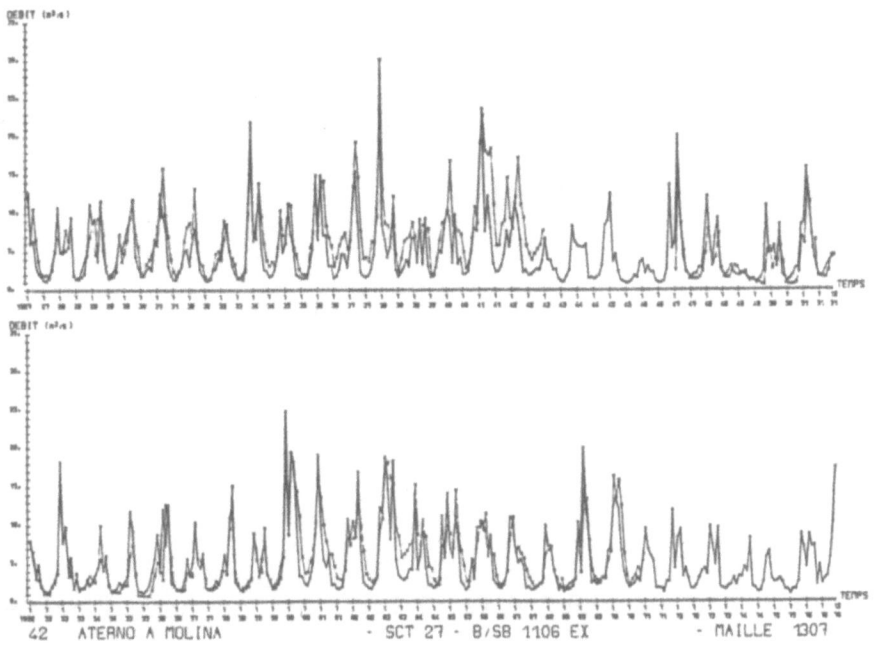

Figure 5. Long Term Hydrologic Records,
Computed and Measured

A similar initiative is underway by Cassa for the Monti Lepini region in Italy. Construction of this new model will incorporate the experience of both the ABRUZZI model and the synthesis on the groundwater basins of France.

Development of these models will permit a comprehensive planning and management of water resources on a regional scale. While art is yet involved, because of the limiations noted above, the science of modeling provides a tool which adds significantly to practice.

8. WATER QUALITY MODELING

David W. Hendricks
Colorado State University

1. INTRODUCTION

The water use locus of a river basin is its stream system. Streams are sources of water supply and, at the same time, they function as sinks for used water. Two categories of used water of interest are: (1) point source discharges of treated waste water, i.e., from municipalities and industries, and (2) non-point source flows, i.e., from irrigated agriculture, urban storm water runoff, etc.

The flows of used water will affect stream water quality through the mixing between two waters of different chemical composition, and through the reactions in the mixed stream. If the resultant stream water quality is to be maintained within the limits prescribed by law, i.e., stream standards, then the flows of used water must be "managed". The key management questions are: To what extent will ambient stream water quality respond to a proposed level of basin-wide effluent discharge control? And conversely: What level of basin-wide effluent discharge control is required to meet proposed stream standards? These questions may be restated in more poignant "cost effectiveness" terms as: (1) For a given expenditure what will be the results?, and (2) For a given result, what will be the cost?

The only way to answer either of these complimentary questions with certainty is to initiate a treatment plant construction program and then monitor stream water quality to empirically ascertain the results. If the results achieved are not as good as desired, then another level of treatment may be imposed. This empirical approach is an important part of the overall management process. But it is not adequate in developing a program initially, when foresight is needed. A water quality simulation model is a tool which can aid in developing such foresight.

2. SIMULATION MODELING

A large number of inputs and diversions may occur within any given stream reach. Further, the reactions of the water quality constituents introduced into the stream may be rather complex. The only feasible way to integrate all of these flows and reactions is through the construction of a simulation model, programmed for a high speed computer.

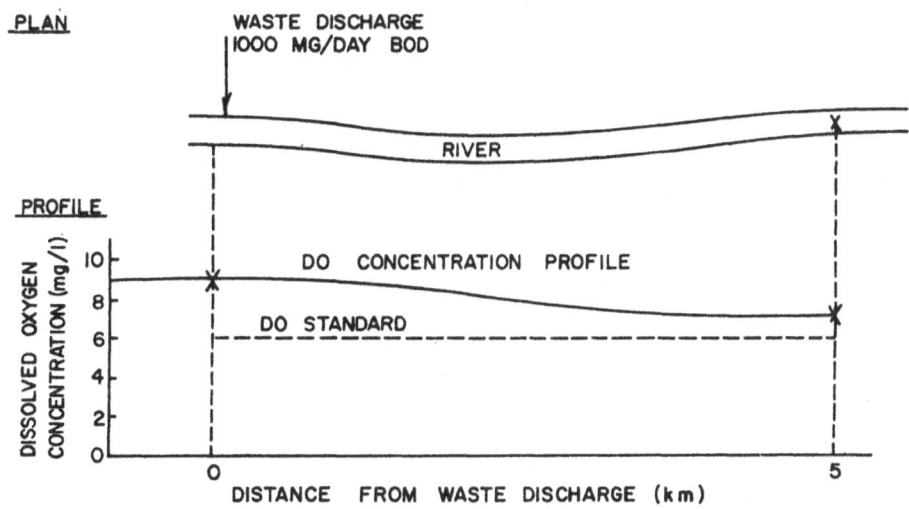

Figure 1. Illustrative of the effect of a wastewater discharge on the dissolved oxygen concentration profile.

A simulation model provides a predictive capability. Thus one can anticipate the results of a new operating mode in terms of changes in stream water quality. For example, if a new waste discharge of 1000 kilograms per day is added to the stream as depicted in Figure 1, what might be the effect on the dissolved oxygen levels, say five kilograms downstream? The dissolved oxygen (D.O.) concentration profile shown could be predicted by a water quality D.O. simulation model. Comparison of such a profile with the D.O. standard, as shown in Figure 1, tells whether a violation of the standard is likely. On the other hand, if a reservoir is located upstream of the waste discharge, one might ask if reservoir releases could be programmed such that stream standards for dissolved oxygen are not violated? A variety of such questions can be answered through simulated modeling.

2.1 The Materials Balance Concept

A rational water quality simulation model must be constructed by a materials balance formulation. The essence of such a model is merely an accounting statement in terms of mass flows in and out of a given fixed element in space, and reactions which occur within.

The fixed spatial element upon which the materials balance, formulation is constructed may be any arbitrary volume, such as a cube, or a slice of a stream reach. Usually in stream modeling, a slice is used as the volume element. The slice implies that the model constructed is one-dimensional. In practical modeling the thickness of the element may be rather large. The reach length, ΔZ, should be short enough that the constituent in question is approximately homogeneous in concentration within the spatial element.

2.2 Mathematics of Materials Balance

Figure 2 is a graphic representation of the materials balance principle applied to a slice element, fixed in space. The

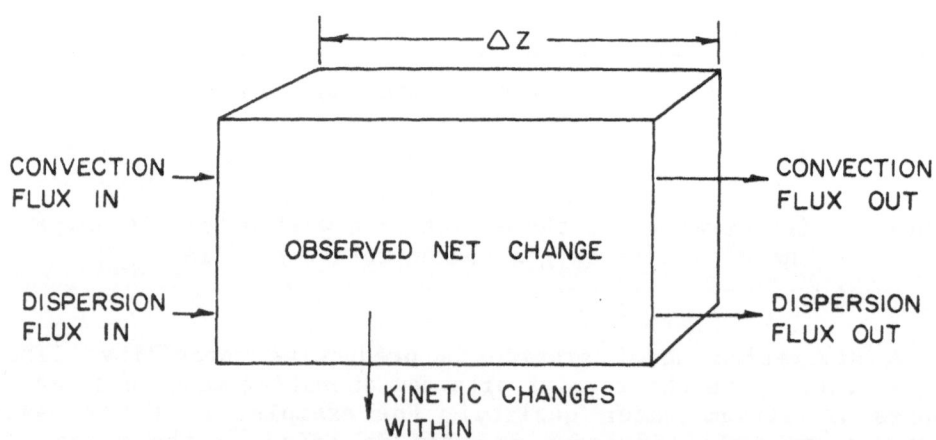

Figure 2.   Inputs, Outputs, and Changes within a
Slice Element of Water

transport terms are labeled <u>convection flux</u> and <u>dispersion flux</u>, respectively. Convection flux includes all material transported by the bulk flow in the stream, which has a mean velocity $\bar{v}$, added or taken from the slice element at the two boundaries. Dispersion flux is that material transport rate superimposed on the convection transport rate by random motion due to the fluid turbulence. The kinetic terms are the rate of change of concentration due to chemical or biological reactions of various sorts. The <u>observed change</u> is the net result of all influences within the element.

The graphic representation of the materials balance principle can be given verbal expression, i.e., the observed changes within the volume element equals the convection flux in minus convection flux out, plus dispersion flux in minus dispersion flux out, plus kinetic changes within. Equation (1) is the corresponding equality statement for a volume element fixed in space, as shown in Figure 2.

$$\begin{array}{l}
\text{OBSERVED} \\
\text{CHANGES}
\end{array} =
\begin{array}{l}
\text{CONVECTION} \\
\text{FLUX IN}
\end{array} -
\begin{array}{l}
\text{CONVECTION} \\
\text{FLUX OUT}
\end{array} +$$

$$\begin{array}{l}
\text{DISPERSION} \\
\text{FLUX IN}
\end{array} -
\begin{array}{l}
\text{DISPERSION} \\
\text{FLUX OUT}
\end{array} +$$

$$\text{KINETIC CHANGES WITHIN} \qquad (1)$$

This statement can be formulated mathematically several ways. The essential assumption, as noted previously, is that the contents of the volume elements are homogeneously distributed. Thus the size of the volume element and the resulting form of the mathematical expression depend upon whether it is necessary that the volume element be finite or infinitesimal for the homogeneity assumption to be valid. If the constituent of interest, c, is changing rapidly with distance, then slice ΔZ must be infinitesimal, and the materials balance expression must be developed in differential form. The mathematical expression for this case is:

$$\left\{ \left[\frac{\partial c}{\partial t} \right]_o = -\bar{v} \frac{\partial c}{\partial z} + D \frac{\partial^2 c}{\partial z^2} + \left[\frac{\partial c}{\partial t} \right]_k \right\}_{i,t} \qquad (2)$$

where: c = concentration of water equality constituent being modeled (M/L^3)

            t = time elapsed from which initial conditions were described (T)

            z = distance downstream from some initial point (L)

            D = coefficient of hydraulic dispersion for stream (L/T^2)

     o = subscript used to designate the <u>observed</u> change rate of the constituent concentration within the volume element

     k = subscript used to designate the <u>kinetic</u> term in the materials balance equation

     i = subscript used to designate a particular volume element.

NOTE: The bracket around the overall equation is subscripted in i, and t; this means that all terms in the equation are unique to that particular i and t designated.

   If, on the other hand, the system can be described macroscopically--that is the water quality changes are relatively small with distance, i.e., Δz is perhaps a <u>reach</u> instead of an infinitesimal <u>slice</u>, then the appropriate mathematical description is:

$$\left\{\left[\frac{\partial(Vc)}{\partial t}\right]_o = Q_e c_e - Q_o c_o + D \frac{\partial^2 c}{\partial z^2} \cdot V + \left[\frac{\partial c}{\partial t}\right]_k\right\}_{i,t} \tag{3}$$

where: V = volume of element (L^3)

    Q_e = flow entering reach volume, i (L^3/T)

    Q_o = flow leaving reach volume, i (L^3/T)

    c_e = concentration of water quality constituent being modeled at point of entering reach i (M/L^3)

    c_o = concentration of water quality constituent at point of exit of reach i (M/L^3)

   Equation (3) is the basis for stream modeling. It can be expanded to incorporate additional terms as needed to fit the situation.

   <u>Expressions in Materials Balance Equation</u>--To appreciate the generality of Equation (3), and to provide a better intuitive understanding of it, each of the differential and product expressions forming the equation should be examined individually. This is done in the following.

$\left[\frac{\partial(Vc)}{\partial t}\right]_o$ - This differential expression can be expanded to give $V\frac{\partial c}{\partial t} + c\frac{\partial V}{\partial t}$. The first differential term is the observed rate of change of c in the volume as a result of all influences on the right side of the equation. The second term in this expansion accounts for volume changes during the time increment dt. The subscript o means <u>observed</u>.

$Q_e c_e$ - The product expression for convective transport can be considered a collective term for all inputs into the stream reach ΔZ. It can be expanded into several terms to include hydrologic inputs, discrete waste discharges, and diffuse return flows along the reach ΔZ. These individual terms all are "lumped" to occur at the beginning of the reach.

$Q_o c_o$ - As with the input flux term, the output flux term can be expanded to include stream outflow from the reach, diversions, and channel losses. All of these terms are lumped to occur at the end of the reach.

$D \dfrac{\partial^2 c}{\partial z^2}$ - This is the dispersion transport term caused by the existence of a velocity profile in the channel cross section, and the random motion of turbulence. The dispersion coefficient, D, must be determined by in-situ field tests.

$\left[\dfrac{\partial c}{\partial t}\right]_k$ - The rate of change of c due to reaction is given by the kinetic term. The subscript k is used as the designation for the <u>kinetic</u> term. If several reactions are involved, all must be delineated. For example if dissolved oxygen is the constituent in question then kinetic equations associated with biochemical reactions, respiration, photosynthesis, and reaeration, must be included. An expression for the photosynthesis term can be quite complex because of the continuous change during daylight hours due to sunshine intensity variation.

All of the terms in Equation (2) and Equation (3) are time varying. The flow Q_e is the result of natural hydrologic influences plus diversions of the upstream i+1 reach, and returns of the i reach upon which the calculation is being performed. Figure 3 shows how these various flow inputs and flow outputs are aggregated at the head and tail ends of a given reach.

The concentration of c in the reach i is computed as the flow weighted average concentration. Input varies due to mixing from flow after all waste discharges, return flows, and diversions are added to or taken from the streamflow entering reach i. If the constituent is conservative, e.g., the chloride ion, then its concentration in reach i is determined by the simple mixing model. However, if the constituent is reactive, e.g., a sugar compound, then its mass is changed within reach i by reaction.

In either case a new concentration, $c_{i,t+\Delta t}$, is calculated for reach i. This calculation utilizes Equation 3 operating over a time period Δt. Equation (3) is then applied to reach i-1, which becomes reach i. The calculation proceeds in the downstream direction to obtain $t+\Delta t$ concentration values at each i. These values are stored for use as $c_{i,t}$ values during the next iteration i.e., after time is increased by Δt. This general computation scheme is outlined schematically in Figure 4.

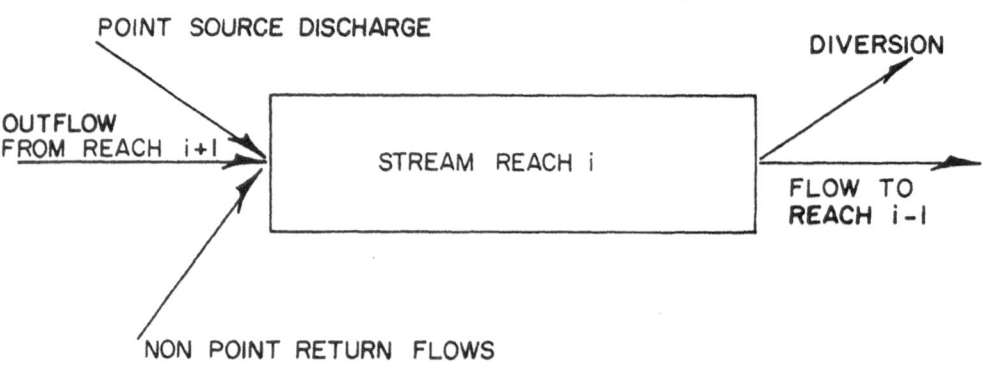

Figure 3. Aggregation of flows at ends of stream reach i.

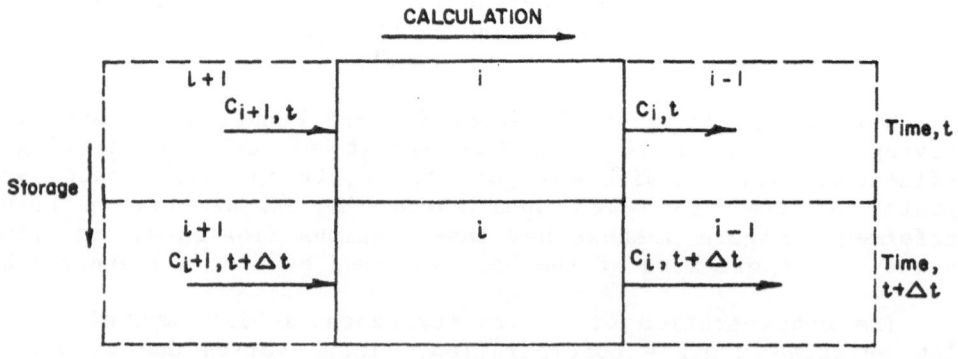

Figure 4. The flow enters slice i over time Δt, the reactive constituents react over time, Δt, and a new concentration $c_{i,t+\Delta t}$ is calculated.

2.3 Reaction Kinetics

The kinetic term $\left[\dfrac{\partial c}{\partial t}\right]_k$ for reach i describes the resultant rate of change of concentration, c, in reach i, due to the aggregate of all reactions within reach i, which are going on simultaneously. For example if dissolved oxygen is the constituent in question the complete description of the reactions and processes involved add up to a complex picture, such as depicted in Figure 5. Figure 5 depicts schematically the addition of oxygen to water by reaeration from the atmosphere and by photosynthesis. At the same time, oxygen is depleted from the water by aerobic biochemical reactions and by respiration. Further, if the water is supersaturated with oxygen it will be lost to the atmosphere by the same exchange process which causes reaeration in oxygen deficient waters.

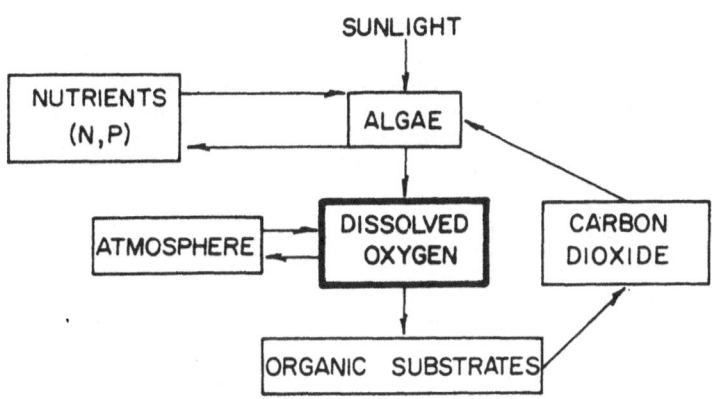

Figure 5. Partial representation of reactions and processes affecting dissolved oxygen level in a stream.

The kinetic statement, corresponding to the reactions is:

$$\left[\frac{dc}{dt}\right]_k = \left[\frac{dD.O.}{dt}\right]_{photosynthesis}$$

$$+ \left[\frac{dD.O.}{dt}\right]_{\substack{atmosphere \\ exchange}} + \left[\frac{dD.O.}{dt}\right]_{\substack{biochemical \\ reaction}} \tag{4}$$

Equation (4) says merely that the net rate of change of oxygen concentration in a given volume element is the sum of the rates of all other reactions and processes which are occurring simultaneously. The individual terms are represented by kinetic models.

2.4 The Steady State Approximation

Equations (2) and (3) are not amenable to analytic solution; they are too complex mathematically. This complexity is accentuated by a variety of boundary conditions (e.g., diversions, tributary streams, waste discharges, etc.), which are different from reach to reach. In order to handle the mathematical complexity, the unique boundary conditions at each reach i, and the possible time varying nature of certain boundary conditions, such as flow, a numerical solution is necessary. This, of course, requires the use of a digital computer. Equation (3) can be made less formidable mathematically, however, through the use of Lagrangian coordinates. With this approach the mathematical model is constructed about a volume element which moves with the stream at the velocity of the streamflow. Thus there can be no flow across the boundaries. This assumption ignores the hydraulic dispersion phenomenon. Thus, with no mass fluxes across the boundaries of the element Equation (3) can be simplified to:

$$\left\{ \left[\frac{\partial c}{\partial t} \right]_o = \left[\frac{\partial c}{\partial t} \right]_k \right\}_{i,t} \tag{5}$$

If the more general "c" term is replaced by "D.O." for dissolved oxygen, in Equation 5, it becomes the basis for the well known Streeter-Phelps equation for oxygen sag. Usually, it is expressed in terms of oxygen deficit. The differential equation is, in terms of the formulation of Equation (5):

$$\frac{dD}{dt} = k_1 L - k_2 D \tag{6}$$

where: D = oxygen deficit, i.e., $[D.O.]_s - [D.O.]$ (mg/L),

L = biochemical oxygen demand (mg/L),

k_1 = reaction coefficient (sec^{-1}),

k_2 = reaeration coefficient (sec^{-2}).

Equation (6) is an ordinary linear differential equation, which has a well-known analytical solution. This solution is not important for the purposes herein, however. The important point is to recognize that Equations (5) and (6) are simplifications of

Equation (3). In the former equation one deals with a fixed spatial element, while in the latter equation the orientation is on a <u>parcel</u>, or a volume element, which moves with the stream. The former permits flow across the boundaries of the spatial element while the latter does not. The left side of Equation (6) is equivalent to the left side of Equation (5), and so forth for the right sides. It should be noted that the Streeter-Phelps model leaves out the photosynthesis kinetic term, i.e., it ignores oxygen changes due to photosynthesis. The Equation (6) solution is merely the time history of a closed reactor as it floats downstream.

2.5 Computation Scheme

Equation (3) is completely general and has no restrictions. It is especially applicable to situations having variable boundary conditions (e.g., diversions or return flows which may vary over time). Because of the necessity to recompute all the variable terms (e.g., concentrations, flows, dispersion coefficient, kinetic coefficients, etc.) at each reach and to keep track of variable boundary conditions, it requires more effort to set up and more computer time to run the corresponding computer program.

Equation (5) is applicable only with the assumption of the steady state approximation, i.e., all flows and waste loadings are held constant with time. Because of this it uses less computer time. The flows and other boundary conditions must be changed to fit new steady state conditions. The computation proceeds reach by reach as in the general model, but in the steady state model no iterations with respect to time are required.

The application of either Equation (3) or Equation (5) requires that the boundary conditions be delineated and that a computation scheme be devised. Figure 6 shows a general layout of a stream system set up for mathematical simulation. The iteration starts at the top of the main stem and then proceeds reach by reach downstream. When a tributary is reached the program shifts to the top of the tributary. A reservoir is handled according to whatever instructions are programmed.

If the model is based upon Equation (5) then there is only one pass through the whole system. Equation (5) will compute the concentration-distance profile reach by reach along a sequence of n reaches as noted above.

If the model is based upon Equation (3) the computation algorithm is more complex. The computation begins at the top of the main stem and handles tributaries as described above. Equation (3) is applied at each reach i, in turn. However, the computation has an iteration, or a "do loop", with respect to

130

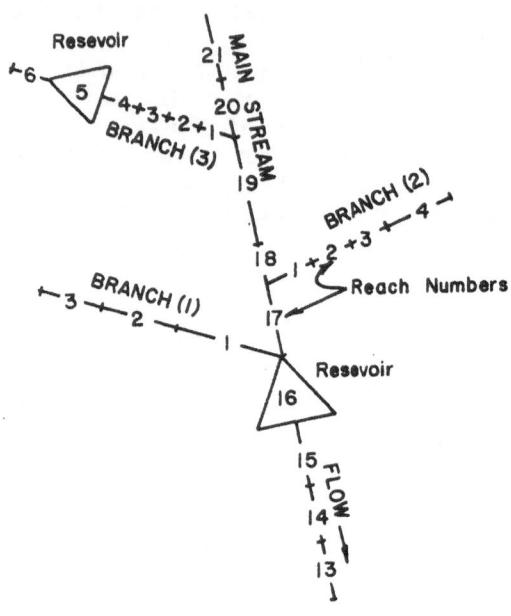

Figure 6. Computation scheme for application of materials in
balance equation (Waddel et al., 1974).

time. It computes, the concentration in reach i at time t+Δt,
i.e., $C_{i,t+\Delta t}$, by Equation (3). This value is stored for use in
the next pass. The computation then proceeds to reach i-1. The
computation can be seen by noting that the left side of Equation
(3) can be expressed:

$$\frac{C_{i,t+\Delta t} - C_{i,t}}{\Delta t}$$

The output of this computation algorithm is a three-
dimensional solution which may be expressed as, $C(Z)_t$, i.e.,
concentration distance profile for a given time; or as $C(t)_Z$,
i.e., concentration-time history at a given point on the stream.
The steady state model output provides only a concentration-
distance profile.

An example of a steady state model based upon Equation (5) is the PIONEER I, reported by Waddel et al. (1974). The model can be used for a large number of water quality constituents, since it has programmed in it the commensurate kinetic expressions.

2.6 Determination of Model Coefficients

The "model coefficients" include the various kinetic coefficients, and a dispersion coefficient. These coefficients can be determined by utilizing either of two approaches: (1) empirical "calibration" of the set of model coefficients as a whole, or (2) independent determination of each coefficient. The "calibration" approach is a trial and error process which involves parametrically assuming values for these coefficients, as a set, until finally there is a match between the concentration profile as measured in the field and the computed profile. This trial and error calibration is then repeated for another set of boundary conditions (i.e., low steam flows vis a vis high flows), which utilizes another corresponding field measured concentration profile. If a set of coefficients can be found such that two or more model computed concentration profiles for different boundary conditions approximate the field measurements, one can say the model is "calibrated." These coefficients can be used then for any other boundary conditions which one may want to impose, i.e., for the purpose of exploring the effects of a proposed effluent standards program on the concentration profile.

Figure 7 shows the results of several runs for the PIONEER I model, by Waddel et al. (1975), comparing model computed concentration profiles with measured field data. These "runs" for the two constituents shown, i.e., dissolved oxygen and ammonia, involved two different sets of boundary conditions. The runs shown were for the purpose of "verifying" the model. This means that with the coefficient set determined, the model should predict concentration profiles for other boundary conditions. Evidently the model was succesful in this test (from an engineering point of view) and so it can be said to be "verified." With the model thus calibrated and verified, new boundary conditions can be imposed as desired.

The other general approach for determining the model coefficients involves independent evaluation of each. The dispersion coefficient, the reaeration coefficient, and the reaction rate coefficient, have been the subject of a considerable amount of research, while there has been little effort to determine the photosynthesis coefficient. The latter is dependent upon sunlight and so is subject to hourly variations; this means that if photosynthesis is considered a considerable amount of added complexity is introduced into the model. Even though one may be able to find information on some of these coefficients from the

132

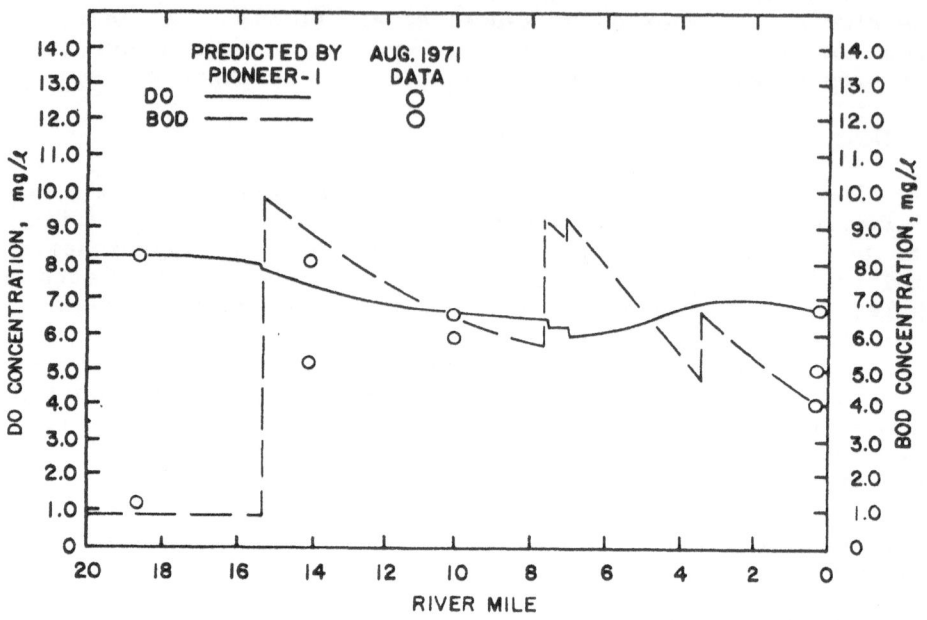

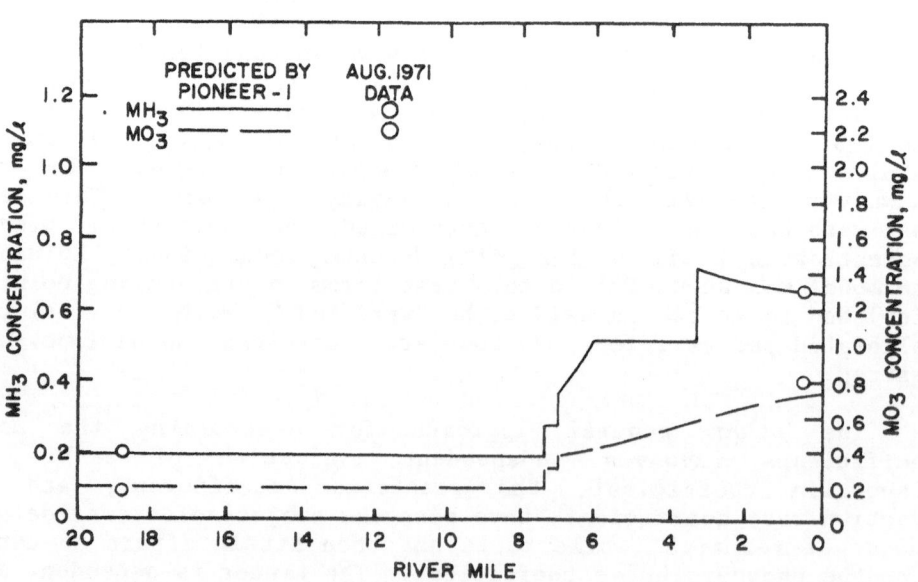

Figure 7.   Sample outputs for the PIONEER I model (Waddel et al., 1975).

literature it is advisable to determine them by in-situ field tests. Dixon et al. (1970) developed a model which incorporated the results of continuous 24-hour field measurements over an annual cycle.

2.7 Development of Boundary Conditions for an Actual Water Quality Model

The model is really no better than its boundary conditions. An example of how boundary conditions are depicted for a stream system is seen in Figures 8 for the South Platte River near Denver. Figure 8 shows the river layout near Denver depicting the tributaries and the reaches within both the main stem and the tributaries. Such schematics must be constructed for other inputs and outputs of water also.

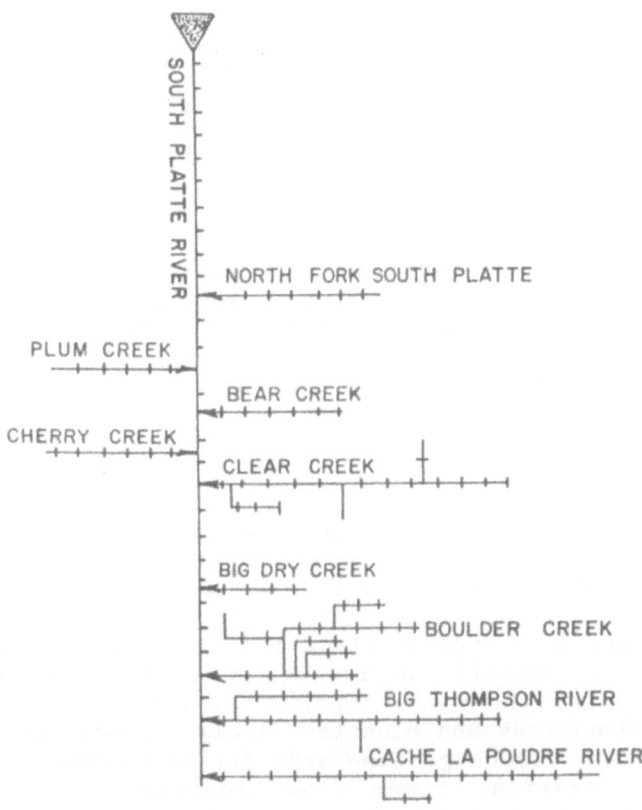

Figure 8. The river layout for the South Platte Basin showing tributaries and reaches (Waddel et al., 1975).

3. USING A WATER QUALITY MODEL

Once the system is described in terms of its boundary conditions, new ones can be imposed to ascertain their effect on stream water quality. Some of the questons which might be explored include:

1. If proposed new effluent discharge standards are imposed on all present and projected point source discharges in the South Platte Basin what will be the corresponding concentration profiles for various water quality constituents for average river flows, and for the ten year low flows?

2. Is there an optimum location for a new regional wastewater treatment plant with respect to its impact on basin water quality?

3. What is the effect of different stream flow levels from reservoir relases on stream water quality?

Many such questions can be formulated. They involve merely the imposition of corresponding boundary conditions on the model.

3.1 Samples of Output

The first question above was asked in a 197% nationwide study by the National Commission on Water Quality pursuant to PL 92-500. Contracts were let for 50 such river basin studies throughout the United States. One of the water quality studies authorized by the Commission is reported by Hendricks and Bluestein (1976). The main objective of the study was to determine the effect of imposing 1977, 1983, 1985 effluent discharge standards on stream concentration profiles of various water quality constituents.

A steady state model is adequate to handle the question posed in the above study. The non-steady state model, i.e., Equation (3), would have more resolution than is needed.

The PIONEER I, a steady state model, developed for the South Platte River by Battelle Northwest in 1973-74 (Waddel et al., 1975) was used in the study reported by Hendricks and Bluestein (1976). The Hendricks and Bluestein study imposed the 1977, 1983, and 1985 efluent discharge standards for the South Platte River, and also the critical river flow conditions. The hydrologic conditions were imposed for the lower quartile year flows for two representative seasons; August and December conditions were deemed most critical. The seven-day ten-year low flow was used also as an even more critical flow boundary condition for the model. The water quality constituents of interest included: temperature, orthophosphate, nitrate, ammonia, BOD, and dissolved oxygen. Each

of these constituents was modeled for the entire length of the
South Platte from above Denver to the Nebraska state line. Some
thirty runs were made to cover the various combinations of
effluent limitations and hydrologic conditions. Examples of
output from the PIONEER I model are given in Figures 9 and 10.

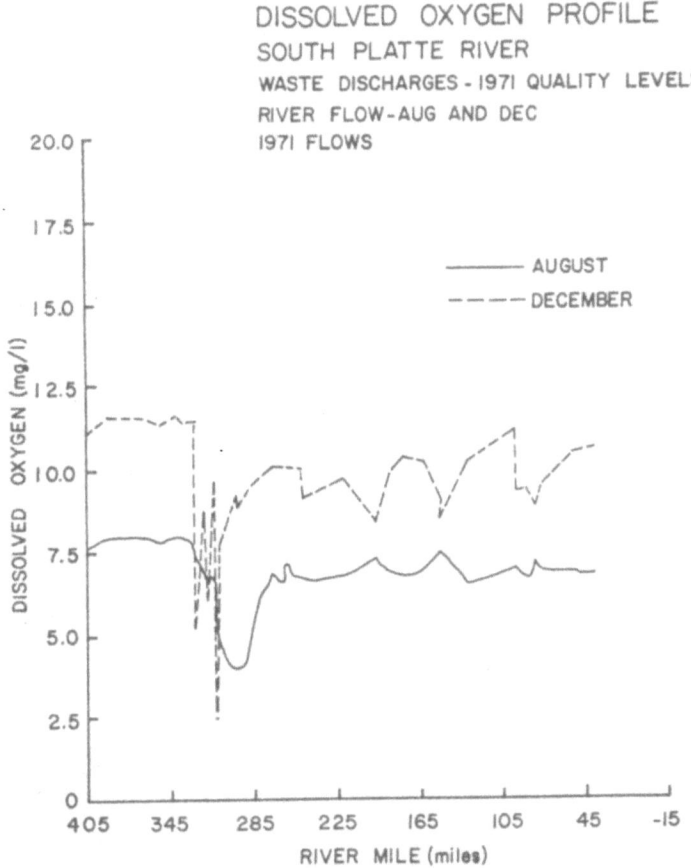

Figure 9. Comparison in dissolved oxygen concentration profile
           for August and December, South Platte River (Bluestein
           and Hendricks, 1975).

Figure 9 shows the comparison in the dissolved oxygen
profiles for August and December. The lower temperature in
December is one of the major factors accounting for the higher
dissolved oxygen levels for that month. Figure 10 is a composite
of outputs from several different runs, showing how the different

136

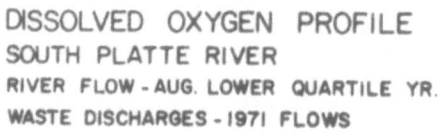

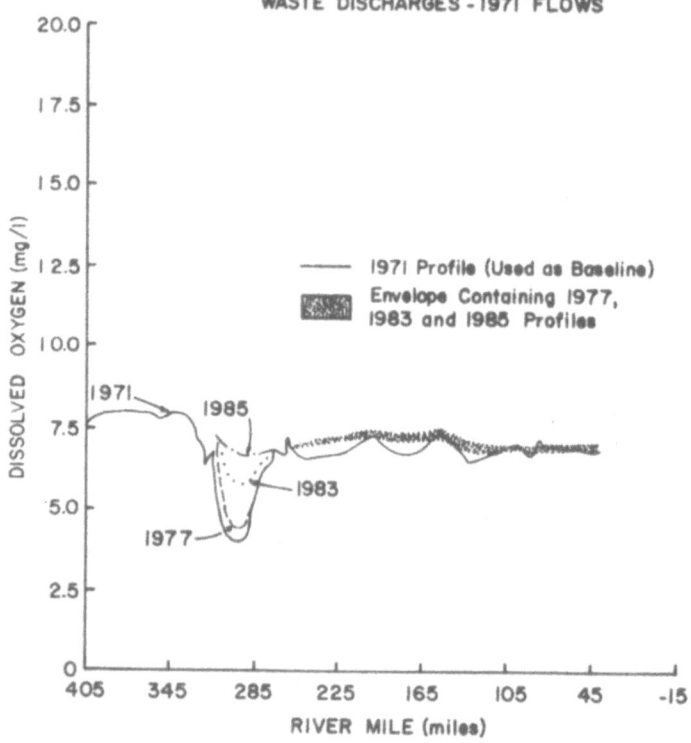

Figure 10. Response of dissolved oxygen concentration profile, South Platte River, to 1977, 1983, 1985 effluent discharge limitations (Bluestein and Hendricks, 1975).

basin wide pollution control policies result in different levels of response in the dissolved oxygen concentration profile. The 1977 profile for secondary treatment shows not too much improvement over 1971 conditions, while a tertiary treatment program in 1983 would result in the dissolved oxygen level meeting standards throughout the stream. The 1985 zero discharge of pollution was programmed into the model literally and the result is seen in the 1985 curve.

3.2 Conclusions

Water quality models can represent complex situations. They incorporate in their construction the hydrologic behavior of a stream system, the kinetic behavior of the various water quality constituents modeled, the mechanics of behavior of the stream, and the structure of the man-made sets of withdrawals and effluent discharges. A large amount of empirical information must be collected and incorporated into the model structure. The model integrates all of this complexity through a set of instructions, i.e., a program, handled by a digital computer.

A variety of studies are necessary in order to delineate the various components comprising the model (e.g., hydrology, reaction kinetics, etc.). Most of these studies could be extensive, consuming much time and money. Thus many approximations must be made in determining the structure of the model, its coefficients, and its range of simulation.

The output of the model is the integrated result of many mathematical operations which approximate the internal complexity of the real system. It can be used to ascertain the water quality response of different management plans.

ACKNOWLEDGEMENTS

This paper is based upon a video tape lecture developed for a short course on water resources planning produced by Colorado State University, in collaboration with the University of Catania, Sicily. It was developed further for the Symposium on River Mechanics at Colorado State University in 1977, and published in Modeling of Rivers, H. W. Shen, editor (1979).

REFERENCES

Dixon, N. P., Hendricks, D. W., Huber, A. L., and Bagley, J. M., 1970. Developing a hydro-quality simulation model. Utah Water Research Laboratory, Report PRWG67-1, Utah State University, Logan, Utah.

Hendricks, D. W., 1979. Application of Water Quality Models, in Modeling of Rivers, edited by H. W. Shen, John Wiley and Sons, New York.

Hendricks, D. W. and Bluestein, M. H., 1976. Response of the South Platte to effluent limitations. Journal of the Environmental Engineering Division, No. EE4, American Society of Civil Engineers, August.

Kartchner, A. D., Dixon, N. P. and D. W. Hendricks, 1969. Modeling diurnal fluctuations in stream temperature and dissolved oxygen. Proceedings of the 24th Annual Purdue Industrial Waste Conference.

Waddel, W. W., Cole, C. R., and Baca, R. G., 1974. A water quality model for the South Platte River Basin. Documentation Report, Battelle Pacific Northwest Laboratories, Richland, Washington, April.

9. DESIGN OF OBJECTIVE FUNCTIONS FOR WATER RESOURCE SYSTEMS

Vujica Yevjevich
George Washington University, Washington, D.C.

1. INTRODUCTION

Specific problems in operation of complex water resources systems are basically two: (1) decisions on water releases, and (2) satisfaction of various levels of service. The first problem relates to how much water will be released from storage in a given time interval. The second problem relates to the distribution of positive or negative effects of these releases among the water related interests.

Many effects result from water release decisions. They include the impact characteristics of releases, and are controlled by constraints and independent actions of parties related to operation of water resources systems. According to Westgate (1980), the releases of water and the resulting levels of service or disservice result in outcomes that are:

(1) Desirable - Undesirable
(2) Satisfactory - Unsatisfactory
(3) Disastrous - Unimportant
(4) Risky - Safe
(5) Require corrections - Do not require corrections

In order to improve the decision-making process, the analysis of water resources systems is usually made with the intent of sharpening the decision techniques, so that the above five dilemmas are answered in some rational ways. Most of these techniques relate to specification of objectives and design of objective functions. The design of mathematical functions relates the objectives to decision variables. The use of the associated optimization provides the quantitative information that enables more meaningful decisions on water releases than will be otherwise feasible. Before the design of objective functions (or as an alternative term, the functions of objectives) is discussed, definitions of various terms are first covered.

2. DEFINITION OF TERMS

Water resource planners and systems analysts make clear distinctions between the key terms in the field, such as goal, objective, purpose, decision variable, objective function, constraint, levels of service, alternative strategy, performance measure, optimization, trade-off, etc. These terms are defined here in the context of water resources systems. There is no unanimity on these definitions. Therefore, they are reviewed for the sake of common understanding of the terms and as a basis for the design of objective functions.

2.1 Goal, Objective, Purpose and Decision Variable

Goals provide the basic reasons why society undertakes water projects. Usually there are more than one goal and they are rarely controversial. For a public water project they are most often defined in terms of increase in national welfare, attainment of regional economic development, solution of social problems (unemployment, social peace, retention of population in case of ongoing migration), and by similar expectations. For a specific water project of nonpublic nature, the goals are most often defined in terms of satisfying the needs in energy, water, water quality, and of the other needs for industry, energy systems, mining, irrigation, etc. Directly or indirectly, they attain some of the same goals as the public water projects. The basic characteristic of goals of water projects is the difficulty to mathematically define (model) how they will be attained in terms of operational decision variables. To do that, the definition and description of project objectives and objective functions are needed.

The objectives of a project are those aspects of the goals that are identifiable, specifiable and quantifiable in terms of purposes and decision variables. They may be controversial because of differences of interests within a variety of affected groups, individuals, institutions and decision-makers. Major objectives are: (a) economic efficiency, (b) equity in distribution of project outputs (positive or negative by the criteria of any affected party), (c) environment quality, (d) level of service (employment, satisfaction of need for various services), and the other similar but usually minor objectives. Two schools or approaches to definition of objectives are: (a) economic efficiency is basically a sufficient objective, with all the other objectives reduceable to these terms; and (b) the other noneconomic objectives are as important as the economic efficiency, and they are considered as nonreduceable to economic efficiency (Marglin, 1967; Major, 1969; Cohon and Marks, 1975).

Economic efficiency is most often measured in monetary terms. Sometimes it is expressed as physical measures (such as cubic meters of water, kilowatt-hours of electricity). The equity criteria, as measured by economic differences between regions or social classes, is translated by the first school in terms of constraints on the objective function describing economic efficiency. It is, however, treated by the second school as an objective. The case is analogous with the objective of satisfying the enviornmental quality of life and the level of service that are either translated into constraints according to the approach of the first school (say a minimum release of water, aesthetic concerns, or the creation of a minimum number of jobs), or they are considered as separate objectives by the approach and criteria of the second school.

Using the economic efficience approach, the next step is the design of an objective function, usually with several constraints superimposed. The school advocating the use of several objectives simultaneously must rely upon application of multiple objective decision making methodologies. The positions of decision-makers are, however, the important factors in selecting either approach, i.e., the single (consolidated) objective, or the multiple objective.

The identification of decision-makers may not be simple in the case of complex water resources systems. This is true especially in democratic societies. In some cases, it is caused by strong influence on decision making by those who are not even conceived by planners and analysts as the decision-makers (say those who have indirect influence on political decisions on public projects). This factor may be equivalent in reality to additional decision-makers. Likely the systems analysts (say in the United States) may have often overplayed the search for versatility and variety of decision-makers. Identification of decision-makers is needed, however, in order to define the specifiable, quantifiable and manageable objectives.

Purposes in water planning have to do with water related interests. The common purposes include water supply, hydropower, irrigation, navigation, flood control, pollution abatement, fish and wildlife enhancement, recreation, etc. They may represent the terms in objective functions. A purpose as the variable should be a function of the major decision variables for the analytical treatment of objective functions through optimization. The

142

decision variables may be water release, water pumping discharge, diversion flow, and similar physical variables. Therefore, the purpose of hydropower production can be expressed in terms of water release.

The objective function is an equation composed of terms that describe the economic performance of quantifiable variables of purposes of a water project, measured either in monetary or in physical terms. This equation is most often supplemented by a set of inequalities, that normally represent goals, legal criteria, or other constraints which the decision making process must respect. When each purpose is expressed as a function of decision variables, the objective function becomes a relationship of the major objective (say the economic efficiency) in terms of the major decision variable (say the water release).

2.2 Constraints, Level of Services, Alternative Strategies

A constraint is either a physically imposed limit, legally prescribed value, or it results from the decision made in advance, namely that the value of a decision variable during a time interval must be either greater or smaller than a specified limit, or that it should fall within a prescribed range of values. Often, a trade-off can be made between the number of decision variables and the number of constraints. The trend among the system analysts is to replace as many decision variables by the constraints as is feasible, because the number of decision variables significantly influence the potential application of most relevant optimization techniques.

The level of service is prescribed delivery of water or other service, often synonymous with constraint; however, this is not necessarily in all cases. The equity in allocation of project positive and negative outputs may be translated into the distribution criteria for the level of service, either among the regions and cities, or among the social strata. These distributions may be further translated into the constraints by some objective criteria, all of it in order to minimize the number of decision variables.

An "alternative strategy" represents another concept or set of premises needed in planning and in design of objective functions. A number of alternative strategies enables a planner or a system analyst to pursue different courses of investigation, and for each alternative through its objective function, purposes, decision variables and the application of appropriate optimization techniques.

2.3 Performance Measure, Optimization, Trade-off

Four fundamental aspects of any water project are: benefit, cost, technology, and risks and uncertainties. All may serve as the most general performance measures of water projects. The classical water resource economics reduces all these measures to only one, usually either B/C (benefit-cost ratio), or (B-C) or (B-C)/C, and B and C the expected annual benefit and cost over the economic life of a project, respectively. Other performance measures may be specified, however, for specific aspects of the project.

A technology is selected prior to the analysis of the B/C ratio, or several technological alternative strategies are investigated, each with its own B/C ratio. Risk and uncertainty as performance measures are not yet commonly introduced in a direct way into the decision making process, though they may affect the decisions or may be separately used, because of various difficulties involved. These include identification and quantification, lack of reliable trade-off techniques, tendency to avoid public statements concerning risk, etc.

Optimization techniques, i.e., minimizing or maximizing of objective functions, are selected according to the problem at hand and the analyst's ability to realistically portray and quantify the objective function. Forcing an optimization technique onto the uncertain or faulty objective function is substituting form for substance.

Trade-off techniques represent in practice the power to translate the effects of selected technology or the assessed risks and uncertainties into the increased or decreased benefits and costs. They permit a derivation of the unique performance measure, such as B/C or (B-C). Trade-offs help reduce the number of performance measures, or even enable their condensation into one such measure only.

3. FUNDAMENTAL PERFORMANCE MEASURES

The four general performance measures of water projects: (a) selected technology, (b) risks and uncertainties, (c) benefits and (d) costs, are mutually interactive and subject to trade-offs. Figure 1 depicts this concept schematically. The objective function should be able to incorporate these performance measures with preference through an aggregation into one consolidated measure. There is nothing especially difficult with the use of the B/C ratio, provided it is realistically applied as the general measure that properly represents all four performance measures.

Complexities in using the above four performance measures simultaneously impose a sequential optimization in practice.

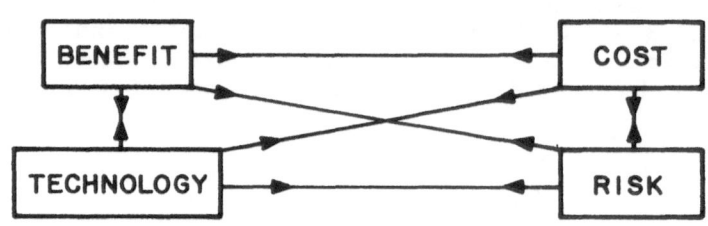

Figure 1.   Interactive actions in project performance.

First, the technology is selected in this sequential optimiation by taking into account the related risks, benefits and costs.   Second, a trade-off often can be made between the benefit and cost on one side and the risk and uncertainty on the other. Third, the resulting benefit and cost, realistically conceived as random variables, are integrated into one variable (benefit-cost ratio, or their difference -- absolute or relative).   Then the objective function is designed to optimize this unique variable of performance measure.   When all the measures cannot be reduced to one unique measure, the multiple objective decision making may be applied by using both the priced and unpriced services of a system.

The future decision making process will likely consist of these aspects: (i) identification and quantitative description of benefits and costs, both conceived as random variables; (ii) selection of "appropriate technology"; and (iii) identification and quantification of all sources of risks and uncertainties, with their effects on the decision making process.   Methods for deter-mining relations between the four fundamental performance measures of water resources systems are needed, if the future decision making process is to depart from the ongoing simple but less realistic approach.   The constant benefit-cost ratio used at present as a measure of water project feasibility, will likely be supplemented in the future by the B/C ratio conceived as a random variable.   The analysis of already realized benefit-cost ratios of existing water projects show always to be different than those constant B/C ratios, used in the decision making to implement a project.

4. DESIGN OF OBJECTIVE FUNCTIONS

4.1 Economic Efficiency as the Major Objective

The most fundamental approach in the design of objective functions is the use of economic efficiency E as the basic objective of a water project. This efficiency may be conceived as the B/C ratio, or as the benefit only, assuming that the resulting cost will not change the optimal B/C ratio, obtained as the optimal benefit divided by the cost. This is likely the case in the operation of water projects, though it may not be the case in their planning phase. In its most general form, the objective function of economic efficiency is defined by

$$E = F (X_1, X_2, \ldots, X_k),\tag{1}$$

with $X_1, \ldots, X_k$ the k quantifiable variables of purposes or subpurposes of water projects. These variables are the quantity of water delivered, hydropower produced, amount of service rendered, navigational depth assured, etc. In the simple linear for, Equation (1) can be written as

$$E = a_1 X_1 + a_2 X_2 + \ldots + a_k X_k,\tag{2}$$

with a_1, $a_2, \ldots$, a_k the coefficients in $/[X], where $[X]$ represents the dimension of any X_i, so that in this case E is expressed in monetary terms. The coefficients a_i are either

positive in case of a positive economic effect, or negative in case of damage, loss, or negative impact by water (floods, drainage-related problems, resulting expensive work in a river channel, etc.). Equation (2) is usually linear in the composition of terms, but the terms themselves may be nonlinear because X_i may have powers different from unity.

The ideal case is when X_1, $X_2, \ldots$, X_k variables of purposes or subpurposes are expressed in terms of only one decision variable, say the water release at a given place and for a given time interval (reservoir release, pumping discharge from an aquifer or a river, water diverted, etc.). If the system is simple (say one reservoir, one aquifer, one river diversion), the desision variable may be only one, say the released, pumped or diverted discharge, Q. Then Equation (1), for

$$X_i = f_i (Q)\tag{3}$$

becomes

$$E = F [f_1 (Q), f_2 (Q), \ldots, f_k (Q)].\tag{4}$$

In case of additive effects of various project purposes and subpurposes, Equation (2) becomes

$$E = a_1 f_1(Q) + a_2 f_2(Q) + \ldots + a_k f_k(Q), \tag{5}$$

subject to various constraints either for each term of Equation (3), or jointly for several or all terms of Equation (5), or for Q itself. In that case, Equations (4) and (5) are reduced to a simple function

$$E = F_0(Q) \tag{6}$$

with constraints translated into a set of inequalities, such as

$$Q \geq Q_{c,1}; \quad Q \geq Q_{c,2}; \quad \ldots \tag{7}$$

or similarly with $Q_{c,1}$ and $Q_{c,2}$ resulting as the limits from the constraints involved.

The simplified case of Equations (4) through (7) is applicable only to a simple subsystem, and usually not to a complex system. Further decision related aspects require the extension of the number of decision variables. Basically, these new decision variables result from extension of the system in space and time, and in distribution of allocated output to subpurposes. The three dimensions of the number of decision variables are as follows:

(1) More than one dicision variable is involved, say n decision variables Q_1, Q_2, ..., Q_n for n subsytems of a complex system, each with the water release Q_j, for j = 1, 2,..., n; this is an increase of the number of decision variables because of the space composition of complex systems by well-defined subsystems (reservoirs, aquifers, rivers, lakes, conveyance subsystems, etc.).

(2) The application of the objective function of Eq. (1) is extended over many time intervals, usually along an optimization horizon of h intervals; this is an increase in the number of decision variables because the impacts of future decisions affect the decisions at present and vice versa.

(3) The distribution of system equity among regions or among social strata require each or some of the X_i-variables of Equation (1) or of particular purposes to be composed of two or more subpurposes, say as $X_{i,1}$; $X_{i,2}$;...; $X_{i,s}$,

for X_i with s = the number of subdivision of an X_i into the subpurpose variables (conceived as the parts of a purpose in water or power allocation); this is an increase in the number of decision variables because of the distrubution of outputs of the project or system to subpurposes of a given purpose.

4.2 Increased Number of Decision Variables

In the case of an increase in the number of decision variables because of complexity of a water system, Equation (3) becomes a function of several decision variables, so that each purpose X_i may be expressed in function of any number of a total of n decision variables, namely

$$X_i = f_1(Q_1, Q_2, \ldots, Q_n). \tag{8}$$

If Q_i = the total water deliver to an irrigational area, Q_1 may be the pumping from one river, Q_2 from another, Q_3 from an aquifer and Q_4 and Q_5 water releases from two surface reservoirs, etc. Each X_i of Equations (1) or (2) may be a function of one or more decision variables Q_j of Equation (8), with j = 1, 2,..., n. Or, for X_i being the amount of the total production of hydroelectric energy or energy needed in water pumping in a time interval, the quantity may be a function of releases or pumped discharges Q_j of several subsystems, but not necessarily all of them.

The increase in the number of decision variables from one to n represents then the multiple variable decision making process. Two approaches may be used in treating this multi-dimensionality: (a) hierarchiacal decision approach, and (b) decision by solving complex problems imposed on optimization by this multi-dimensionality. Each approach has positive and negative aspects.

The hierarchical approach starts first with the general dicision, related to all or most Q_j's. This is either the sum of Q_j, or a function of Q_j, or a representative value, conceived as one variable decision, say Q_0. Then, by the specially designed method or procedure the top hierarchical decision, Q_0, is redistributed to several lower hierarchy values, at one, two or more levels of sub-hierarchical decision making, as $Q_{0,i}$ values, i = 1, 2,..., m. Figure 2 illustrates for the lowest sub-hierarchical level, the decision variables $Q_{0,i}$ of the next higher level are redistributed to Q_j variables, j = 1, 2,..., n.

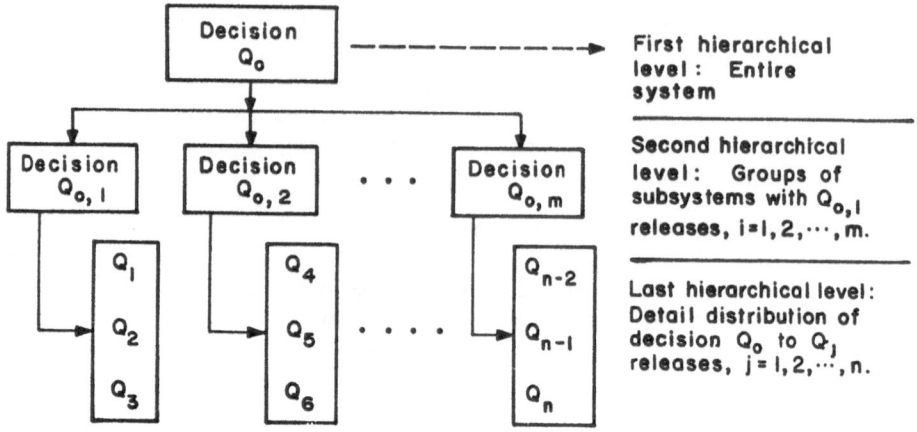

Figure 2.   Scheme of hierarchical distribution of general system decision release Q_0, at time t_0 for a time interval, into the subreleases of subsystems.

The approach of multi-dimensional decision making requires the optimization over all of the n decision variables. This may be feasible only for one time interval decision. It is difficult, and computer time limiting, if decisions are optimized along the horizon of h time intervals at each time interval decision along with the n decision variables.

Here is the case in which the state-of-the-art of optimization techniques and the computer potential and cost limit the treatment of complexity and reality in the design of objective functions. Then the efforts are directed either at simplification of objective functions so that some standard optimization algorithms can be used, or the efforts are made to find new methods of optimization, by new solutions to complex optimization problems. Likely, future efforts in both directions will lead to improvement in the application of optimization techniques to realistically designed objective functions involving a large number of decision variables.

4.3 Optimization Over a Horizon

The basic aspect of any water storage facility in a water system is an optimization between the present benefit and satisfaction of demand, control and protection, and the risks of future shortage, deficit, damage, lack of control and protection, with the resulting penalties, loss of benefit, increase of cost and outright failures. The trend among the water users and system beneficiaries is to get <u>now</u> as much of benefits as possible, and hope that the <u>worst</u> will not come in the immediate future. The optimal operation of storage facilities, however, requires that the satisfaction of present and future demands ·be balanced in terms of probabilities of future occurrence of both supply and demand. Therefore, the need is to make an optimal decision at the time t_0 for the forthcoming time interval under the condition that the decision made does not significantly increase the risk of future shortages and associated losses or to lead to waste of water.

The coefficients in Equation (2) or (5) may be disigned in such a way, that they are also functions of Q. A simple example is the case of hydropower. The value of energy per kilowatt-hour may be high for a low Q, say for $Q < Q_a$, and small for $Q > Q_a$, or for X_i, the hydropower quantity variable, may change with Q either in steps or as a continuous function, depending how the energy is priced. A very low value of a_i for $Q_i > Q_a$ will automatically force the release discharge to an optimal upper limit, even for the immediately forth-coming time interval. In the case of "distribution of water shortage" over the period of future deficit, with the total regulated supply smaller than the total demand for the horizon h of time intervals, this is not the case, since $Q_i < Q_a$ may be valid for all the intervals.

The horizon h, after which any release of water in the future does not affect (or affect in a neglible way) the dicision for the present time interval, is a random variable and not a constant, as usually either assumed or claimed. This fact complicates the process of optimizing decisions over the h future time intervals. Therefore, Equations (4) and (8) lead to a sum

$$E = \sum_{t=1}^{h} F[f_1(Q_1, \ldots, Q_n) + \ldots + f_k(Q_1, \ldots, Q_n)], \qquad (9)$$

with h either arbitrarily selected or determined in a proper way from its probability distribution. Equation (9) then incorporates both the n subsystem decision variables and the h horizon dicision variables, with a total or $r = nh$ decision variables.

Bacause r increases fast with an increase either of n or of h, the hierarchical decision Q_0 for $n = 1$ reduces r to h, which permits the use of dynamic programming in order to optimize dicisions over the time horizon h. This is a support for the application of dynamic programming by using the hierarchical decision making. The determination of the optimal future horizon is still in search of reliable, theoretical or practical solutions (Anderberg, 1979).

4.4 Treatment of Economic Equity

In case of distributions of benefits and costs, resulting from the operation of a complex water resources system, to regions or social groups, the additional distribution related decision variables may be needed. A purpose of irrigation water supply may have a subpurpose to assure the quantity of water to small farmers at the reduced prices, or water to small rural communities under more favorable conditions than to large urban communities. These distribution decisions represent the new variables in Equation (8), say adding the new decision variables, $Q_{n+1} + \ldots + Q_{n+s}$, so that now the number of variables is $(n+s)$, where s = the number of all additional decision variables that are subdivisions of the n decision variables, $Q_1, \ldots, Q_n$.

The imposed distributions of equity type (as the results of project goals and objectives) can be accomplished by one of the two alternatives of treatment of the objective functions: (a) the use of an increased number of decision ariables in the objective function, say from r = nh to r = (n+s)h, which definitively is not a desirable approach; and (b) the translation of the basic project equity objectives into the constraints or distribution criteria. An example is the use of a percentage of any release Q_i for irrigation by small farmers or for water supply to communities, allocated at a fixed or proportional price in comparison with the other farmers or with urban communities, respectively. The latter approach is simpler than the former, in terms of complexities in developing the objective functions and applying the optimization techniques (Brill, 1972).

4.5 Treatment of Environmental Quality Objectives

The treatment of objectives in solving the environmental problems, especially for those aspects of water-related problems that are not quantifiable either as benefits for their positive effects or as costs for their negative effects, is somewhat similar to the treatment of the equity objective. Many purposes of environmental enhancement may be quantifiable in terms of economic efficiency, or as the constraints to objective functions. However, some can not. The aesthetic problems of a project, however, are never measurable either quantitatively or even objectively (what is beautiful is highly controversial). Simply, such problems of environmental enhancements will be solved either by the position taken by the most influential decision-makers, or they will result from concessions to most vocal social groups, to "compromise in order to implement the project."

5. BASIC CONTROVERSIES INVOLVING OBJECTIVE FUNCTIONS

Proposals on how to make basic decisions on water projects have resulted in controversies that also affect the design and use of objective functions. Some of them are outlined briefly in the following text.

5.1 Decision Making in Planning Size and in Operation of Reservoirs

Two extreme positions may be discerned among planners and decision-makers of how large a crucial reservoir should be, or what should be the total capacity of all reservoirs in a water system. Similarly, extreme positions may be found on how best to make decisions in their operations.

In deciding the size of a storage reservoir, an extreme position is "to have it as large as one can get by without a significant opposition." It is based on four fundamental premises:

(1) The number of sites for storage reservoirs is limited and each should be used to its maximum potential;

(2) The need seems always to exist for more water storage space;

(3) The economic calculations often neglect increases in values of storage capacities with time, as the demand pressure on water resources increases; and

(4) Rarely complaints are heard of too much storage, in comparison with more frequent complaints of underdesigned storage capacities.

In this extreme position, the use of objectives through objective functions is only a small imput into the process of balancing various interests. Many other nonquantifiable aspects may prevail.

The second extreme position is to design storage capacities by the objective of economic efficiency, in using the present-day interests, economic criteria, political perceptions, and visions of future needs. This is obtained mostly by satisfying special, short-range, local or regional interests. In this case, the design of objectives and objective functions, and the connected exercise of optimization bear heavily on the final decisions.

All transitions exist between the above two extreme planning positions. Basically, the stop point of the pendulum between these extremes depends whether the values and interests of the present generation (of interested parties as well as of engineers, economists, analysts and politicians) dominate the decision making process or their modesty in perception and the care for future generations leave them the flexibility and opportunity to expand, adjust, amplify or modify the inherited water resources systems without a prohibitive cost. As in all the other economic and political decisions related to public investments, the swings between the present interests or the vision of all the future interests, will be the determinants of final decisions on water resources projects and systems. In this basic dilemma, the objective functions play a smaller role in decision making, when the long-range interests prevail, than the case is when the present-day problems, interests and narrow concepts of the self-centered generation predominate.

5.2 Quantifiable versus Nonquantifiable Objectives

The major differences in positions by planners of water resources systems involve the quantification of objectives in physical, monetary or ranking terms. At one extreme the position is that the large majority of water project outputs, effects and impacts can be quantified if only sufficient efforts, data collection and specific evaluations, even by using assumptions, are undertaken. What cannot be quantified is of the secondary importance, and can be taken care of by constraints or criteria.

The opposite extreme is that the significant proportion of project outputs, effects and impacts cannot be realistically quantified. They should be treated in different ways than by the use of objective functions. The examples are indirect benefits and costs (nonmeasurable impacts on society in general), aesthetic value judgements, values related to quality of life that are not measurable in monetary or physical terms. This view may extend to the extreme that humans should not temper with nature, because they can only damage it but not improve it by their water projects.

The history of quantification of project objectives and impacts gives a support to the first position of quantification. For example, recreational benefits, project impacts on fish and wildlife, in-stream increased flows, propagation of project outputs through the regional or national economy, all have been considered as nonquantifiable objectives many years ago. But presently they are considered quantifiable, which gives this position an outlook for continuous progress with time in quantification of objectives and of effects and impacts of water projects. On the other hand, some strong positions and historic experience of negative impacts in the interaction of humans and water environments, point out at many remaining difficulties in that final push, if it ever arrives, of quantifying the sensitive, elusive and value-prone aspects associated with water projects.

All intermediate positions between the above two extremes have been advocated in the past. What seems to be the state-of-the-art on quantification of objectives of water projects, is that objectives should be quantified in the form of outputs, effects and impacts, as much as pratically feasible. The remaining effects and impacts can then be sorted in two categories: (a) quantifiable but based on some general assumptions that may not be acceptable to all interested parties, and (b) nonquantifiable but identifiable effects and impacts, that are subject only to value judgments. The first category, subject to errrors, may be included with the well-quantifiable objectives; however, with proper taking into account of these errors. In that case, two objective functions may be used, one with fully quantifiable objectives, and the other with additional objectives that need various error-prone assumptions in quantification. The optimization of the two functions would give information on the effects of this "semiquantifiable" objectives. The second category usually can be treated only through value judgements, constraints, criteria and the political process at all levels of the review and inputs by the general public.

5.3 Solutions of Water Problems versus Solutions of Employment of System Analysis

In the discussion of applications of objective functions and optimization techniques, it is often difficult to pinpoint clearly whether some claims or counterclaims of benefits of these applications are unbiased positions to further the solutions of problems in planning and operation of water projects, or they are advanced primarily to increase the employment of the new breed of specialists, who identify themselves as systems analysts. This is a known phenomenon in any emerging technology, subdiscipline, or scientific advancements. It is often encountered in the applications of differential equations (differential equations analysts), or statistics (statistics analysts), or probability (probability analysts), in various phases of planning, design and operation of water projects.

History has shown, however, that all these specialists are the needed collaborators (often advisors only) in the complex process and team work of various phases of an important water project. Also, it shows that the judgment of the usefulness of various methods of solving the problems with optimization of objective functions rest definitely with the planners and decision-makers.

6. TIME FACTOR AND OBJECTIVE FUNCTIONS

An aspect of objectve functions seems to be nearly always neglected, namely the continuous or stepwise evolution of objectives, purposes and decision variables with time. This is particularly important in the planning phase, when optimizations and performance measures stretch over the entire project life. It is less important in the operational phase, because the evolution in objectives is taken into account by the change of objective functions (say, the change in objectives, purposes, decision variables, coefficients, forecast, etc.).

One will be able to give examples of water systems that have significantly changed their objectives, and several times, during the project economic life. Therefore, the solidified position on objectives in the planning stage often looks only as the game of numbers, rather than a feasible approach of most likely full set of evolutions in the project life as well as beyond it.

These reflections on effects of the time factor on objectives and objective functions, are a decisive determinant in the final, resulting benefit-cost ration at the end of the economic project life, and have important impact on basic planning strategies. It gives a heavy weight to long-range goals and interests in comparison with the present-day interests, concepts and economic measures of performance.

ACKNOWLEDGEMENT

This text is a contribution by the project "Investigation of Objective Functions and Operational Rules of Storage Reservoirs," conducted at Colorado State University, Fort Collins, Colorado, U.S.A., and sponsored by the Office of Water Research and Technology, U.S. Department of Interior, through the Water Resources Institute at Colorado State University, as Project B-195-COLO. Studies were partially sponsored by the Colorado State University Experiment Station, ES-114. The studies leading to this paper were also financially supported by the cooperative U.S.-Spanish research project on "Conjunctive Use of Various Sources of Water," sponsored by the Joint U.S.-Spanish Committee via Research Institute of Colorado, Fort Collins, Colorado.

REFERENCES

Anderberg, Lars A. W., 1979, "The anticipatied decision influence period in real time reservoir operation," Ph.D. Dissertation, Colorado State University, Department of Civil Engineering, Fort Collins, Colorado.

Brill, E. D., 1972, "Economic efficiency and equity in water quality management," Ph.D. Dissertation, The Johns Hopkins University, Department of Geography and Environmental Engineering, Baltimore, Maryland.

Cockrane, J. L. and M. Zeleny (eds.), 1973, Multiple Criteria Decision Making, University of South Carolina Press.

Cohon, J. L. and D. H. Marks, 1975, "A review and evaluation of multiobjective programming techniques," Water Resources Research Journal, Vol. 11, No. 2, April, p. 208-220.

Goicoechea, A., D. R. Hansen and L. Duckstein, Multiobjective Decision Analysis with Engineering and Business Applications (to be published by Wiley and Sons in November, 1981.)

Haimes, Y. Y., W. H. Hall, and H. T. Freedman, 1975, Multiobjective Optimization in Water Resources Systems, Elsevier.

Major, D. C., 1969, "Benefit-cost ratios for projects in multiple objective investment programs," Water Resources Research Journal, Vol. 5, No. 6, December, p. 1174-1178.

Marglin, S. A., Public Investment Criteria, 1967, M.I.T. Press.

Westgate, John T., 1980, "Design of objective functions for reservoir operations," M.S. Dissertation, Colorado State University, Department of Civil Engineering, Fort Collins, Colorado.

10. WATER QUALITY MODELING OF GROUNDWATER SYSTEMS

Marcello Benedini
Water Research Institute
National Research Council, Rome

1. INTRODUCTION

Groundwater aquifers serve both as reservoirs and as conduits, thus making water readily available over large areas of land for any convenient purpose. A survey sponsored by the Commission of European Communities (1) has shown that groundwater is a source of supply for a variety of economic activities. In Denmark, for instance, groundwater satisfies 100 percent of the water requirements for agricultural uses, 98 percent for domestic uses, and a large proportion of industrial uses. In some countries it is used even as a coolant for conventional power stations. France alone estimates that it will be using annually some 1000 mcm (million cubic meters) of groundwater by the year 2000. Potential groundwater resources in Italy amount to some 13,000 mcm (2). Most of this is located in the north and only a small fraction, i.e., 22 percent, occurs in the south and on the islands. A recent Italy-wide survey indicated that current withdrawal from wells is about 8700 mcm per annum.

With such pervasive dependence, protection of the quality of groundwater resources is of the utmost importance. The aquifer is passive, i.e., control is not possible, and so pollution is virtually irreversible.

It is only within the past two decades, that there has been much interest in the problem of groundwater quality. This has been stimulated by the general need to protect resources of vital interest to the economies of entire regions, and by the rapid spread of groundwater pollution. The problem must be viewed within a broad framework involving numerous factors, such as

157

geology, hydrogeology, surface and groundwater exchange, withdrawals, discharges, etc., and discipline interests, such as science, engineering, law, economics, politics, etc.

For the above reasons it would appear desirable to treat groundwater protection in the broader context of water resources management. Such management should utilize the scientific methods of systems analysis that have been found so effective in many other equally complex fields.

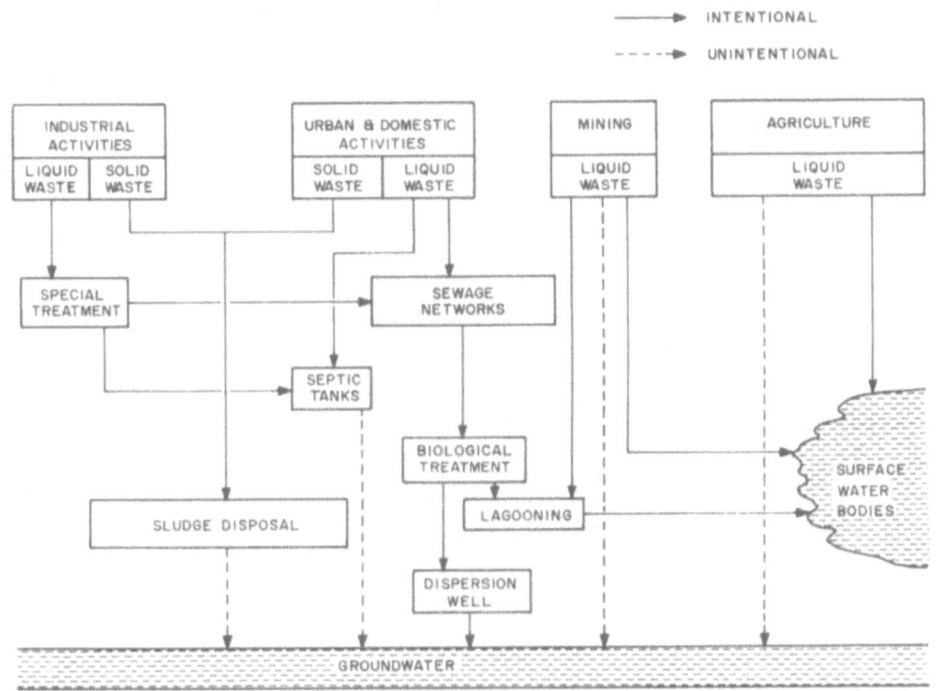

Figure 1.   Sources of Groundwater Pollutants

2. SOURCES OF GROUNDWATER CONTAMINATION

Pollutants can reach an aquifer from a variety of sources. Figure 1 illustrates, showing a range possible origins. In most cases groundwater pollution is the result of passive events. For example, a liquid waste may seep through the subsoil due to leakage from sources not intended to leak; sewers are one such source. Surface water also, particularly rainwater, may have contact with a deposit from which contaminants are leached; they are then carried to an aquifer. In the case of seepage from the surface the unsaturated layer also is affected; the contaminants in water undergo filtration, adsorption, and other changes in the

porous media constituting both the upper unsaturated strata and
the aquifer itself. Some of this retention may be a continuing
source of pollution in that the substances orginally retained may
be gradually released. In many cases, however, contaminant laden
water is introduced directly into the aquifer by injection wells
as a method of wastewater disposal.

3. CONTAMINATION PATTERNS

Figure 2 illustrates more specifically some of the ways in
which groundwater pollution may occur. As outlined by the Food &
Agriculture Organization of the U.N., (5) the pollutant typology
is a function of the cause and the way an aquifer is reached. In
urban waste discharges, organic matter and bacteria are the main

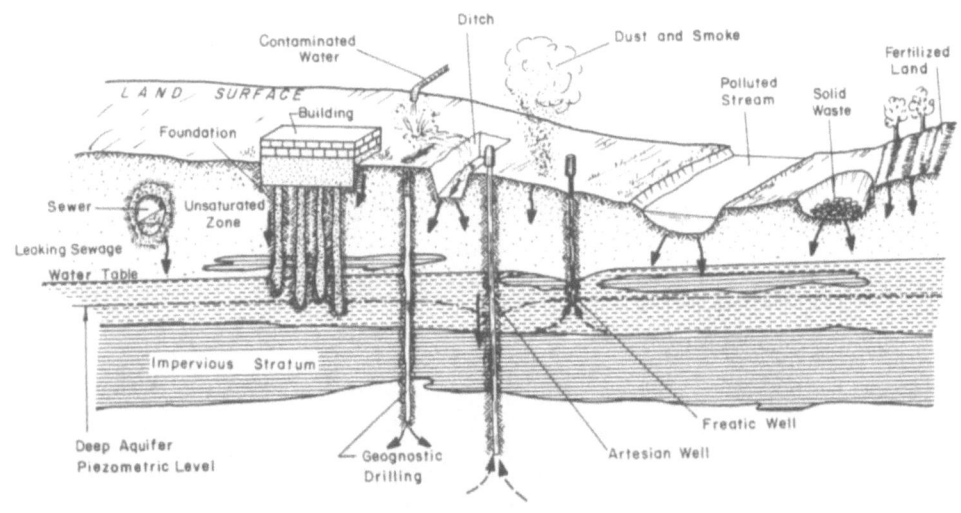

Figure 2. Examples of how Groundwater may become Polluted

concerns. The organic matter is subject to decomposition and
mineralization by bacterial reactions. Aerobic bacterial
reactions will occur in the upper unsaturated soil strata. At some
depth, however, the system becomes anerobic due to depletion of
the oxygen. Attention must be given also to the hazards of
bacteria and virus travel through the porous media. This is
especially a problem with fissured aquifers.

Leaching of solid wastes and treatment plant sludges by
rainwater is another category of pollution hazard. In large urban

centers the presence of hydrocarbons and lead due to motor vehicles and the gradual air and soil pollution resulting is also a hazard to groundwater. Pollution by radioactive substances from hospitals and clinics has become important recently.

Another major source of pollution is agriculture. The widespread use of both natural and chemical fertilizers is not easily controlled. They are distributed uniformly over entire areas. The spreading of fertilizers over the ground results in the appearance of several nitrogen compounds, such as urea, ammonia, nitrites and nitrates. Phosphorus compounds occur also, especially orthophosphates. On contact with the soil and the oxygen in the air, the nitrogen and phosphorus compounds are involved in various chemical reactions. The polluting effects of agriculture are spread also by dusting and spraying with pesticides, fungicides, and herbicides, which often are persistent in the environment. Observations carried out on one of the most common insecticides, DDT, have confirmed that it can persist in the soil for more than 20 years. The postulated duration for another pesticide, dieldrin, is even longer. The persistence of these substances can be explained by their low solubility. However, it must be admitted that this acts also as a safety factor by slowing down the rate of aquifer penetration. Nevertheless, this process of release in tiny doses, of the order of $\mu g/l$ does occur. The effect on ecosystems has yet to be evaluated. Nitrogen compounds are also released naturally by the vegetation in the course of its own chemical and biological activity.

Equally disturbing is the industrial pollution, the most distinctive characteristic of which is the presence of metals and organic compounds. These may end up in the aquifer after being transported in liquid discharges, or as a result of solid wastes being leached by rain water. A special role is played in this respect by waste from certain small-scale industrial activities such as the processing of agricultural products (olive pressing, winemaking) or zootechnical products, and from small workshops and service stations. As these activites are located in areas with no sewage systems, they affect the aquifers directly by discharging into natural fissures or quarries lacking the natural protective action of the topsoil.

Mining is another source of groundwater pollution. Mine drainage and leaching of tailings piles are usual sources of polluted water from mining.

4. SEAWATER INTRUSION

Pollution of coastal aquifers due to saltwater intrusion is a common occurrence. The hydraulic mechanism associated with the problem is illustrated in Figure 3. It becomes an acute problem when water is pumped out of the aquifer, since the resulting intrusion cones cause subsequent pollution of a larger portion of

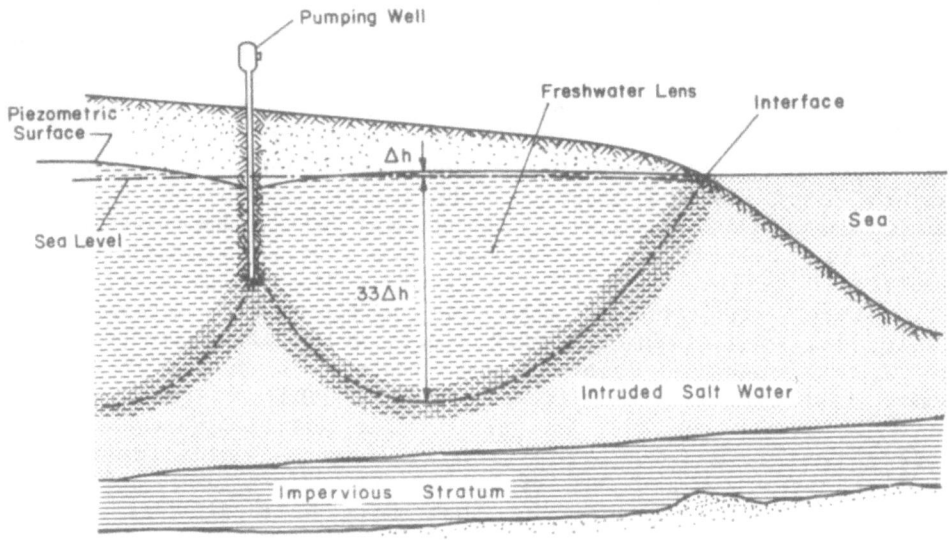

Figure 3. Hydraulic Mechanisms Associated with
         Seawater Intrusion

the aquifer. In the case of a coastal aquifer the presence of a saltwater layer at lower levels of an aquifer is due to the great hydrostatic pressure exerted by the sea and the higher specific gravity of seawater. Lying above these layers is that of the surface and rainwater, whose motion depends the piezometric head caused by the natural conditions. The interface formed between the two layers, although conventionally represented as a single surface, actually consists of a comparatively thick transition layer inside which the saline concentration varies rapidly and continuously from that of the sea to that of fresh water. Pumping freshwater from a well will upset this equilibrium. As the density of seawater is about 1030 kg/m^3 and that of fresh water is about 1000 kg/m^3, a variation of Δh_f in the level of the upper, freshwater layer must be offset by an opposite variation in the interface level of

$$\Delta h_s \overset{\sim}{=} 33\ \Delta h_f$$

This theoretical law, which is known as the Ghiben-Hertzberg relation, (6), may be considered as generally verified (7), albeit with some approximation. The main problem in application is the difficulty involved in correctly positioning the interface, which has already been described as a layer of brackish water of variable density. The thickness of this layer is affected by changes in the level of the free surface of the freshwater layer, i.e., by changes which cause amplified variations in the theoretical interface level.

5. BEHAVIOR OF POLLUTANTS IN AQUIFERS

The behavior of the substances entering groundwater may vary considerably since they may interact with mineral particles

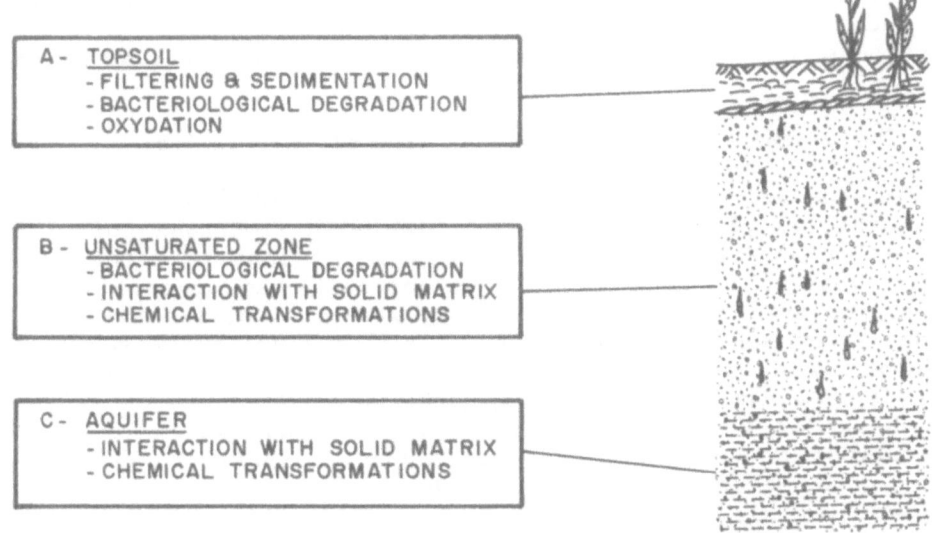

Figure 4.   Contaminant Removal Processes Associated
with Groundwater Percolation Zones

comprising the porous media of the aquifer. Figure 4 illustrates the different removal processes which occur in the different percolation zones of the porous media. In the simplest case, interaction with the mineral matrix may involve adsorption only, leading to an increased persistence of the pollutant in the aquifer, even when the incoming water has regained its previous purity. Under these circumstances, as has been mentioned, the process of desorption of pollutants from the solid matrix will result in continued release of small quantities of these substances over very long periods and thus recontamination of the water in the aquifer will occur.

More aggressive chemical reactions also may occur between the pollutant and the matrix, leading to the formation of compounds which ultimately will be carried away by the water or else they may be bonded to the matrix. Generally speaking, the upper portion of the mineral matrix near the surface plays a protective role from these effects. In addition to its mechanical filtering action, it is also the locale of intense chemical and biological activity which eliminates much of the organic substances and the bacteria and modifies certain inorganic compounds in the water as well.

This protection is no longer afforded when the water flows directly into the aquifer through injection wells or tunnels, or through natural fissures. In this case, the filtering action is exerted only by the mass of solid material in direct contact with the incoming flow. Within the aquifer the pollutant will spread. The extent depends on the characteristics of the granular mineral matrix.

The chemical nature of the mineral matrix is, in fact, responsible for the kind of chemical reaction that may take place between the matrix and the substances contained in the liquid. A vital role is played by pH, particularly in limestone soils (8). The action on organic pollutants and metal ions is more complex (9). Grain size and shape, together with the degree of grain compaction and cementation are of importance also, insofar as they favour water motion and pollution transport. It is interesting to observe how a liquid waste having solid matter in suspension can cause gradual reduction in soil porosity as a result of sedimentation due to the very slow velocity. This frequently occurs in agriculture when untreated liquid waste is spread directly over the ground, thus quickly making the surface soil layers almost impermeable.

6. POLLUTION IN ARTIFICIAL RECHARGE

The qualitative aspects of artificial recharge of an aquifer are of special significance. Artificial recharge has become a practice which permits storage of water in aquifers. Thus it is necessary that such water meet designated standards of quality right from the start.

In some cases, however, the aim of the recharge is not to increase the amount of water available in storage but to modify the natural conditions of the motion in the aquifer. In coastal aquifers, for example, the piezometric surface is altered to cause a gradient toward the sea. Figure 5 illustrates the two kinds of situations, e.g., recharge for storage and recharge to control the hydraulic gradient in the a and b diagrams, respectively.

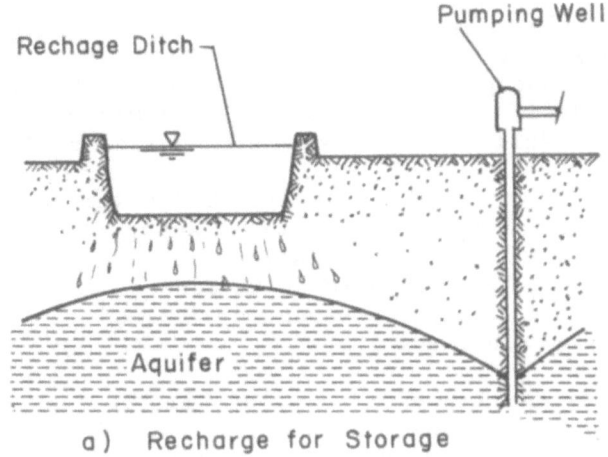

Rechage Ditch

Pumping Well

Aquifer

a) Recharge for Storage

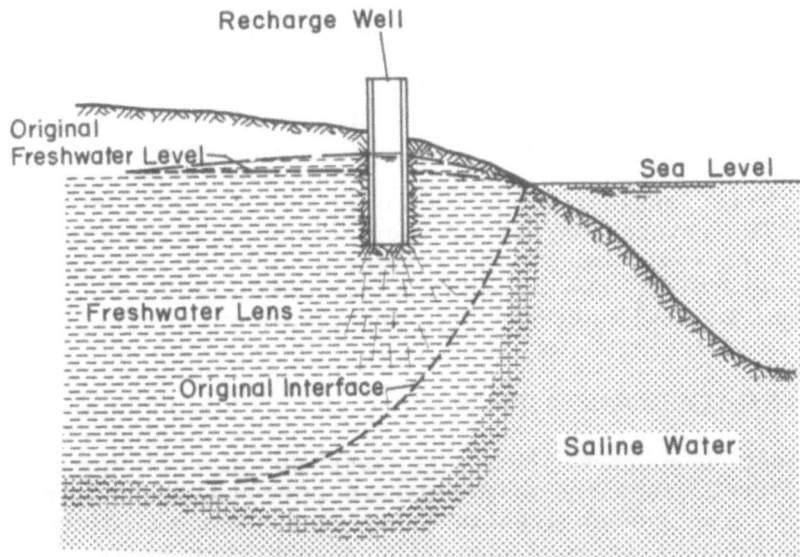

Recharge Well

Original
Freshwater Level

Sea Level

Freshwater Lens

Original Interface

Saline Water

b) Recharge to Control Hydraulic Gradient

Figure 5. Recharge of Groundwater

In some well known cases the recharge water used consists of urban effluent that has undergone a sophisticated tertiary treatment, consisting at least of coagulation, filtration, activated carbon adsorption, and chlorination. The product water meets water standards. The cost of such treatment is high. But the soil system itself is capable of attenuating small quantities of contaminants even further (10). For instance, it has been ascertained (11) that by allowing polluted water to flow into a sandy subsoil, and by subsequently withdrawing it from a pumping well about 140 m away after a retention period estimated at about 100 days, drinking water standards with respect to phosphates (92 percent elimination) and suspended solids (100 percent elimination) may be achieved.

Even smaller distances, e.g., as little as 5 m between introduction and withdrawal, would allow the initial BOD to be reduced by 75 percent and the COD by 50 percent. In order to ensure the complete removal of all viruses and bacteria, larger distances would be required, e.g., more than 450 m.

On the basis of these considerations less advanced treatment could be adequate. The conditions required to obtain a satisfactory effluent also could be attained by lagooning of the effluent prior to injection. The water subsequently withdrawn should be disinfected. In any case there would be a considerable saving in plant and operating costs.

All these considerations do not, however, take into account the possible "saturation" of the soil's purification power or of clogging which would reduce the filtering capacity of the recharge zone. In practice, the actual implementation of these measures must be done with all attention to these factors.

7. LEGAL AND ECONOMIC ASPECTS

In most countries there are laws to protect groundwater. Usually, direct discharge of pollutants into underground strata is prohibited. The recent Italian legislation bans the discharge of effluents, even if treated, into the ground when there is chance of the underlying aquifer being affected. Guidelines are given to help determine whether contamination is likely. Also, special care must be taken when authorizing irrigation with wastewater effluents. In addition, there must be safeguards to prevent contamination due to accidental spills or other discharge on the ground surface.

Water quality standards for groundwater have not been established. This is based upon the premise that surface water quality standards are adequate and that any surface water infiltration into groundwater will undergo attenuation with respect to contaminants. But in the case of accidental contamination, for instance when a sludge pile is eroded by rainwater, it is no longer possible to speak of limits of acceptability and everything must be done to control the channelling of the run off waters and their destination with respect to the various ways in which they could reach the water table.

The economic aspect of this legislation and the associated regulations is initially focussed on the cost of effluent appropriate to the surface watercourses. The need to protect aquifers against accidental contamination on the other hand entails special costs for construction of physical separation facilities (e.g., mud containment sumps, rainwater collection basins, underground barriers, etc.), for operation, monitoring, and for treatment of the collected water. The economic evaluation of aquifer management must also take account of the self-purifying capacity of the soil layers, which may be utilized in lieu of treatment.

8. MODELING GROUNDWATER POLLUTION

Models are needed to study the behavior of a groundwater system. Models may be of any kind, i.e., physical, analog, or mathematical. Physical models have not been applied widely in pollutant transport and spread. Probably this is due to the complexity of reproducing the real situation. The best known analog models are the Hele-Shaw and electric analogs. The Hele-Shaw analog has been used in the study of the motion of a fresh-water layer floating on an intruded saltwater layer (12) under Ghiben-Herzberg conditions and without mixing. A difficulty in analog modeling consists of finding a hydrodynamic, electrical, or other kind of physical phenomenon which obey the same laws as those representing the pollutant transport process in a liquid medium. The use of semiconducting papers, as described by De Wriest (13), are used widely.

The most widely used models are mathematical, in which a numerical solution is sought. In order to use a mathematical simulation model it is necessary to know the basic relationships governing the motion of liquids and the pollutants contained in them. In the case of seawater intrusion, attempts have been made to make a formal approach based on relationships of equilibrium between the two layers (14). A more realistic treatment, however, is that which takes into account the motion of both the freshwater layer and the underlying salt water layer. This is done by transcribing the fundamental equation of groundwater hydrodynamics twice and assuming an equilibrium for the interface (15). With reference to Figure 6, the following may be written for the freshwater layer,

$$\frac{\partial}{\partial x}\left(K_x^f \frac{\partial h^f}{\partial x}\right) + \frac{\partial}{\partial y}\left(K_y^f \frac{\partial h^f}{\partial y}\right) + \frac{\partial}{\partial z}\left(K_z^f \frac{\partial h^f}{\partial z}\right) = q^f + S^f \frac{\partial h^f}{\partial t}$$

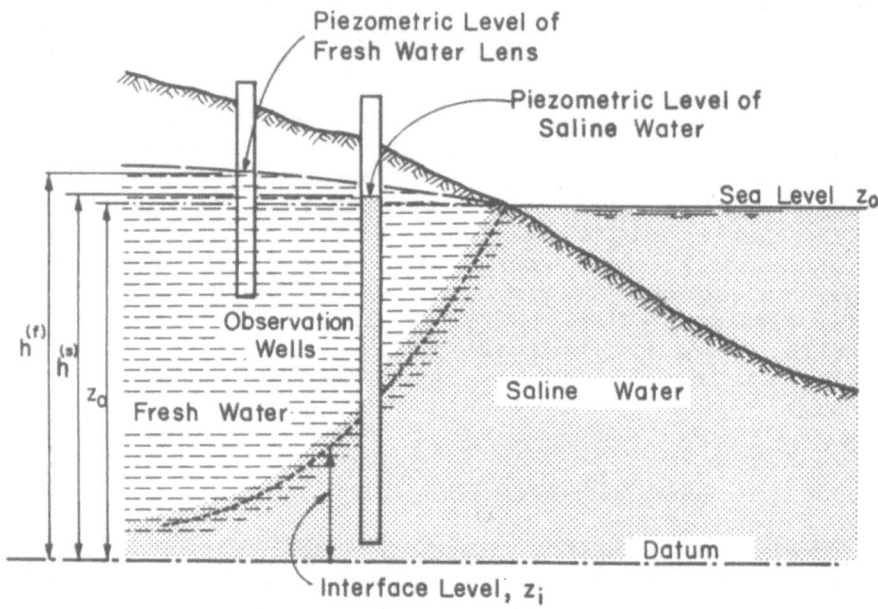

Figure 6. Problem Set-up for Modeling Seawater Intrusion

and for the saltwater layer,

$$\frac{\partial}{\partial x}(K_x^s \frac{\partial h^s}{\partial x}) + \frac{\partial}{\partial y}(K_y^s \frac{\partial h^s}{\partial y}) + \frac{\partial}{\partial z}(K_x^s \frac{\partial h^s}{\partial z}) = q^s + S^s \frac{\partial R^s}{\partial t}$$

and for the interface,

$$z_i = \frac{\rho^s h^s - \rho^f h^f}{\rho^s - \rho^f}$$

The numerical integration of these two equations makes it possible to trace out the interface position as a result of given withdrawing conditions that may be expressed in terms of the functions q^s and q^f. Of greater complexity is the aspect pertaining to pollution transport inside the liquid mass of the aquifer, as could occur in the case of penetration of the upper soil layers by polluted surface waters.

The phenomenon of groundwater pollution may be represented schematically by introducing a foreign species into the aquifer; the concentration is given as $C = C(x,y,z,t)$, i.e., mass of pollutant per unit volume of water. According to the well-known Fick's law, if between two points a distance dx apart in the aquifer a variation dC exists in the concentration, a pollutant transport flux, J, is set up between the point with the higher concentration and that with the lower. This can be expressed as:

$$J = -D \frac{dC}{dx}$$

where D is a coefficient of proportionality. This process is called diffusion.

A first factor in this process is the effect due to the random motion of the molecules, which gives rise to the "molecular diffusion," formally expressed by $D = D_M$, the coefficient of molecular diffusion (for the mixing of seawater with fresh water its value if of the order of $10^{-5} cm^2/s$). A second factor which must be considered are the strong local concentration gradients occurring in the liquid mass when the polluted liquid flows along the winding paths through the grains in the solid matrix. This also is accounted for by means of another proportionality factor D_T known as the coefficient of mechanidal diffusion. The two above-mentioned pehonmena can be accounted for by

$$D = D_M + D_T$$

All these effects are gathered together in the coefficient of dispersion D, which theoretically depends on numerous physical variables. The most important is the velocity, v, and the intrinsic permeability of the porous medium.

If U is the module of velocity, the coefficient of dispersion may be taken as

$$D_i = D_0 + \alpha U$$

The term, α is a constant of proportionality which depends on the porous medium. The velocity is obtained by solving the hydrodynamic problem, while the parameter α_i may be obtained experimentally from field measurements as described by Molinari and Rochon (16). Furthermore, under the effect of the liquid motion, part of the pollutants are transported by the convection process, is taken into account by referring to the range of velocity as represented by the variations (in terms of velocity components v_x, v_y, v_z),

$$\frac{\partial}{\partial x}\left(v_x C\right) \quad ; \quad \frac{\partial}{\partial y}\left(V_y C\right) \quad ; \quad \frac{\partial}{\partial z}\left(v_z C\right) \ .$$

In a rapidly moving liquid convection predominates over dispersion.

As a final note, some of the reaction phenomena are difficult to depict mathematically, i.e., change in bacterial counts, or reactions with the soil matrix. A synthetic expression, $f(x,y,z,t)$, is used, which must be added to the expressions describing the phenomena dealt with previously, using a positive sign for a source, or a negative one for a sink. The one-dimensional mathematical expression for pollutant transport in the aquifer may be given as,

$$\frac{\partial}{\partial x}\left(D_i \frac{\partial}{\partial x} C\right) - \frac{\partial}{\partial x}\left(V_x C\right) = \frac{\partial}{\partial t}\left(nC\right) \pm f(x,t).$$

in which the term n represents the porosity.

9. APPLICATION OF MODELS

Application of the mathematical models described in the preceding section is complex in that the same equations must be rewritten many times for each of the elementary configurations into which the aquifer is divided. This calls for a high degree of skill and "sensitivity" that can be acquired by the operators only after appropriate training. Furthermore, suitable data must be available on the nature of the aquifer, pollutant characteristics and the ways in which the water withdrawn is to be used.

Although formally very complicated, the procedure for integration of the equation has been simplified considerably by the advances made in large computer science and technology, as well as by the availablity of suitable software. Considerable advantages are offered by the "finite elements" procedure which is replacing the nevertheless effective "finite differences" procedure.

Output may be obtained from the model in the form of readily readable and usable printouts and graphs. Computer generated plots such as those in Figure 7 are particularly useful in visualizing the results of the computation.

Mathematical modeling does not, however, stand alone. The examination of results, however presented, is a task requiring great effort and considerable knowledge of the real situation. At this stage, on the basis of their experience, the operators must decide on the "correctness" of the results. This is done by

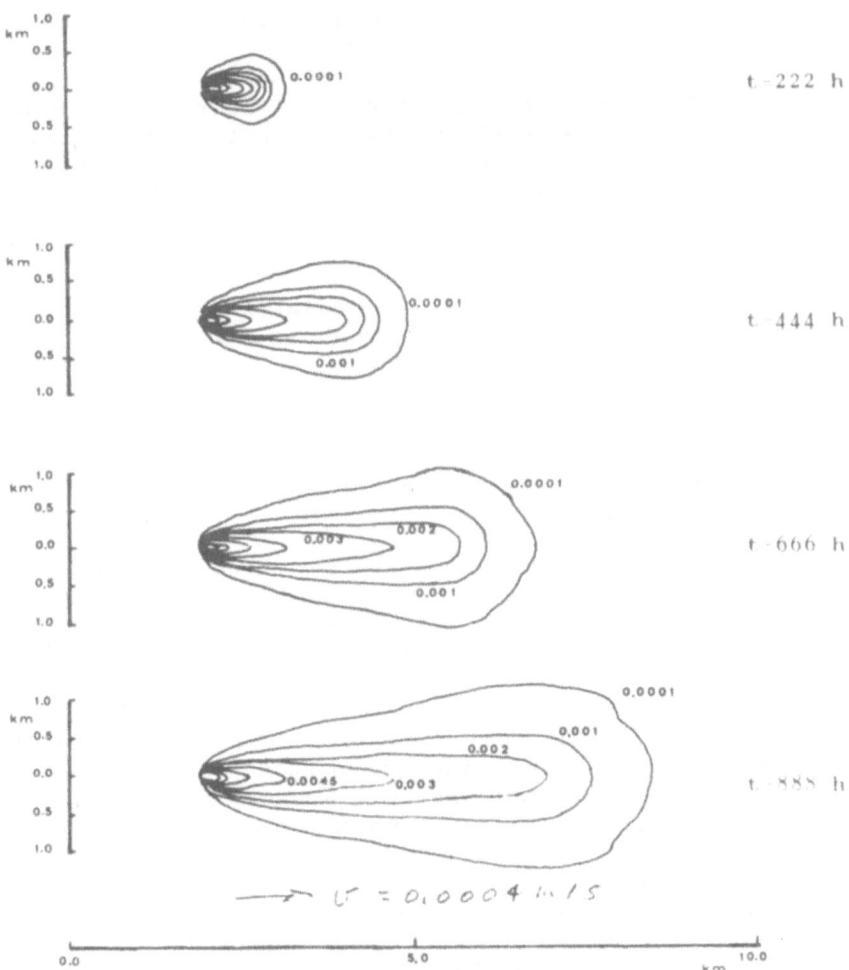

Figure 7.   Computer Plots of Isoconcentriation Curves in kg/m^3 for a 0.1 kg/s Tracer Injection (17)

comparing the output with field data, or by means of new model applications.

The use of models can allow determination of the predictable state of pollution inside the aquifer as a function of known conditions of groundwater exploitation.   At the same time it is necessary to solve the hydrodynamical problem in order to evaluate velocity distribution and piezometric pressure behavior inside the aquifer.

As seen in the preceding chapters, water velocity must be known also in order to evaluate and quantify the phenomenon of convection transport. However, the hydrodynamic field may also be influenced by the presence of some pollutants whose physical characteristics (e.g., specific weight, viscosity) may be quite different from those of water, for instance, hydrocarbons.

The literature outlines the principles governing motion of groundwater. From this three key points should be noted:

1. In groundwater flows convection usually predominates over diffusion

2. Molecular diffusion phenomena become important when long periods of time are considered,

3. Mechanical diffusion phenomena become important under turbulent flow conditions inside the small canals in the porous medium. In this case, the Reynold's number of the filtration does not necessarily lie outside the range of validity of Darcy's law.

Mention must also be made of the attempt to find a qualitative expression for the phenomenon of matrix-pollutant-interaction so that the above mentioned expression $f(x,y,z,t)$ is meaningful even if its inclusion in the fundamental equation can lead to much complexity.

The pollutant attenuation phenomenon is controlled by the concentration of the pollutant in the liquid phase, (C), and by the quantity, (S), of pollutant already absorbed or released by the solid phase (18), according to the expression,

$$\frac{\partial S}{\partial t} = k_1 G^n - k_2 S$$

The terms n, k_1 and k_2 stand for the overall multiple effects of a physical, chemical and fluid nature, and must be determined experimentally.

REFERENCES

1. Commission of European Communities, "Evaluation of water resources in the European community," Brussels, 1977.

2. Conferenza Nazionale delle Acque, "I problemi delle acque in Italia," Roma, 1972.

3. Bear, J., Zaslovsky, D. E. and Irmay, S., "Physical principles of water percolation and sepage," UNESCO, New York, 1968.

4. Fried, J., "Groundwater pollution," Elenco Publ., 1975.

5. Food and Agriculture Organization of the U. N., "Groundwater pollution," F.A.O. Irrigation Papers, n. 31, Rome, 1979.

6. Cotecchia, V., "Studies and investigations on apulian groundwater and intruding seawater (Salento Peninsula)," Quaderni Istituto Ricerca sulle Acque, n. 20, Roma, 1977.

7. Cotecchia, V., "Direct and laboratory observations of the mixing phenomena of fresh and salt water in coastal groundwater. Particular case of fracutred media," Proceedings of the Symposium on Hydrodynamic Diffusion and Dispersion in Porous Media. I.A.H.P., Pavia, Italy, 1977.

8. Hall, E. S., "Some chemical principles of groundwater pollution," Proceeding of Conference on Water, Water Research Association, Reading, Sept. 1972.

9. Quaderni P. F., "L'inquinamento del suolo," Ambiente, 1979.

10. Cohen, P. and Durfor, C. N., "Artificial recharge experiments utilizing renovated sewage plant effluent, a feasibility study at Bay Park, New york, U.S.A.," I.A.S.H. Symposium of Haifa, Israel, 1967.

11. Frankel, R., "Economics of artificial recharge for municipal water supply," I.A.S.H. Synposium of Haifa, Israel, 1967.

12. Zanovello, A., "Scritti in onore del prof. Mazzolo," Padova, 1962.

13. De Wriest, R. J. M., Geohydrology, John Wiley, New York, 1965.

14. Troisi, S., "Modelli di inquinamento marino di falde acquifere costiere," Atti del Convegno su Metodologie Numeriche per la soluzione di Equazioni Differenziali del l'Idraulica," Bressanone, 1978.

15. Sirangelo, B. and Troisi, S., "Studio metodologico di nuove tecniche di gestione delle risorse idriche sotterranee: 1 - Modello di sumulazione idrodinamica, "IDROSIM," Quaderni IRSA, n. 50, 1980.

16. Molinari, J. and Rochon, J., "Mesure des parametres de transport de l'eau et des substances en solution en zone saturèe," La Houille Blanche, numero special 3/4, 1976.

17. Gambolati, G., Troisi, S. and Volpi, G., "Un modello di inquinamento agli elementi finiti in sistemi filtranti." Ropp. Interno IRSA, 1981.

18. Boochs, P. W. and Barovic, G., "Numerical dispersion and sorption of pullutauts in groundwater," Proceedings of Symposium of Hydrodynamic Diffusion and Dispersion in Porous Media, Pavia, 1977.

PART III: DECISION MAKING

11. HYDROLOGIC FORECASTING FOR OPERATION

Vujica Yevjevich
George Washington University, Washington, D.C.

1. INTRODUCTION

This chapter is based on the premise that both water supply and water demand of a complex water resources system contain compositions and variations of their time series, that make them conducive to hydrologic forecast. Because demand is often highly affected by climatic and hydrologic variables (precipitation, temperature, soil moisture, snowmelt, etc.), it is subject also to hydrologic forecasting techniques, similar but not identical to water supply forecast. The ensuing discussion is not concerned with the details of forecast methods; rather it stresses the interaction of the power and utility of forecasting in providing benefits for the operation of water resources systems.

1.1 Definition of Operational Forecast

Figure 1 serves to define the idea of an operational hydrologic forecast. At the time t_0, the forecast is made for both supply, S, and demand, D, for the forthcoming time interval, $\Delta t = t_2 - t_1$. The realized supply and demand during the interval, Δt, are rarely identical with forecast values. Differences $\Delta S = S_f - S_r$ and $\Delta D = D_f - D_r$ are forecast deviations, for supply and demand, respectively. In Figure 1, the forecast of water surplus $(S_f - D_f)$ is different than the realized surplus $(S_f - D_f)$. This is due to errors in both forecasts. Any forecast method of water supply and water demand may be beneficial in the operation of complex water resources systems, provided the method gives smaller deviations $S_f - S_r$ and $D_f - D_r$, on the average, than

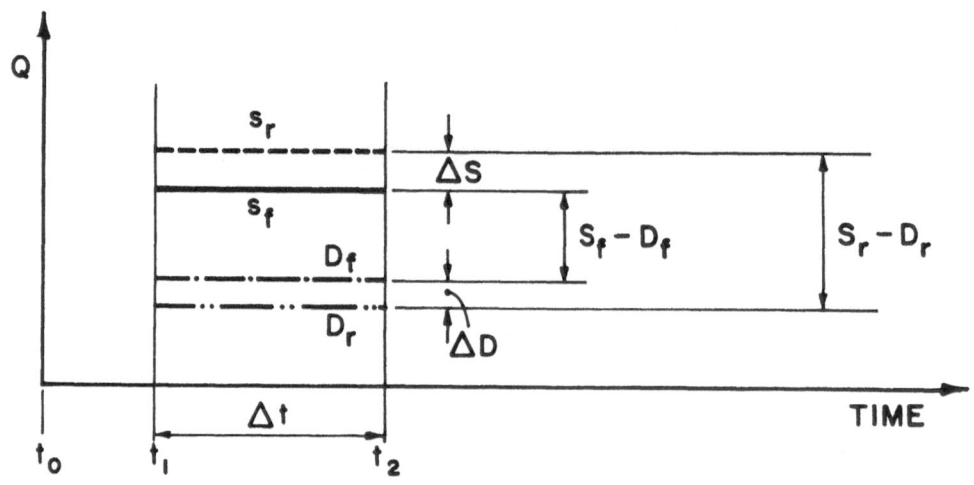

Figure 1.   Definition of Operational Hydrologic Forecast

the simple generation of potential future realizations of supply and demand (or the Monte-Carlo simulation).

The generation of time series samples assumes that only past information is used, and no effects are taken into account of variables that in one way or another precede the occurrence of, and affect the forthcoming outcome of supply and demand. To illustrate, let us designate the discharge of water supply or water demand of the future interval, Δt, by Q_i. Further assume that these flows are autocorrelated and normalized, say by $\log Q_i$. Then an example of the generated future discharge values, in forms of $\log Q_i$, are

$$\log Q_i = \sum_{j=1}^{m} a_j \log Q_{i-j} + \sigma x_i, \tag{1}$$

with x_i = independent values of standard normal variable a_j = autoregressive coefficients, σ = standard deviation of residuals, and $\log Q_{i-j}$ = preceding flows.   Only the past information of observed flows is used.   Forecasts should produce a more accurate anticipation of S and D than Equation (1).

1.2 Forecast of Supply and Demand

Operation forecasts uses various methods to forecast supply and demand. They are the anticipated system inputs and outputs for the given time horizon. The time horizon is set of forthcoming time intervals over which it is feasible to make a forecast. The forecast horizon determines its range and designation. Examples include short-range forecasts measured in hours and days (especially in flood control and day-to-day operational decisions), middle-range forecasts measured in weeks and months (especially in the operation or seasonal storage capacities), and long-range forecasts measured in months and eventually years (mainly for operation of very large over-the-year regulation reservoirs or the conjunctive water use of large surface reservoirs and aquifers).

The accuracy of a forecast decreases as the forecast interval, Δt, and the horizon of h intervals, each of Δt length, increase. The time series of supply and demand may have all or only some of the four basic components of time series, i.e., trend, periodicity, intermittency and stochasticity. A water supply time seris may not contain trend (no trend in their parameters in the long run). Also, it may not be intermittent (say, sufficiently large rivers with no zero flows). But water demand definitely does contain trends. It may be also intermittent (such cases are demand or irrigational water, for navigational and pollution abatement waters, recreation water, etc.). The unavoidable structural components in both supply and demand are periodicity in parameters (seasonality) and chance variation (stochasticity). The stochastic variation about the periodicity-governed fluctuations in parameters is the major factor that requires forecast in water supply, and the fluctuation in all components but especially the stochasticity component require forecast in water demand.

1.3 Goodness of Forecast

Two well-defined performance measures in the operational hydrologic forecast are needed: (1) a measure of the goodness of forecast itself, and (2) a measure of operational benefit of the forecast. The first measure was given more weight in the past than the second one, because: (1) it is easier to design and compute, and (2) the forecast agencies need to demonstrate the goodness of forecast in order to justify it.

The goodness of forecast is a crucial factor in the assessment of its application value in operation. It is given either as an absolute or as a relative performance measure. The best absolute measure of forecast, that is of the same dimension as the forecast water supply variable, is

$$M_s = [\frac{1}{N} \sum_{i=1}^{N} (S_r - S_f)^2]^{\frac{1}{2}} , \tag{2}$$

in which S_r and S_f are defined in Figure 1; N = number of forecast values used in the derivation of the measure. Replacing S_r and S_f in Equation (2) by D_r and D_f, the absolute measure M_d of the goodness of forecast of water demand is obtained.

The relative measure is needed if forecasts of supply and demand, or between supply and demand, are to be compared. It is

$$R_s = [\frac{1}{N} \sum_{i=1}^{N} (\frac{S_r - S_f}{S_r})^2]^{\frac{1}{2}} , \tag{3}$$

in which for each value of $i = 1, ..., N$ is the ratio $\Delta S/S_r$, of the error in forecast ΔS over the realized supply value of S_r. The measure R_d for demand is Equation (3) with S_r and S_f replaced by D_r and D_f. The forecast measure of Equation (3) avoids a significant impact of very large values of $S_r - S_f$ or $D_r - D_f$ of Equation (2) for the high values of S_r or D_r. The R_s and R_f values then serve these purposes: (1) they measure how good are methods of forecast in terms of errors; (2) they indicate how forecasts compare at different places and conditions; and (3) they compare supply and demand forecasts.

2. BENEFIT OF OPERATIONAL FORECAST

2.1 Differences in Benefit

The real measure of the benefit of water supply-demand forecast is the difference in benefits with and without such forecasts. This measure may be given in absolute terms as,

$$\Delta B = B_f - B \tag{4}$$

and in relative terms as,

$$b = \frac{\Delta B}{B} = \frac{B_f}{B} - 1, \tag{5}$$

in which B_f = the benefit with forecast, and B = the benefit without forecast, as the relative measure in terms of the benefit without forecast.

Equations (4) and (5) each may be subdivided into two components, B_s and B_d, and b_s and b_d, respectively, as the increased benefits from the supply-demand forecast. Both are needed if to justify economically the costs of the forecasts.

The use of Equation (4), or Equation (5), permits calculation of a benefit-cost ratio, i.e., $\Delta B/C$, or $\Delta B_s/C_s$, and $\Delta B_d/C_d$. The terms C, C_s and C_d are the costs of forecasts and their implementation. Equation (5) can be used in the same manner.

Thus, Equations (4) and (5) have a role in justifying the application of forecasting to the operation of water systems. Forecasting methods are not unique and a search for improved methods must be related to their potential to increases in operational benefits.

2.2 Maximum and Minimum

The minimum benefit from a forecast, B_{min}, occurs when there is no forecast. In this case, the manager knows nothing about the future supply and demand except past history. The operating decisions are made by a standard rule. The maximum benefit, B_{max} is produced if the forecast is perfect. In this case, both the supply and demand time series are assumed known for the entire economic life of the project. Operation decisions are made in such way that waste is avoided unless it is in excess of system requirements.

The maximum benefit of forecasting is given by the expression:

$$\Delta B_{max} = B_{max} - B_{min} \tag{6}$$

The same idea in relative terms is:

$$b_{max} = \frac{B_{max} - B_{min}}{B_{min}} , \tag{7}$$

The benefits term, B_{min}, for no forecast situation is a reference. Procedures for calculation of B_{min} and B_{max} are given in the following sections.

2.3 Determination of Minimum Benefit of Operation Without Forecast

Techniques are operational already for modeling the time series of supply and demand. They are based on past samples of these series together with adjustments for expected trends. They permit the generation of a set of m samples of water supply and m corresponding samples of water demand, for the selected life of N time intervals of the system. The generated S- and D- series are used as pairs of sequences because they are two crossing processes, often also mutually correlated, say as the j-series, with j = 1, 2,..., m. To operate any system with storage, either by using an ongoing operational rule, or by designing the operational rule for specific study, the generated bivariate samples of (S,D)-series are the input data.

There are two basic conditions to generate these S- and D- time series. First, the operation manager does not know at an initial time, t_0, the supply, S_i, and the demand, D_i, in the forthcoming time intervals, except by using the preceding values S_{i-j} and D_{i-j}, j = 1, 2,..., s, and the application of Equation (1). At time t_0, the values of $\log Q_i$ for $E(x_i) = 0$ of Equation (1), and then Q_i, are computed for the next time for both supply and demand. Second, once the decision is made for the system operation in the interval i by the operational rule, the decision-maker is provided with the "realized" values of S_i and D_i (in this case the generated values of samples), to be used in Equation (1) for the next interval, i+1, and beyond it.

This approach is feasible only because computers have made this approach cost-effective. The generation of a large set of m samples of given size, N, for the operational decision time interval, Δt (hours or days for flood control reservoirs), the application of operational rules, and the recording of results, can be programmed easily and accomplished economically.

Let us assume now that the benefit of the i-th interval decision results in $B_{j,i}$, as the elementary benefit for sample j and interval i. The total minimum benefit for the j-th bivariate sample of supply and demand, over the N time intervals of the systems operational economic life, is then

$$B_{j,min} = \sum_{i=1}^{N} B_{j,i} \, . \tag{8}$$

For m samples, there are m values of $B_{j,min}$ in Equation (8). Their probability distribution is defined by the mean and standard deviation as follows:

$$\bar{B}_{min} = \frac{1}{m} \sum_{j=1}^{m} (\sum_{i=1}^{N} B_{j,i}) = \frac{1}{m} \sum_{j=1}^{m} B_{j,min} \qquad (9)$$

and

$$s_b = \{ \frac{1}{m} \sum_{j=1}^{m} \sum_{i=1}^{N} (B_{j,min} - \bar{B}_{min})^2 \}^{\frac{1}{2}} . \qquad (10)$$

Figure 2 show these terms graphically.

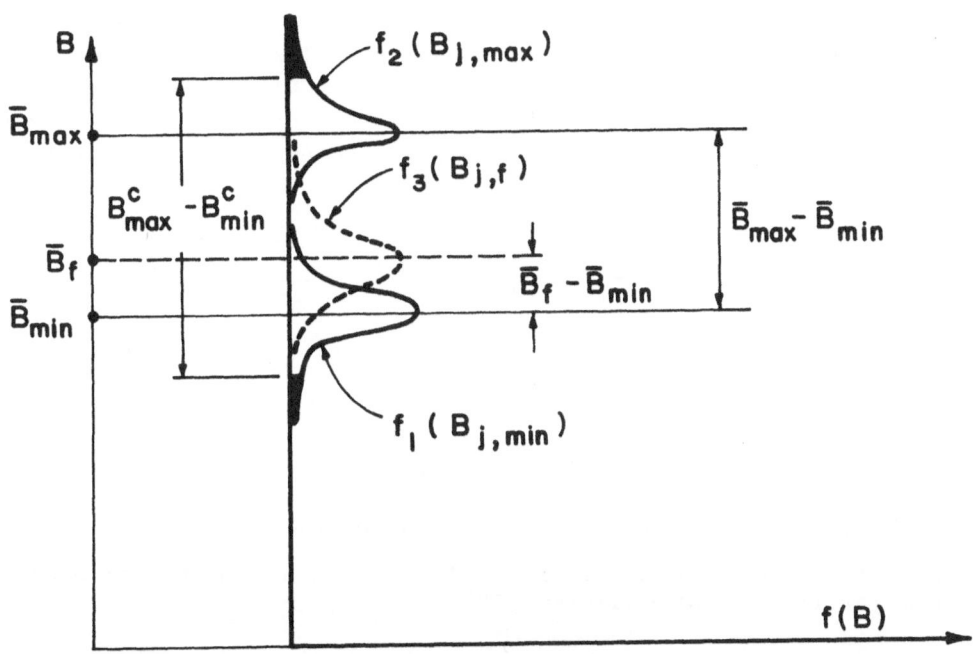

Figure 2. Distributions of B_f, and B_{min}, and B_f and the Range of Potential Benefit Increases Resulting from the Forecasts of Water Supply and Water Demand

An alternative in defining the minimum benefit is by equating it with the mean of Equation (9), i.e.,

$$B_{min} = \bar{B}_{min} \quad . \tag{11}$$

Another alternative for the minimum benefit is to use the lower confidence limit of $\bar{B}_{min}$ with s_b of Eq. (10) and a selected probability level for the minimum benefit of the sample not to be smaller.

2.4 Determination of Maximum Benefit of Operation with Known Future

The maximum benefit of operation of a water system with storage will be obtained if the supply and demand series are completely known over the N intervals of the economic life of the system. The adjustment of demand in deficit periods, the anticipation for storage of water surplus without water spillovers, and taking into account the value of water according t the time of release and the demand release discharge, will produce the maximum possible benefit $B_{j,max}$ for each generated sample j of supply and demand. This is merely the sum of the benefits over the N time intervals, i.e.,

$$B_{j,max} = \sum_{i=1}^{N} B_{j,i} \tag{12}$$

Then for $B_{j,i}$ the benefit of any time interval i, is,

$$\bar{B}_{max} = \frac{1}{m} \sum_{j=1}^{m} B_{j,max} = \frac{1}{m} \sum_{j=1}^{m} (\sum_{i=1}^{N} B_{j,i}) \quad . \tag{13}$$

Similarly as for B_{min} m samples give values of $B_{j,max}$ of Equation (12), the mean of Equation (13), the standard deviation similar to that of Equation (10).

The mean, $\bar{B}_{max}$ is assumed to be B_{max} for comparison purposes. Similarly as for B_{min} an upper confidence limit can be assigned as the B_{max} as indicated in Figure 2.

2.5 Determination of Forecast Benefit

The computation of operational benefit, B_f, of forecasts of supply and demand may be obtained by first accomplishing the following tasks:

(1) The series of forecast input variables are generated first for water supply (say a set of precipitation series over an area, snowmelt series, and series of the other variables used in forecast), and then for water demand;

(2) The forecast output series of supply and demand are produced by using the forecast method, to which the forecast errors are superimposed to produce the "realized" series; and

(3) The resulting supply and demand series are the same, namely m bivariate samples, each of size N, as those used for the computation of B_{min} and B_{max}.

To accomplish the above tasks, the historical data are used for the three aspects of forecast: (a) input series, (b) output supply and demand series, and (c) series of differences S_r-S_f and D_r-D_f (of realized and forecast values). Then, the three types of series are analyzed and modeled, by preserving their time structure and dependence, as well as their log cross correlations.

Two basic approaches may be used in order for the bivariate samples of supply and demand to be the same as those used for the computations of B_{min}, B_{max} and B_f. They are outlned as follows:

(1) The forecast input series are generated. Then the forecasts of the series S_f and D_f are performed as the forecast methods specify. Finally differences, S_r-S_f, and, D_r-D_f are added in order to produce the "realized" series S_r and D_r. These "realized" series are used for the computation of B_{min} and B_{max}.

(2) The samples of bivariate supply and demand series, S_r and D_r, are generated. Then the generated samples of differences, S_r-S_f and, D_r-D_f are subtracted from S_r and D_r series, with the forecast series S_f and

D_f obtained. This second approach does not require the generation of forecast input variables and the application of forecast methods, if the historical series of S_r and D_r and of difference S_r-S_f and D_r-D_f are sufficiently long and reliable. In this case the operation of the system is made by using the i-th values of the forecast series S_f and D_f any time a decision is made for water releases in the i-th time interval. A similar procedure is used if the forecasts are made for the of h time intervals.

Either approach will produce the unit benefits, B_i, so that the benefit for the j-th sample is

$$B_{j,f} = \sum_{i=1}^{N} B_{j,i} \tag{14}$$

The frequency distribution of m values of $B_{j,f}$ for m samples can then be obtained, with its mean

$$\bar{B}_f = \frac{1}{m} \sum_{j=1}^{m} B_{j,f} = \sum_{j=1}^{m} (\sum_{i=1}^{N} B_{j,i}), \tag{15}$$

and the corresponding standard deviation.

Instead of using the seem of N unit benefits $B_{j,i}$ in all previous benefit equations, Equation (14) can be divided by n, which is the number of years in the systems economic life of N decision intervals. Then $\sum_{i=1}^{N} B_{j,i}/n$ represents the average annual benefit of a sample as the estimate of the expected annual benefit.

The measure of the benefit of forecast can be either B_f of Equation (15) , or a confidence limit may be used, similarly as for the values of B_{min} and B_{max}.

2.6 Relative Measure of Operational Benefit of Forecast

Once B_{min}, B_{max} and B_f are computed, the relative measure of operational benefit of forecast may be

$$b_f = \frac{B_f-B_{min}}{B_{max}-B_{min}} \tag{16}$$

with B_f = the benefit of forecast, in case of the same samples of supply and demand as used for B_{min} and B_{max}. For $B_f = B_{min}$, or no value of forecast, $b_f = 0$. For $B_f = B_{max}$, then $b_f = 1$, or no further improvement in forecast is feasable. Figure 2 shows the distributions of B_f, B_{min} and B_{max}, permitting their comparison.

3. HYDROLOGIC FORECAST METHODS

It is not simple to classify the hydrologic forecast methods, because of feasibility to use many classifications such as mathematical type, physically based or not, accuracy, benefit, cost, data input, time horizon, etc. From the point of view of relationships of forecast output variables to the preceding, forecast input variables, the three broad groups of forecast of water supply (and to some extent also of water demand) are:

(1) Simple <u>analytical relationships</u> of output to input forecast variables are generally either deterministic equations or statistically derived relationships.

(2) <u>Physical mathematical models</u> using the unit hydrograph approach on one extreme, to the approach of the most elaborate physical modeling on the other extreme. The latter would account for water in all the important phases of the water cycle. For the supply forecast the input variables are precipitation and snowmelt. For water demand they are the hydrologic and climatic variables which affect the demand side.

(3) <u>Stochastic mathematical models</u> use forecast input variables that represent both the "memory" of a natural system (such state variables as storage, antecedent moisture, preceding flows, etc.), and the recently observed precipitation or completed snowmelt.

Hundreds of forecast methods are described in the literature. They differ in basic characteristics, such as: (a) accuracy, (b) needed input data, (c) computational sophistication, (d) processing time , (e) benefit, (f) cost, (g) feasibility of use in operation, etc. Many subjective preferences affect the selection of a forecast method.

If the average annual cost of forecast and implementation of a method is C_f, and the average annual forecast benefit is $B_f - B_{min}$, the result of its application is best measured by a benefit-cost ratio.

$$\frac{B}{C} = \frac{B_f - B_{min}}{C_f} \qquad (17)$$

Equation (17) is the real measure of usefulness of forecast. Problems may be encountered, however, in the evaluation of the three variables of Equation (17). Because governmental forecasting services often neglect some aspects of the cost (say, the cost of operation of the general gauging network that is used along with the special forecast network), the cost C_f may be underestimated. Furthermore B_f and B_{min}, as the above analysis has pointed out, may require special efforts for an accurage evaluation.

The major criterion in the selection of a forecasting method among a variety of available methods is the minimum mean square error (MSE) of Equation (2), or its relative value using Equation (3). The assumption is that this method also will produce the maximum in differences of benefits, $B_f - B_{min}$ or its relative expression, Equation (16). The minimization of Equation (3) does not necessarily mean the maximization of Equation (16), because accuracies in the forecast of low, medium and high flows do not affect Equations (3) and (16) identically.

There is general tendency among the forecasting agencies to develop and use for a river one forecasting method only for all the purposes of forecast, though a different method for each purpose may be the better approach. The trend is toward the development of an elaborate forecast method, computer oriented, which will be recalibrated from time to time as new data are accumulated.

4. FORECASTING HORIZON

The forecast of precipitation is a difficult task. Though the outlook for precipitation in the days to come may be well predicted, there is a significant difficulty in forecasting closely the amount of precipitation. Many weather forecasting services give their precipitation forecast in terms of probability of rain and snow, even for the next day. Forecast is usually given for several days in advance, with a weather outlook for a week or so. The monthly precipitation forecast, in form of "normal", "above normal" and "below normal" precipitation (and temperature) seems to have been until now just a little better than the simple probabilistic statements.

Long-range forecasting of precipitation is not feasible at present. All attempts of the past to develop such a forecast have failed. As the hydrologic forecasting is at the receiving end of the weather forecast, the best judgment on forecast feasibility of precipitation is by the study of its potential in hydrologic forecast.

A quantative forecast really begins with the observed precipitation, snowmelt and moisture conditions in river basin. The weather forecast is good mainly for warning purposes, say to alert the forecast service and the users of forecast output, that some chances exist for the extreme precipitation and weather conditions to occur.

A general pattern in hydrologic forecast is that the larger a river basin, the longer is the horizon of forecast for a given accuracy of forecast. The reason is simple. The larger a river basin, the longer is the water travel time from the most distant parts of the basin to the river position of forecast. Furthermore, the larger a river basin, the smaller on the average may be its overall average slope, and smaller is the average water velocity along the drainage system.

5. CONCLUSIONS

Hydrologic forecasting for operational purposes is basically of a short-range type, say of days or up to a couple of weeks, except in case of forecasting based on accumlated snow, which extends to the entire smowmelt season. The value of a forecasting method is best measured by the mean square error (MSE) of the differences of realized and forecast values. In measuring forecasting benefits, the modeling of time series of supply and demand, and the forecast, permit the computation of average maximum benefits, average minimum benefits, and average benefit of operation. The range of forecast benefits between maximum and minimum gives two results: (1) the amount of benefit from forecast; and (2) the benefit potential from an improved forecast.

ACKNOWLEDGEMENT

This text is a contribution by the project "Investigation of Objective Functions and Operational Rules of Storage Reservoirs," sponsored at Colorado State University, Fort Collins, Colorado, USA, by the Office of Water Research and Technology, U.S. Department of Interior (through the Water Resources Institute at Colorado State Unversity), Project B-195-COLO. Studies were partially sponsored by the Colorado State University Experiment Station ES-114. The studies leading to this paper were also financially supported by the cooperative U.S.-Spanish research project on "Conjunctive Use of Various Sources of Water," sponsored by the Joint U.S.-Spanish Committee via Research Institute of Colorado, Fort Collins, Colorado.

REFERENCES

Anderson, A. Eric, "Streamflow simulation models for use on snow covered watersheds," Proceedings of Snow Cover Runoff, U.S. Army Cold Regions Reasearch and Engineering Laboratory, Hanover, NewH ampshire, 26-28, September 1978.

Brasil, L. E., and M. D. Hudlow, Calibration Procedures Used with the National Weather Service River Forecast System, IFAC Symposium on Water and Related Land Resource Systems, Pergamon Press, Oxford and New York, 1981.

Curtis, C. David, and G. F. Smith, "The National Weather Service river forecast system - update 1976," Proceedings Int. Seminar on Organization and Operation of Hydrological Services (Fifth Session of WMO Commission for Hydrology, Ottawa, Canada, July 15-16, 1976).

Ostrowski, T. Joseph, "National Weather Service produces useful for reservoir regulation," Report, Hydrologic Research Lab, National Weather Service, NOAA, Silver Spring, Maryland, USA, 1980.

Peck, E. L., E. R. Johnson, K. M. Drouse, T. R. Carroll and J. C. Schaake, Jr., "Hydrological update techniques used by the U.S. National Weather Service," Proceedings of the Oxford Symposium, IAHS-Publication No. 129, April 1980.

Sittner, W. T., "Determination of flood forecast effectiveness by the use of mean forecast lead time," NOAA Technical Memo NWS Hydro-36, U.S. Dept. of Commerce, Silver Spring, Maryland, USA, 21 pp., 1977.

U.S. Dept. of Commerce, "Operations of the National Weather Service," Silver Spring, Marylad, USA, 303 pp., 1978.

WMO, "Intercomparison of conceptual models used in operational hydrological forecasting," WMO Operational Hydrology Report No. 7, WMO - No. 429, Geneva, Switzerland, 172 pp., 1975.

12. WATER DEMAND FORECASTING

Luis Veiga da Cunha
Laboratorio Nacional de Engenharia Civil, Lisbon

1. INTRODUCTION

The purpose of water demand forecasting to achieve at the same time, an optimum use of water, and the economical programming of facilities investments. An adequate national and regional water demand forecast enables comparison of investments in water resources with other categories of investments, and also to determine on appropriate regional distribution of the investments.

Projections of national water demand must also account for regional diversity and consider the effects that the projections may have on regional planning. To understand regional influences it is useful to compare forecasting methods used in several countries (1).

A water demand forecast is not an end, but an input for decision making. In choosing models for demand forecasting, it is therefore necessary always to take care that they meet the objectives established and that the economic value of the information obtained is satisfactory.

In many countries, including developing countries, the fact that water has been considered a free good implied that the effort devoted to anticipating water demand was comparatively weak. For this reason the only elementary methods have been used to forecast water demand.

Until recently the location of water consuming activities in most countries has had little consideration toward an overall strategy for regional water utilization. Private enterprise together with decision-makers, i.e., politicians, led the process of distributing people and social and economic activities throughout a territory. Only later were the engineers asked to provide the water. The axiom has been that water must be available in the required quantity and quality. This scene has been changing in many regions of the world to one in which management of water demand has evolved. Water is considered a production factor whose costs condition the decisions in land-use planning. The planning of water resources is thus no longer centered on water itself, but rather on the needs that water use is intended to meet. In this changing scene the definition of future water demand has a role in the choice of the water demand forcast method.

In the past, the process of planning water resources was carried out by engineers, almost exclusively. They had sufficient training for assessing and comparing the tangible benefits of the solutions proposed. More recently, however, it has been recognized that the engineering disciplines are sufficient to adequately consider the diversity of problems that arise. Those relating to the environment, quality of life, demographic balance, etc., require multidisciplinary teams of engineers, economists, sociologists, jurists and other professionals.

This paper deals mainly with the problem of water demand forecasting, with focus on developed countries. The concepts developed hold also for developing countries; but many aspects can be different. For instance, in the poorer of the developing counties the impact of extreme situations, particularly drought, may be felt in a very critical manner, and may have serious effects on social aspects relating to health, and even survival. In the developed countries, the economic impacts predominate.

Furthermore, the impacts of water related investments are far more important in developing countries than in developed countries. The benefits of well planned investments therefore may aid significantly in the development of those countries. Badly planned investments, on the other hand, may have economically disastrous consequences.

The transfer of technologies associated with water uses from developed countries having no critical water problems to developing countries, which often are situated in arid and semi-arid regions, may cause the water demand in such countries to increase to an untenable extent. Thus, it is necessary to use "appropriate technologies" in the developing countries.

In summary, although the increases in future water demand in the developing countries may incorporate technologies appropriate for efficient use of their water resources, that is not the crux of the problem (1). Rather, the crucial concerns have to do with the basic water requirements connected with fundamental aspects of public health, e.g., water supply and sewerage, and of development, e.g., power production and irrigation, that are crucial.

2. THE NEED FOR WATER DEMAND FORECASTING

In this chapter the term water demand is used in its economic sense, meaning the relationship between the quantities of water used as a function of the price of the supply. Further, the general conclusion resulting from various water demand studies is that above a certain minimum for meeting the basic needs corresponding to drinking, cooking and personal hygiene, water demand behaves in the same way as the demand of other consumer goods and services; that is, demand decreases as the price increases. This behavior is valid both for domestic uses (2, 3) and for the other uses such as industrial (17).

For cases in which water is a factor of production such as industry and agriculture, increases in water prices means that the product price increases. Usually this leads to a reduction in sales and/or a reduction in water consumption by the user, provided another production factor can be substituted for water.

It is useful at this stage to mention some fundamental problems relating to the identification of water demand.

First, water demand is a derived demand, since water is a production factor of various products intended to meet a demand by the consumer. Second, water demand is related to water quality. This means that water pollution control policies can condition water demand. Third, when speaking of water demand a distinction must be made between water withdrawal, and water consumption.

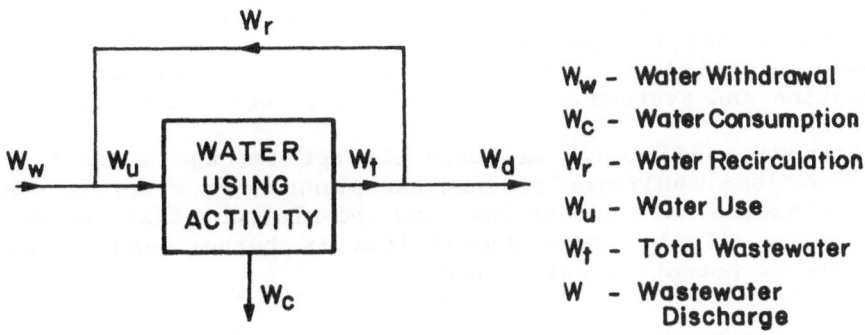

Figure 1.   Definition of Basic Concepts Related to Water Demand

Figure 1 defines these concepts as well as those of water use, total wastewater and wastewater discharge, and water recirculation.

In order to forcast water demand it is necessary to classify the various water uses, as proposed by a United Nations publication (4). This is done according to two general criteria. They are:

1. purpose of use, such as municipal and rural, agricultural, industrial and infrastructural water use (such as transportation, recreation, wetland habitats, and conservation of estuaries).

2. forms of use: water withdrawal, in-stream use, and on-site use. Table 1 shows how the general classification scheme is related to specific water uses (transformed from Water Series No. 3 (4)).

When water supply is insufficient to meet demand, it is necessary to resort to a mechanism for rationing water. This is not accomplished by either of the water rights doctrines, i.e., the riparian rights doctrine and the appropriative rights doctrine. Both have proved to be inefficient as automatic water management instruments when water is scarce; they do not take into account the marginal productivity of the water. Thus, various technical, economic and social measures have been implemented in several countries to influence water demand.

The water saving measures are well known and a description is beyond the scope of this chapter. They may involve reduction of water withdrawal, reduction of losses, or both. Improvement of the quality of wastewater discharged can also be considered as a water saving technique.

Economic instruments can create economic incentives to reduce water demand and the pollution discharge in water bodies. These economic incentives can consist of pricing for water withdrawal or consumption and systems of charges for wastewater disposal.

Economic and social measures may act through a modification of the national policies of land use planning in order to achieve a distribution of production and population that ensures a distribution of the water demand that is better adapted to the available or potential water supply.

Table 1. Categories of Water Demand

	Municipal and Rural	Agricultural Demand	Industrial Demand	Infrastructural demand
Drinking	W			
Domestic uses	W			
Public uses in settlement	W			
Livestock	W	W		
Fish and wildlife				
Flood loss management	I-O-W	I-O-W		I-O-W
Drainage	O-W	O-W		
Swamps and wetland habitat		O		
Utilization of estuaries		I-O		
Irrigation	W	W		
Navigation				I
Hydropower				I
Soil moisture conservation		O		
Mining			W	
Steam power	W		W	
Cooling	W		W	
Processing	W		W	
Boiling	W		W	
Waste disposal	I	I	I	
Recreation	I			I
Water sports	I			I
Aesthetic enjoyment	I			I

W-water withdrawal; I-instream use; O-onsite use

3. METHODS FOR WATER DEMAND FORECASTING

An overall water demand forecast may be made by determining a relationship between water use and certain indices of economic development, such as the gross domestic product. More detailed studies may analyze water demand changes in the different consumer sectors. A common forecasting method is trend extrapolation based upon observations of past data. This method, though simple, often leads, however, to unacceptable results, especially in the case of long-term extrapolations. It neither considers any modifications of the social and economic factors that determine the water demand, nor does it take into account the evolution of technologies associated with its use.

To circumvent these difficulties, the current trend is to identify the factors determining water consumption and to establish empirical relationships between them and water demand. Examples of such factors are population, level of agricultural and industrial activities, and technological development (5). However, since application of these methods can not be thoroughly checked, it must be done carefully and critically.

Since many of the variables affecting water demand depend on regional policies and various ill-defined complex relationships, the water demand forecast is marked by intrinsic uncertainty.

There are five types of uncertainty in a water demand forecast. They are:

1. natural hydrologic phenomena, related to the availability of water resources,

2. inadequate hydrologic, economic, social, and environmental data,

3. an ill-defined concept of the future, which tends to increase when the forecast period is increased,

4. incomplete understanding of the natural technologic and economic factors influencing resources problems,

5. imperfect models for analysis of water resources problems.

Analysis of uncertainty involves three steps, i.e., identification of its origins, quantification of the different kinds, and, inclusion of the uncertainty in the analysis and subsequent determination of the solution.

3.1 Scenarios

Water demand forecasts currently may use the technique of alternative scenarios, corresponding to specifications of future conditions of economic and social development, technological progress, life style and public policies. The implications of each alternative scenario must be analyzed independently. Some of the more important variables in the generation of alternative scenarios for water demand forecasting are the following (6):

1.   several assumptions about future population levels,

2.   several economic growth assumptions in various industries,

3.   several assumptions on technological developments as related to water use,

4.   the effects of alternative macro-economic policy alternatives, such as promoting regional as opposed to national economic growth,

5.   the effects of various types of pricing policy for water and other elements of overall water management strategy,

6.   the effects of a policy to promote environmental quality.

Figure 2, taken from reference (4), summarizes the procedure for regional water demand forecasting based on alternative scenarios.

3.2 Regression Techniques

The application multiple regression techniques provides a sound basis for forecasting water demand. These techniques make it possible to correlate the variations in water demand in the past, W_d, with the variables that influence such demand, i.e., X_1, X_2, ... X_n, through equations of the following type:

$$W_d = a_1X_1 + a_2X_2 + \ldots + a_nX_n + e$$

in which a_i are coefficients and e is an error term.

Variables may be classified as endogenous, when determined within the regression model itself, and exogenous when determined externally, independently of the functional relationships described by the model. The multiple regression model may be presented in the form of a reduced model that expresses each exogenous (or dependent variable) as a function of the endogenous (or independent) variables, and of an error term. Application of

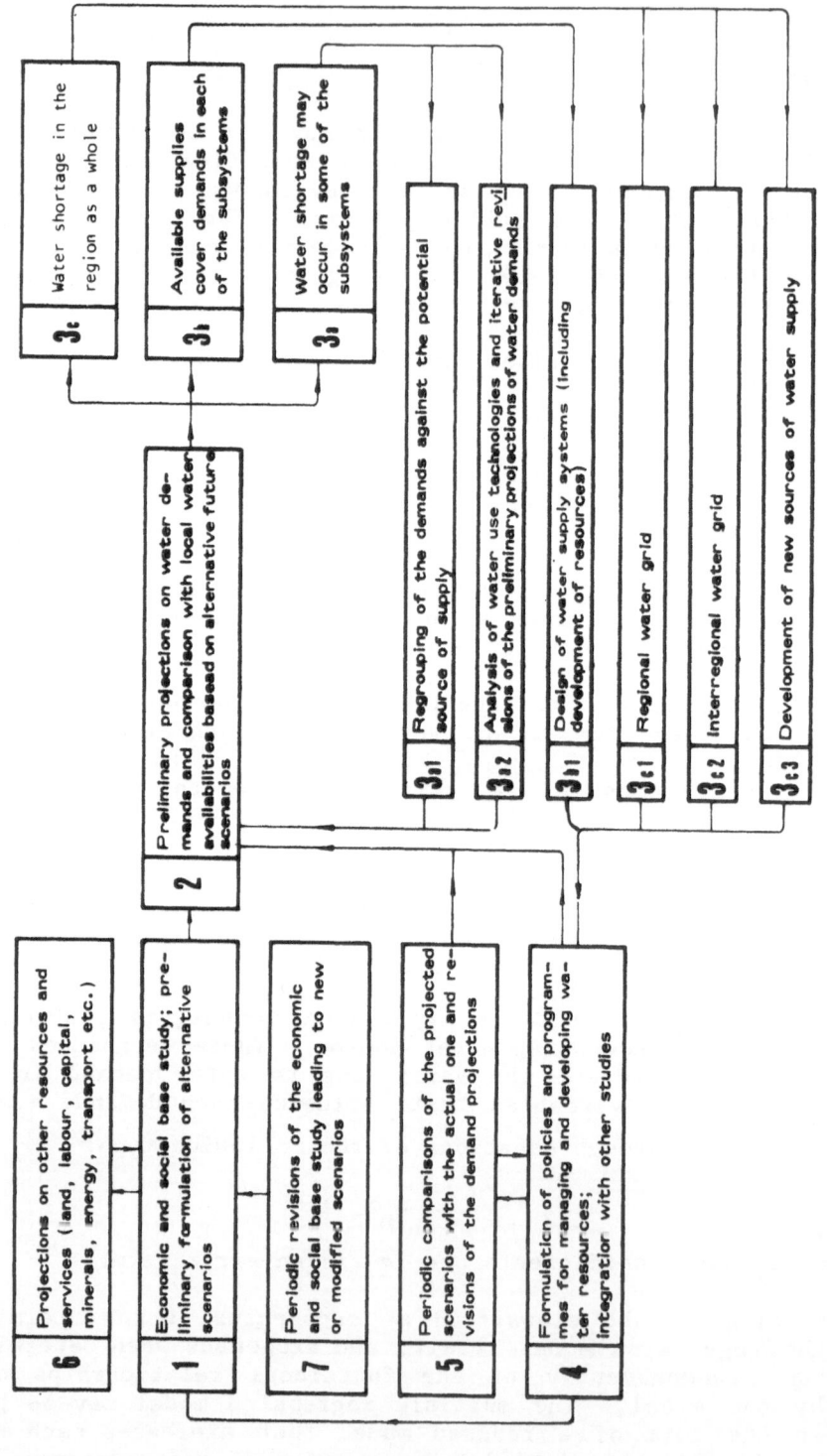

Figure 2. Flow Chart of a Regional Water Demand Projection Based on Alternative Scenarios

these multiple regression techniques to water demand forecasting presupposes that the influence of the explanatory variables will keep to the same pattern in the future, i.e., that the reduced form coefficients remain constant.

To forecast water demand, there is first an estimate of the reduced form coefficients on the variables which explain water consumption in the past, and it is assumed that such coefficients will remain constant in the future. In order to complete the application of the technique it is necessary to project independently the various explanatory variables, supposing that their variation is independent of water use. By applying the regression model to the projected values of the variables, an estimate of the water demand is obtained.

Whittington (7) calls attention to some limitations associated with the application of multiple regression techniques to water demand forecasts. These are essentially of three kinds: the validity of the assumption that the reduced form coefficients are constant; specification problems that may occur if the price of water is not one of the explanatory variables, and may make the estimates of the coefficients biased and inconsistent; and the fact that these reduced form forecasting techniques combine the influences of supply and demand relationships.

The accuracy of the regression techniques decreases as the period of forecast increases.

An example of the regression technique, in which there is only one explanatory variable, is the simple correlation method or the so called water use coefficients approach. This model is often used, and consists of finding a variable in the past has shown high correlation with the water demand, and assumes that the same type of correlation will prevail in the future.

Various explanatory variables may be selected, such as population growth, economic growth, the output product of the activity whose water demand it is intended to forecast, or even, in cases where they are applicable, variables such as the number of employees, the quantities of certain raw materials consumed or the value added per unit of output. Refinements of the method can be separate considerations of water use coefficients associated with different production technologies, or considerations of nonlinear relationships, i.e., the introduction of coefficients whose values vary with time, in accordance with trends observed in the past or foreseeable in future.

The coefficients method corresponds essentially to a form of application of the well known input-output analysis often used in economics for describing the mutual relationships between input

flows and output flows in a given economic system. This application corresponds to considering the flow of water as one of the
elements of the analysis. Although input-output analysis is
usually done on the basis of consideration of constant coefficients, it is also possible to foresee the variation of the coefficients with time. Application of input-output analysis allows
not only a forecast of variatons in water demand but also an
appreciaton of the effects of different forms of water use on the
values of the final outputs that are the object of consideration
in application of the analysis. The principal limitation of
simple or multiple regression methods is due to their not being
supported by any theory capable of explaining the water use. As
Whittington (7) notes "in applying these methods, no behavioral
model is tested."

In fact, the results of the regression analysis do not
explain the mechanisms of water use in a given activity, or the
way such mechanisms can be modified. It is therefore necessary to
examine, from a technical point of view all pertinent aspects
relating to the water use and especially the types of use to which
the water is put, the technical alternatives to such use and the
costs associated with those alternatives. The technological
alternatives to be considered are related not only to possible
different ways of using water, but also to modifications to the
processes involved in the water use, and to actual modelling of
the other production factors occurring in the process, the use of
byproducts or a variation in the nature and quality of the
products. What is aimed at is an identification of the various
technological alternatives envisaged, and the costs of such alternatives, in order to ascertain whether the benefits associated
with the reduction in consumption and in the pollutant discharge
of wastewater justify the costs associated with implementation of
the alternatives envisaged. System analysis techniques provide
the answers to this type of question.

3.3 Methods Based on Prediction of Demand Function Changes

These methods are based on a forecast of the modifications
that water demand is expected to undergo in coming years. The
forecast is made on the basis of the data available on the
variables influencing water demand. An example is the economic
growth forecast which included the foreseeable variation in the
price of water, the water resources management policy to be
implemented, tehcnological modifications influencing water
consumption, and the foreseeable changes in water treatment
technologies.

In practice an effort is made to disaggregate the different
types of demand, then conduct the study for each individual demand
function, and finally aggregate them again to obtain the overall

water demand function. By proceeding in this way, it is expected to obtain overall demand functions for the different types of use referred to previously: agricultural, industrial, municipal and infrastructural.

A very important aspect is consideration of the influence of technological progress in water demand. Technological modifications may affect the demand curves in various ways. There are very many recent examples of modifications in technological processes, particularly in industrial technologies, which determine appreciable changes in water demnad. These technological changes may in principle determine both increases or reductions in water consumption, but from the point of view of water resources management it is obviously the latter that are important, and there has been noteworthy progress in this respect in recent years.

A different type of technological modification likely to determine a reduction in water consumption are those directly relating to water use itself, both in regards to water supply and wastewater discharge, particularly in those aspects concerned with treatment technologies. Sawell and Bower (8) calling attention to the fact that "technological changes most often pertain to industry, although other sectors may be affected as well," mention four types of technological changes: "new products, new processes, new or different raw materials and new methods for handling water." New products may replace or substitute for an established commodity. New processes may be developed which make production of a commodity cheaper or more efficient. New raw materials may be introduced to an established production process. These three changes can generate either decreased or increased water use. New methods for handling water refer to inplant changes such as automatic control on water utilization, the adoption of water recycling, etc., almost always lead to decreased water use."

4. WATER DEMAND FORECASTING FOR DIFFERENT KINDS OF WATER USE

Water demand forecasting can be done at national, regional and local levels, but it is mainly at the last two levels that the forecast is more useful. In spite of this, the analysis of water demand forecasting should first proceed from a national to a regional level and from a regional to a local level, and then in the opposite direction. Thus there must be a disaggregation of the data compiled on a national level, then forecasting of water demand at regional and local level, and finally aggregation of those results for a national picture.

A regional water demand forecast implies criterion for dividing the country into regions. This is difficult baecause characterization with respect to the water supply may not coincide with the administrative or planning regions which are organized by economic and demographic data that are needed for characterization of water demand. The ideal would be for regional disaggregation to be so done that the units considered allow a hydrological or an administrative aggregation in accordance with what is intended.

Concerning certain products, such as the outputs of the agricultural, hydropower and main industrial sectors, it is necessary to define the regional distribution of these activities and carry out a control for the national totals, checking that the assumptions put forward are consistent with the projections of the total population, of the active population, or of the gross domestic product. For activities such as industries, mining, thermoelectric plants or even municipal water uses, it is possible to achieve significant modifications in water demand by water reuse or recycling techniques.

4.1 Water Demand in Human Settlements

The water demand of human settlements can correspond to urban and rural uses. Urban water demand includes domestic uses (drinking, cooking, kitchen and toilet use, yard and garden watering, car washing, etc. and commercial uses (stores, offices, laundries, private, public or municipal works, etc.), public services (street washing, watering of parks and public gardens, transport infrastrucutres, firefighting, public buildings such as schools and hospitals) and recreation and tourist services (places of entertainment, sports facilities, swimming pools, hotels, restaurants, etc.). Sometimes the water demand of industrial units connected to the municipal water supply network is also considered as part of urban water demand, but it seems more reasonable to include it under industrial water demand. Rural water demand includes domestic uses and usually also water for livestock and irrigation in the part that is not usually considered under agricultural water demand. As regards rural water demand should forecast separately livestock and irrigation demand.

Water requirements for urban consumption are usually fixed on the basis of a study of demographic projections and definition of per capita consumption indices depending on various factors, such as climatic characteristics, the level of economic development, income distribution, the size of the human settlement, the types of size of the industries to be installed and the water pricing system that would be used. Efforts are also made, for each area, to define average proportions of domestic, public and commercial use. Domestic consumption is generally influenced by the cost of water, above a certain minimum consumption corresponding to basic needs, particularly when the price is progressive which obviously involves the existence of metering, and is also influenced by family incomes and the number of household appliances. The knowledge of this influence, together with a forecast of the future variation in the number of dwellings, usually provides the basis for water demand forecasts. The water demand related with commercial activities is usually estimated as proportional to domestic demand.

Water demand for public uses is usually considered as proportional to domestic demand, provided that the characteristics of the human settlement are not greatly modified, or it is directly estimated from a knowledge of the future evolution of the number of gardens, public buildings, length of the streets that have to be washed, etc.

In developing countries, rural water demand can involve a very important part of the investments devoted to water resources development, and the experince of the last twenty years in these countries is very useful for forecasting usual water demand and establishing appropriate planning in that field. The experience reported by Widstrand (11) and by Sargent and Sargent (12) are good examples of descriptions of the problems faced.

4.2 Agricultural Water Demand

The term agricultural water demand usually includes water used for crops, for livestock and, in some cases, water used in forestry and aquaculture. The water that is an essential factor in agricultural production may be supplied directly by precipitation (in humid areas) or indirectly by irrigation after transportation from neighboring water bodies or aquifers. A combination of these two forms of water supply is common in semiarid and semihumid regions.

Throughout the world irrigation is regarded as the main water using activity. It is an important contribution to helping solve world food problems and encouraging developmet strategies in developing countries. Unfortunately these expectations have not been fulfilled in many developing countries, as is reported for instance by (13, 14). The main factors that are usually regarded as most important for identifying water requirements for agriculture are the climate, soil characteristics, the kind of crops and their combinations, the crop reciprocal substitution of irrigated farming for dryland farming, the farming techniques used, the quantities of water needed for maintaining or improving the salt balance of the soil, and the degree of efficiency of the water use. Consideration of these factors allows a definition for the irrigated crops of the amount of water per unit of irrigated land and this, associated with definition of the areas that it is expected to irrigate in the future, enables to forecast agricultural water demand.

Social and political factors also condition the forecasting of water demand for agriculture, since the costs of irrigation works are often political or linked to meeting social needs as regards the creation of employment or fixing the population in certain regions. Nevertheless, an effort must be made to state clearly the real costs of the water used in agriculture, so that the investments in agriculture may be compared with those made in other sectors.

In water demand forecasting it is essential to bear in mind the increased efficiency that can be achieved in irrigaton developments. The waste of water used in agriculture in the form of overuse of water, excessive losses in water delivery or non-economic selection of crops, is very common, even in regions with severe water shortages.

Water demand in irrigation aims at adequately meeting the consumptive water use of crops. There remain, however, some difficulties connected with uncertainty concerning the relationship between irrigation water applied and crop yield. The forecast of water demand has to be made through process of disaggregation. In a given region and for a group of farmers who grow similar crops and use similar resources, particularly irrigation methods and equipment, an attempt is made to establish relationships between the most profitable combination of crops and water use. This analysis is made for a certain range of water prices, and the corresponding water production function is thus determined.

Short-range forecasts of water demand for agriculture may be quite reliable when done at a local level with the help of farmers experience. Medium and long-range forecasts and regional and national forecasts have proved particularly difficult, and it is necessary to try and base the forecasts mainly on the evolution of the overall economic situation, the relevant technological evolution and the institutional measures that it is expected to apply. Systems analysis techniques have to be applied by carrying out the sensitivity tests that may prove necessary (15).

4.3 Industrial Water Demand

The uses of water in industry are classified as follows: processing water, which is the water that during the manufacturing process comes into direct contact with the final product or with the intermediate products; cooling water for cooling of the various items of equipment subject to heating in the industrial process; boiler water, which is used in steam generation for power production or for the manufacturing process, and water for general uses, such as cleaning and air conditioning.

In most industries water is used in two or more of the ways mentioned, and in varying proportions, although the use of water for cooling tends generally to be the most important, especially owing to the large amounts of water used in the cooling of thermal power plants. There is also considerable variation in the relationships between water consumption and water withdrawal, according to the kind or industrial use.

Industrial water demand may be correlated with the amount of raw materials processed, the value of the gross product, the value added of the output, or the number of workers. These correlations are subject to variation according to the level of the industrial technologies.

A correct water demand forecast based on variables such as those mentioned, depends on the existence of a sufficiently detailed industrial development plan. Should there be no such plan, the forecast is more difficult and it is usual to establish correlations with the population or the gross national product, which obviously are always very fallible as they do not take into account the evolution of the technology which has recently been having considerable influence on the values of industrial water withdrawal and waste discharge. Water recycling in industrial plants is one of the import ways of reducing water consumption, often with resulting economy in production.

The industries with the highest water consumption include chemicals, pulp and paper, primary metal industries, petroleum industries, besides thermal power production plants that have a high water ocnsumption for cooling purposes, as has already been mentioned. Mining may also involve considerable water consumption. Industrial water demand depends on factors such as the type of industrial process involved, the nature of the inputs and outputs used in the industrial process, the design of the industrial plant, the level of industrial activity and the processing of te industrial wastes.

The disaggregation of industrial water demand must be done according to the regions, size of industrial plants, production technology and product output mix. For each case considered it it necessary to try and establish relationships between water demand and the explanatory variable, always considering the various alternatives, including recirculation, technological changes, waste discharge load reduction and water supply treatment. Another important aspect of the water demand forecast is specification of the quality of water required. References relating to the application of regression analysis to industrial water demand forecasting include the works presenting methods used in Canada, by Tate (6), in U.S.A. by Thompson (16) and in England by Rees (17) and Smith (18).

4.4 Water Demand in Hydroelectric Energy

The use of water for hydroelectric power production does not involve consumption of water, but modification of the flow regime of the watercourses owing to energy production needs may have an effect on other uses. It may, however, be said that the forecast of water requirements for hydroelectric power production is less critical than for other uses. Currently, water requirments for hydroelectric power production are estimated by taking into consideration the forecasts of the evolution of power consumption, and this is done by the authorities responsible for power production bearing in mind the various conditioning factors. This forecast will be done within the framework of the river basin or on a national scale, whenever there is a national network, as now happens in most countries. Forecasts of hydroelectric power consumption must be considered within the overall framework of the power policy of each country or region. It is important, from the point of view of water resources, to explain aspects such as the seasonal distribution of demand whether the hydroelectric developments have reservoirs, are run-of-river, or evolve pump storage.

4.5 Water Demand for Navigation

Water demands for navigation are mainly determined by the need to maintain sufficient depths for navigational purposes. This involves the availability of certain volumes of water which, although it is in practice not consumed, may not be available for other purposes. The evolution of water requirements for navigation will be influenced by the development envisaged for the volume of freight, the type of vessels and the regime of operation of river transport. Moreover, river transport must be considered within the national transport policy.

4.6 Water Demand for Cultural and Recreational Purposes

Water uses intended to meet cultural requirements relating to preservation of the environment or recreational uses, do not normally involve high water consumption and do not, therefore, raise important problems as regards water demand forecasting. In fact, in these cases the conditioning factors are mainly the aspects related to water quality. Recreational uses of water are mainly concerned with bathing and swimming, speedboating, sailing, rowing, waterskiing, fishing and protection of wildlife and the landscape, and also with winter sports when snow occurs.

In order to forecast the use of water associated with recreational activities it is necessary to determine the value ascribed by users of the recreational activities, and this usually raises difficult problems owing to the subjective character of such uses.

In certain cases where use of water resources involves specialized equipment, such as pleasure craft, fishing material or camping tents, criteria have already been tried for determining the benefits to the water users. In such cases, assessment of benefits is based on the study of variation in consumption of such equipment with water quality. Conclusions may also be drawn from an analysis of the amounts received, in the case of areas where access to the recreational uses of the water or visits to natural parks is subject to control and payment. Another possible way is to hold inquiries among the potential users of water recreational areas in order to find out how much such users are prepared to pay for enjoyment of those areas, and what is the incidence of the water quality on the amounts they are willing to pay.

REFERENCES

1. Kindler, J. (Ed.), "Proceedings of a workshop on modelling of water demands," Laxenburg, International Institute for Applied Systems Analysis, 1978.

2. Howe, C. W. and Lineweaver, "The impact of price on residential water demand and its relation to system design and price structure," Water Resources Research, 1st Quarter, p. 13-32, 1970.

3. Grima, A. P., Residential Water Demand - Alternative Choices for Management, Toronto, University of Toronto Press, 1972.

4. Water Series No. 3, "The demand for water: procedures and methodologies for national planning," New York Natural Resources, United Nations, 1976.

5. Cuhna, L. V. et al., "Management and law for water resources," Fort Collins, Colorado, Water Resources Publications, 1977.

6. Tate, D. M., "Water use and demand forecasting in Canada: a review," Laxenburg, International Institute for Applied Systems Analysis, RM - 78-16, 1978.

7. Whittington, D., "Forecasting industrial water use," Laxenburg, International Institute for Applied systems Analysis, 1978.

8. Sewell, W. R. D. and Bower, B. T., "Problems and procedures," in Sewell, W. R. D. and Bower, B. T. (Ed.), Forecasting the Demands for Water, Ottawa Department of Energy, Mines and Resources Policy and Planning Branch, 1968.

9. Russell, C. S., "Industrial water use," in Howe, C. S. et al., "Future water demands. The impacts of technological change, public policies and changing market conditions on the water use patterns of selected sectors of the United States economy: 1970-1990," Resources for the Future, 1979.

10. Penman, A., "The experience with effluent change scheme of the City of Winnipeg," Inland Watery Directorate, Canada Department of the Environment, 1974.

11. Widstrand, C., The Social and Ecological Effects of Water Development in Developing Countries, Oxford, Pergamon Press, 1978.

12. Sargent, F. O. and Sargent, B. P., <u>Rural Water Planning</u>. <u>The Wave of the Future</u>, F. O. Sargent, South Burlington, Vermont, 1979.

13. Carruthers, I., <u>Contentions Issues in Planning Irrigation Schemes</u>, Water Supply and Management vol. 2, Pergamon Press, 1978.

14. Bottrall, A., <u>The Management and Operation of Irrigation Schemes in Less Developed Countries</u>, Water Supply and Management, vol. 2, Pergamon Press, 1978.

15. Maidment, D. R., "Systems analysis applied to agricultural water demand," in Proceedings of a Workshop on Modelling of Water Demands, Laxenburg, International Institute for Applied Systems Analysis, 1978.

16. Thompson, R. G., "Economic demand for water and economic costs of pollution control," in Proceedings of a Workshop on Modelling of Water Demands, Laxenburg, International Institute for Applied Systems Analysis, 1978.

17. Rees, J., <u>Industrial Demand for Water: A Study of Southeast England</u>. London, Weidenfeld and Nicolson, 1969.

18. Smith, R. J., "Modelling of water demand and wastewater discharge in England and Wales," Laxenburg, International Institute for Applied Systems Analysis, 1977.

19. McDonald, D. G. et al., "Water demand. A selected annotated bibliography," Laxenburg, International Institute for Applied Systems Analysis, 1978.

13. OPERATING RULES FOR STORAGE RESERVOIRS

Jose D. Salas and Warren A. Hall
Colorado State University, Fort Collins, Colorado

The problems of the best planning and the best operation of storage reservoirs have occupied the professional men from the time the first modern reservoirs have been built in the second half of the last century. With time, the operational rules in the form of operational curves, tables or equations have been developed. Basically, experience with planning and operation of reservoirs, on a trial and error basis, has lead to the various types of operational rules based on the original reservoir objectives and various constraints. The main purpose of this paper is to examine the various factors involved in developing guidelines and operational rules for reservoir operation. A summary describing purposes, objectives, system operation and related factors for selected reservoir systems is given.

1. CONCEPTS AND DEFINITIONS

Various concepts and definitions related to reservoir operation, which are often used in the literature are reviewed herein. The intention is to set some common understanding and terminology useful in approaching the problem of operating reservoirs.

1.1 Purposes

Reservoirs are designed, built and operated for specific purposes such as irrigation, hydropower, flood control, water supply, recreation, navigation, water quality control and fish and wild life enhancement. There are other ways how these purposes

are often referred to in practice. For instance, water conservation is the purpose of storing (conserving) water for releasing
where it is needed for irrigation, hydropower, water supply or
navigation. Thus reservoir purposes may be listed simply as
conservation, flood control, water quality control and fish and
wildlife enhancement. Some water supply and water quality related
purposes are also referred to as low flow augmentation while
instead of water quality, terms such as salinity control, wastewater dilution or water quality improvement are commonly used.
Other less common reservoir purposes are for sediment control. In
general, reservoir or project purposes are sometimes denoted as
project or reservoir functions and project or reservoir
objectives. Reservoirs built for one project are called "single
purpose" and those with more than one purpose are called
"multipurpose" reservoirs.

While some of the foregoing purposes may be compatible, quite
often they are imcompatible and conflicting in the degree and time
of service. For example, flood control is in conflict with
conservation purposes because while flood control is enhanced to
the degree that a reservoir is emptied, conservation may require
the filling of the reservoir. Other examples can be found of conflicts between recreation and flood control, recreation and water
supply or irrigation and hydropower.

1.2 Operational Objectives

Operational objectives are referred here to those which spell
out more specifically details of the purposes of the reservoir.
For instance, the primary operational objective of a water supply
reservoir may be to improve the assurance of the availability of
water in particular times and places while the primary operational
objective of a flood control reservoir may be to assure the
required flood storage in order to avoid or reduce the downstream
damages.

An operational objective of a flood control reservoir may be
stated as:

"To best meet the seasonal target reservoir levels and
stages at a given control point."

On the other hand an operational objective for a multipurpose
reservoir may be stated as:

"Optimize production of water and hydropower that meets
required levels and releases for flood control and
recreation."

1.3 Objective Function

The specific operational objectives of a reservoir transformed into some form of performance criteria or function is referred to as "objective function."

Objective functions may be in the form of economic measures considering either benefits, costs, or both. An example of such an objective function may be stated as:

"Maximize the return from firm water, firm power, dump water and dump power," or

"minimize the expected present value of thermal generating costs."

Objective functions may be also stated in the form of performance characterisitics or indices which reflects the way in which the reservoir satisfies the expected or target demands. An example of this type of objective function may be stated as:

"To minimize the sum of the expected squared deviations from seasonal target reservoir levels and target reservoir releases."

Finally, objective functions in the form of risks of not meeting the expected target water deliveries during the planning horizon of the reservoir may be formulated. For instance, if a reservoir is operated under a specified policy during a time horizon $t = 1, 2, \ldots, n$ the objective function may be to minimize the sum of some deficit indices (say, $I_t = 1$ if reservoir delivery < demand) throughout the time horizon. Such performance measures may, of course, include some measure of the length and severity of the deficits incurred.

1.4 Operation Constraints

Operating constraints vary depending on each particular reservoir system. Usual constraints include the physical bounds such as maximum and minimum reservoir levels, maximum releases which depend on such factors as the capacities of gates, turbines, penstocks and canals and maximum diversions. Constraints are also stated to meet required or mandatory (minimum) releases in specified target reservoir levels or releases. Often some of the operational objectives of the reservoir are met by including them in the set of constraints and in some cases computational constraints are included to simplify the specified method of analysis.

1.5 Operating Policies and Procedures

A distinction is made here between the terms "operating policy or operating rule" and "operating procedure." Let us first describe the latter.

An "operating procedure" is such a mechanism, set of instructions, equations, tables or simply judgement decisions by which reservoir releases and diversions are determined based on the current and/or forecasted state of the system. In general, the state of the system may be described by reservoir levels or volumes, inflows, stages or flows at control points which act as independent variables to determine the reservoir release dependent variable. An additional independent variable may be a vector variable describing possible consequences of actions or decisions whereby trade-offs are judged which in turn will determine the final decision reservoir release.

The foregoing "operating procedure" quite often is based on some guidelines or rules which help in determining the actual reservoir releases and diversions. Such guidelines or rules are called herein "operating policy or operating rule." Operating policies may take different forms such as target reservoir levels or volumes, commonly called "rule curves," target flows and stages at control points, storage allocation and zones and flow or stage zoning.

A further distinction is made herein between the foregoing definitions or "operating procedure," and "operating policy" and the techniques or algorithms which determine them. One should keep in mind, however, that depending on the technique or algorithm used often the "operating procedures" does not require the operating policy or rules defined above. In such cases such operating procedures are also referred to as operating policy. Likewise, often an operating procedure does not require any further technique or algorithm. It is thereby definition such as the case of the so-called "standard operating rule."

2. TYPES OF OPERATING POLICIES

2.1 Target Flows and Stages

Target flows and stages at specified control points in the system are often defined which serve either as upper or lower bounds (depending on the specific reservoir operational objective) or as guides whereby reservoir operators are expected to maintain such targets as close as possible. For instance, in the Kentucky-Barkley reservoir system (see Table 1) current and forecasted target stages at the Cairo control point for several days in

Table 1. Description of Purposes, Objectives, System Operation
and Related Factors for Selected Reservoir Systems

System and Category	Description
KENTUCKY-BARKLEY (part of the TVA system)	

1. Type of System

Two Reservoirs: Kentucky (K) and
Barkley (B) joined by a floodway

2. Schematic System
Configuration

3. Purpose
·General

Multipurpose

·Specific

·Flood Control
·Navigation
·hydropower

4. Operational
Objectives

Flood Control Objectives:
·Safeguard the Mississippi (Miss.) river
levee system.

·Reduce the frequency and magnitude of
flooding land outside levees along the
lower Ohio & Miss. rivers.

·Reduce the frequency of using the
floodway.

5. Operational Criteria
& Other Factors

·The magnitude of flooding on the lower
Ohio & Miss. rivers is related to the
Cairo control point.

·Drawdown the reserviors in advance of
a flood

6. Operational
Constraints

·Extent of reservoir drawdown is limited
by headwater navigation depths.

·Since K-B hydropower plants are an
integral part of the TVA system full
generation and/or full peaking capacity
is desired at all times.

Table 1.   (continued)

---

7. Operating Policies
    ·Control Points   ·Seasonal target stages at Cairo
                                control point.

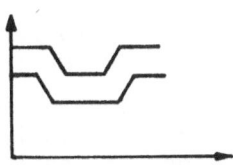

    ·Reservoir Level   ·Seasonal reservoir levels (rule curve)

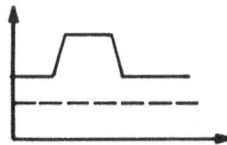

8. Operating   ·Reservoir release decisions are
    Procedures   primarily based on actual and forecasted
                             flood stages at Cairo control point and
                             actual reservoir levels.

                            ·Release decisions are also influenced
                             by other project purposes.

9. Method of Analysis   ·Simulation and engineering judgement.

10. Further Comments   ·Actual system and actual study.

11. Reference   Halsey (1969)

KANSAS RIVER

  1. Type of System   Multireservoir (17 reservoirs)

  2. Schematic System
     Configuration

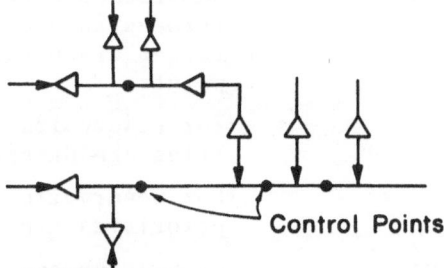

  3. Purpose
     ·General   Multipurpose

     ·Specific   ·Flood Control
                             ·Conservation

214

Table 1.   (continued)

4.	Operational Objectives	Flood Control Objectives: Protect the damage centers
5.	Operational Criteria & Other Factors	•Complete protection for urban developments (Phase III). •Partial protection for agricultural lands (Phase II). •No protection for marginal developments (I). Phases I, II & III are related to river protection levels & storage levels.
6.	Operating Policies •Control Points	•Target protection levels at the control points.

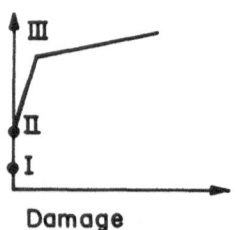

	•Reservoir Level	•Seasonal reservoir flood control capacity.

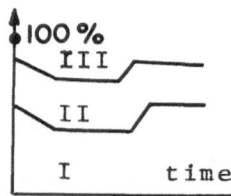

7.	Operating Procedures	•For single reservoirs, releases are determined based on reservoir phase, the forecasted uncontrolled flow and the target discharge at the various control points. •For reservoirs in series release priorities are determined. •For reservoirs in parallel release priorities are determined.
8.	Methods of Analysis	•Simulation and engineering judgement.
9.	Further Comments	•Actual system & actual study
10.	Reference	Coomers (1969)

advance are two of the important variables guiding the releases from upstream reservoirs.

As an example of target flows Loucks (2) refers to the Finger Lakes System where target reservoir releases varying seasonally are specified by the system managers. These target releases represent the most desired reservoir discharges in each month of the year. As another example, the North River system analyzed by Mejia et. al., (5) considered predetermined target flows at a control point downstream from a system of reservoirs.

In addition to, target flows and stages that may be specified at given control points in the system, operating policies are often given in terms of flow and stage zoning varying seasonally (also called ranges or phases). This zoning is done in order to provide more flexibility to the release and diversion decisions in the system. Usually the system operator makes his decision based on both the reservoir and flow (or stage) states.

Figure 1 shows for the Kentucky-Barkley reservoir system the seasonal stage zoning at the Cairo gage located downstream in the

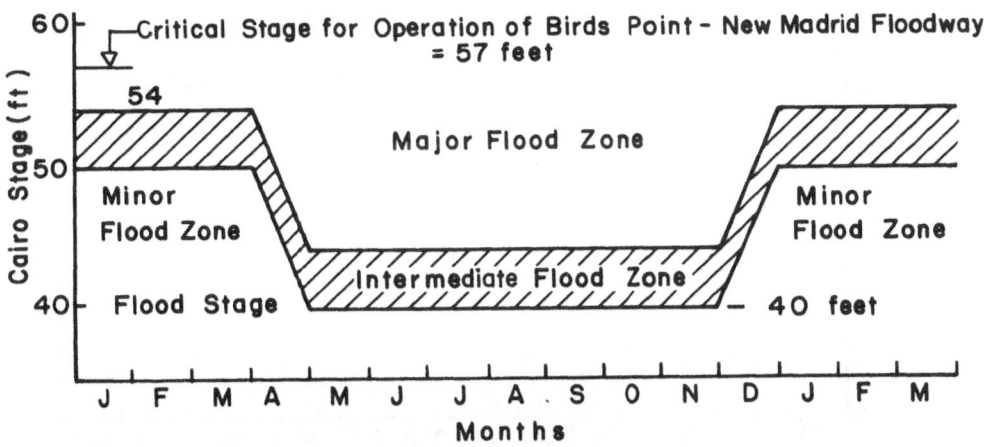

Figure 1. Stage Zoning for the Kentucky-Barkley Reservoir System at the Cairo Stage, Ohio River (10)

Ohio River. Observe that the flood state occurs when the Cairo stage H reaches 40 ft. When H > 40 ft the state of flooding is further divided into 3 zones which varies with the season.

2.2 Target Storage Levels or Volumes (Rule Curves)

Target storage levels or volumes are specifications of seasonal state conditions which are expected or planned for a reservoir system. They serve as a guide for seasonal reservoir operation and they are usually called "guide curves" or "rule curves." They can be either single or multiple curves depending on various possible conditions of say, possible inflows for the planned year. Furthermore, multiple rule curves may in turn define zones within the storage.

Figure 2 illustrates the 1978 annual operating plan (rule curve) for the Seminoe Reservoir of the North Platte River System;

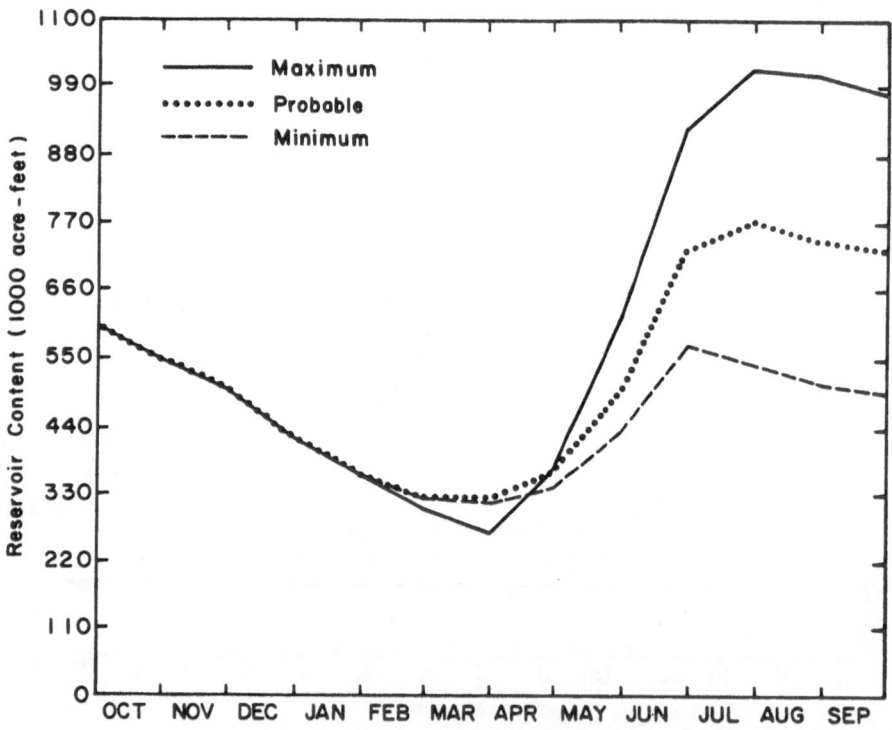

Figure 2.   1978 Annual Operating Plan for the
Seminoe Reservoir (9)

Wyoming, U.S.A. The system is operated for water conservation, flood control, recreation and fish and wildlife. The plan consists of three curves: the "reasonable maximum," the "reasonable minimum" and the "most probable" conditions of inflows into the reservoir. Such plan is restudied and revised from time to time during the year whenever new information or changing conditions indicate this to be desirable (9).

Figure 3 illustrates the typical operating guide (rule curve) for a tributary reservoir of the TVA system providing flood

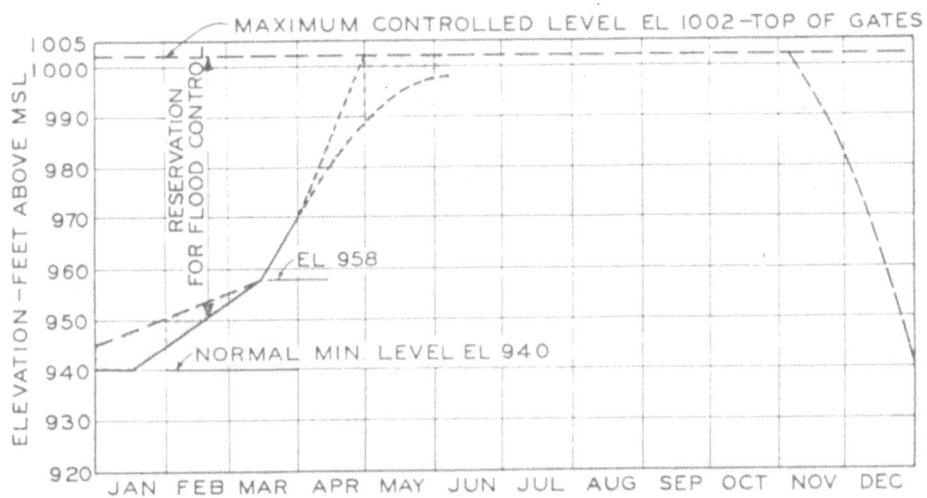

Figure 3. Typical Operating Guide Curve for a Tributary Multiple-Purpose Reservoir (6)

control and conservation storage for hydropower and navigation. The guide curve shows the normal reservoir levels to be expected throughout the year, the storage allocated for flood control and the maximum and minimum levels. The greater flood storage reservation in January 1, gives assurance that the March 15, reservation will be available in the case a series of floods make it difficult to drawdown the reservoir to the March 15, level (6). Drawdown of the reservoir before January 1, provides water for navigaton and hydropower during the drier months. The lesser reservation after March 31, makes allowance for the decreased chance of floods near the end of the flood season.

Figure 4, on the other hand, illustrates the typical operating guide for a reservoir located at the main Tennessee River which provides flood control and conservation storage for hydropower and navigation, but in addition provides a permanent

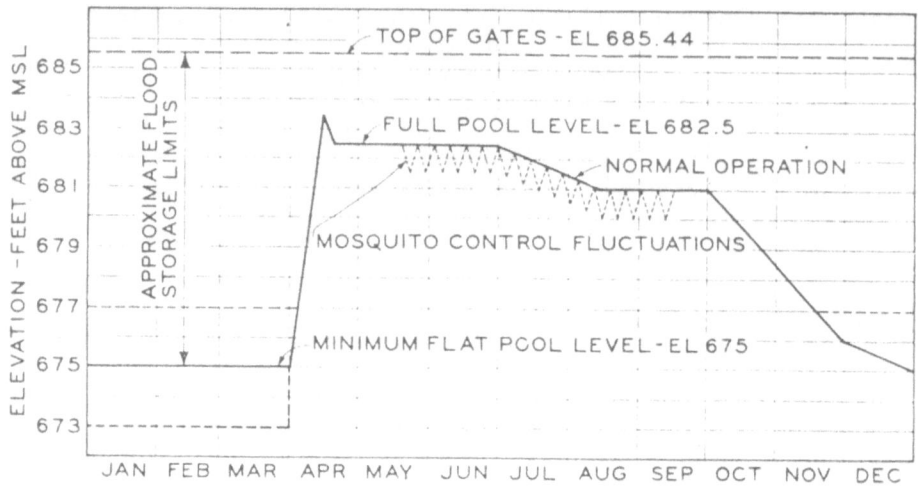

Figure 4. Typical Operating Guide Curve for a Main Tennessee
River Multiple-Purpose Reservoir (6)

pool for navigation. The guide curve shows the normal reservoir
levels for every month and maximum and minimum levels. The curve
also shows the 273 level or 2 ft down the minimum level as
permitted level in order to retain storage capacity for flood
control provided that navigation depths are maintained throughout
the reservoir (6). After March 31, the reservoir level reaches
the 682.5 elevation and the zone between elevation 682.5 and
685.44 is the flood storage reservation during the summer. The
guide also shows that lowering begins during the summer and
accelerates during the fall preparing the system for the new flood
season.

Some of the foregoing examples have shown that the rule
curves vary seasonally and essentially divide the storage into
storage zones for different purposes. For instance, Figures 3 and
4 show that the rule curves divide the total storage into two
zones: the one above the curve which we may call it the flood
control zone, and the one below the curve called the conservation
zone. The foregoing two zones may be further subdivided so that
in general we may divide the storage zones into:

(1)   the inactive storage which is the storage at the bottom of the reservoir and it is usually limited by some required minimum levels due to say allowable navigation depths, location of gates and/or the estimated sediment reserve volume. Extreme dry periods may cause the reservoir to reach this inactive zone in which case reservoir releases may not be feasible.

(2)   the conservation storage which is the storage zone above the inactive storage from which water requirements for irrigation, hydropower, navigation, recreation, etc. are satisfied. In order to give more flexibility to the operating decisions this conservation storage may be further subdivided into a buffer conservation storage and one or more storage zones above the buffer storage. Dry periods may situate the reservoir into the buffer storage in which case the reservoir operator usually takes some rationing measures to reduce reservoir releases. Above the buffer storage there may be either one conservation storage as used for the Oswego River System, New York (8), two conservation storages as used in the Trent River System, Canada (7) or two to five conservation storages as used by the U.S. Army Corps of Engineers.

(3)   the flood control storage which is the storage zone located above the conservation storage in order to provide storage space during the occurrence of floods. In some cases this storage has been further subdivided into two, one immediately above the conservation storage and the other on top of the first. The latter is called spill or surcharge zone (7).

In addition to the division of storages mentioned above, operating policies have been suggested in the form of rule curves within a specified storage zone (usually within the conservation zone). Furthermore, the concept of storage zones has been applied for both single and multireservoir systems. In the latter case, the reservoirs are operated to maintain all reservoirs in the same zone so far as possible.

2.3 Policies for Decision in Space

Policies are necessary for allocation of storages, releases and diversions spatially in a multireservoir system as well as for determining the release priorities among the several reservoirs. Although such policies depend very much on the particular characteristics and configuration of the system at hand, as well as on the objectives, we will give some typical examples.

(1) For a multireservoir system where the objective is to minimize the water wastage the following policies may be used (3, 7):

   a. for a reservoir system in series, the downstream reservoirs are depleted before using upstream reservoirs water to meet downstream demands,

   b. for a reservoir system in parallel, one procedure may be to release water first from reservoirs having the largest drainage area to reservoir storage capacity ratio. This policy is valid if the runoff per unit area is essentailly the same in each reservoir's watershed.

(2) For a system where the objective is the equitable distribution of water among the various users, a reasonable policy may be (4):

   a. if the flow in the river is insufficient to cover the total demand, the deficit in the water supply will be shared by all the system units in proportion to their respective demands. That is, at a given diversion i, the amount of water diverted would be given by

$$W(i) = c(i)\,D(i)$$

$$c(i) = \frac{Q(i)}{D(i)+T(i)}$$

where $D(i)$ = demand at the diversion i, $Q(i)$ = available flow in the river and $T(i)$ = net demand downstream from i.

(3) In a system of reservoirs whose objective is to meet a target water supply level downstream at a control point, as long as water levels in the reservoirs used for recreation are not violated, a priority reservoir policy may be to drawdown the reservoirs sequentially in the inverse order of their recreational utility per unit volume of stored water (5). This policy of course has the shortcoming of possible waste of water in future time periods by spilling at downstream reservoirs.

3. TYPES OF OPERATING PROCEDURES

Operating procedures are given in different forms and various levels of completeness. Operating procedures in general are designed to suit the system operating objectives, to conform with the given operating policies and system characteristics and constraints. Operating procedures may be in the form of precise definition of say reservoir releases as a function of certain state variables such as reservoir levels or they may require additional judgement on the part of the reservoir operator to react to events developing in real time. In any event, two types of operating procedures are important. Those dealing with decisions in time and those dealing with decisions in space.

Decisions to be made in real time operations of a reservoir of a system of reservoirs, can largely be characterized by the quantity of water to be released from a reservoir in the immediate following period of time. Naturally these releases depend on whether the reservoir is of single purpose (say flood control or conservation) or multipurpose (say flood control and conservation). If the reservoir is for flood control and conservation, generally the release procedures and decisions will vary depending on where the reservoir state is at, i.e., the flood control state or at the conservation state. However, in each case one could say that the releases consist of two parts:

(1)   releases directed to meet predetermined "rule curves" or some form of required target levels of assured level of service or avoidance of failures and

(2)   the optimal release corresponding to supplemental energy, water or flood control reservation, etc., which will enhance the various purposes of the reservoir system.

In general one could say that an operating procedure or release instructions for a reservoir system can be expressed mathematically as

$$Y_t = f_t (S_t, \hat{X}_t, Q_{t-1}, \hat{Q}_t, \underline{T}_t, \ldots)$$

where Y_t = releases during time period t, S_t = state of the reservoir at the beginning of time period t, $\hat{X}_t$ = forecasted reservoir inflow during time period t, Q_{t-1} = vector of flow or stage states at control points in the system during time periods $t-1$ (beginning of time period t), $\hat{Q}_t$ = vector of forecasted states (flow or stages) at control points in the system and $\underline{T}_t$ = vector of target or desired releases for the time period t.

For example, for a reservoir for conservation (or when the reservoir state is in the conservation zone) the bench mark procedure for operating such reservoir is the so-called "standard operating rule" which defines uniquely the reservoir releases as a function of the available water in the system (see Figure 5).

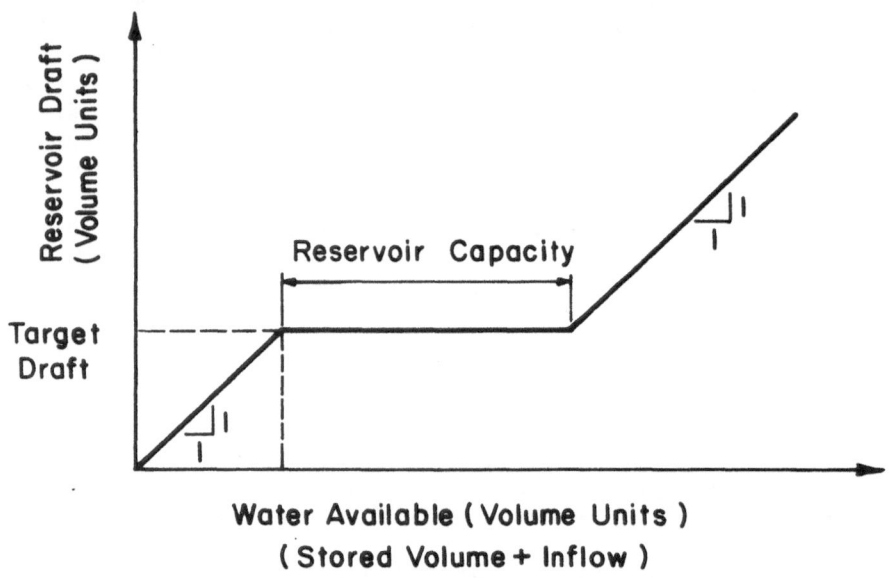

Figure 5. Standard Operating Policy

Mathematically this release rule is expressed as

$$y_t = S_{t-1} + x_t \text{ , if } S_{t-1} + x_t \leq T_t \text{ , therefore } D_t^- = T_t - S_{t-1} - x_t$$

$$y_t = T_t, \text{ if } T_t < S_{t-1} + x_t \leq T_t + V, \text{ therefore } D_t^- = 0, D_t^+ = 0, \text{ and}$$

$$y_t = S_{t-1} + x_t - V, \text{ if } S_{t-1} + x_t > T_t + V, \text{ therefore } D_t^+ = S_{t-1} + x_t - V - T_t,$$

where D_t^- and D_t^+ are deficits and surpluses during time t, respectively. The shortcoming of the rule is that it does not provide a mechanism for rationing water whenever there is insufficient water nor does it have a mechanism for releasing more water when there is plenty available. However, modifications can be made to improve the performance of the standard rule. For instance a rationing criteria may be that whenever the available water supply is smaller than the target release, to let such available supply become a new smaller target and the actual reservoir release will thus become reduced proportionately. Bower et. al., (1) suggested a hedging rule in order to accept small current deficit in output so as to decrease the probability of a more severe water or energy shortage later in the drawdown-refill cycle. This is especially justified when the proposed uses of water have nonlinear loss functions. Bower et. al., (1) also suggested the pack rule for stepping up releases toward the end of the drawdown-refill cycle to free reservoir space for predicted inflows that would otherwise spill. Several other forms of release procedures either in the format of mathematical equations or tables has been used or proposed not only for defining such release procedures in time but for spatial reservoir release allocation as well.

Recently Anderberg (13) suggested a new concept called "the anticipated decision influence period (ADIP)" whereby reservoir operations can be improved by considering decisions only during relatively short time periods of drawdown or refill which are critical for the reservoir operation. Likewise, Ballesteros (12) used the ADIP concept to derive operational procedures where the risk of not meeting specified demands was included.

4. SUMMARY AND CONCLUSIONS

Guidelines and operational rules of reservoir systems have been examined and presented within the context of the overall process of water reservoir management. Some emphasis was placed on classifying and categorizing definitions and concepts commonly used or proposed in reservoir operations. Reservoir operations can be made by setting up an operating policy or rule and by corresponding operating procedures. The several forms of such policies and procedures have been reviewed including examples and some discussion of advantages and limitations.

ACKNOWLEDGEMENT

The support of the projects "Conjunctive Water Uses of Complex Surface and Groundwater Systems" Bilateral U.S.--Spanish Project and "Investigation of Objective Functions and Operational Rules of Storage Reservoirs," OWRT Project B-195-COLO are gratefull acknowledged.

REFERENCES

1. Bower, B. T.; Hufschmidt, M. M. and Reedy, W., "Operating procedures: Their role in the design of water reservoir systems by simulaion analysis" in <u>Design of Water Resources Systems</u> by Maass et. al. (Harvard University Press, 1962).

2. Loucks, D. P., "Computer models for reservoir regulation" <u>Jour. of Sanitary Eng. Div. ASCE</u>, SAO, (1968) 657-669.

3. Loucks, D. P. and Sigvaldason, O. T., "Multiple reservoir operations in North America" (paper presented at the 1979 Poland Symposium on Reservoir Operations, 1979).

4. McBean, E. A., Lenton, R. L., Vicens, G. J. and Schaake, J. C., "A general purpose simulation model for analysis of surface water allocation using large time steps" (Ralph M. Parsons Lab. for Water Resources and Hydrodynamics, Dept. of Civil Engineering, M.I.T., Technical Report No. 160, 1972).

5. Mejia, J. M., Egli, P. and Leclerc, A., "Evaluating multi-reservoir operational rules" <u>Water Resource Ressearch</u>, Vol. 10, No. 6 (1974) 1090-1098.

6. Shelton, R. A., "Management of the TVA reservoir system" (paper presented at the ASCE National Workshop on Reservoir Systems Operations, Boulder, Colorado, Auguast 13-17, 1979).

7. Sigvaldason, O. T., "A simulation model for operating a multipurpose multireservoir system" <u>Water Ressource Research</u>, Vol. 12, No. 2 (1976) 263-278.

8. Tedrow, A. C.; Liu, C. S.; Halton, D. B. and Hiney, R. A., "The use of system analysis in the development of water resources management plans for New York State" (report to the U.S. Department of Interior, OWRT, Albany, New York, 1970).

9. U.S.B.R., "Annual operating plan of the western division, Pick-Sloan Missouri Basin program" (Lower Missouri Region, Denver, Colorado, 1978).

10. Halsey, D. H., "Flexibility in water resource management as related to reservoirs" (proceedings of a seminar on reservoir systems analysis, the Hydraulic Engineering Center, U.S. Corps of Engineers (1969) paper 2).

11. Coomes, R. T., "Flood regulation of Kansas River basin reservoirs" (proceedings of a seminar on Reservoir Systems Analysis, The Hydrologic Engineering Center, U.S. Corps of Engineers (1969) paper 3).

12. Ballestero, T. P., "Seasonal risk-based reservoir operating rules" (Ph.D. Dissertation, Department of Civil Engieering, Colorado State University, (1981)).

13. Anderberg, L.A.W., "The anticipated decision influence period in real time reservoir operation" (Ph.D. Dissertation, Department of Civil Engineering, Colorado State University (1979)).

14. APPLICATION OF DECISION THEORY TO OPERATION

Lucien Duckstein
University of Arizona, Tucson

1. INTRODUCTION

The purpose of this chapter is to introduce a framework for the application of statistical decision theory to the operation of large-scale and complex water resources projects. In plain words, the chapter deals with a model that describes the process of data gathering, information generation based on those data, decision-making and evaluation of the consequences of the decisions. By providing measures of performance, the approach points the way to improving any hydrologic system that involves both an information and a response subsystem.

The model emphasizes that information may be perfect or imperfect and decisions may be optimal or nonoptimal. Four cases are thus defined; the case of imperfect information with non-optimal design appears to have received little attention in the literature and is thus investigated in greater detail than the other three cases. The approach demonstrates that it is important to consider not only how the information is generated but also how that information is used in decision making. Decision theory has traditionally been applied to small-scale problems, such as the design of a flood levee (3), that of a drainage system (14), or the operation of a reservoir (11). For complex systems such as the network of water sources and users considered in the "identification of subsystems" report, a general framework is necessary. We propose to use, for such a framework, the Information Response (IR) system first developed in Sniedovich and Davis (17) and Krzysztofowicz et al. (12), for flood-forecasting-response systems, then generalized to various hydrologic problems in Davis et al. (2).

The IR concept accounts for the possibility of imperfection or nonoptimality in both information provided by sample data and decisions made on the basis of this information. System performance thus depends on both quality of information and quality of response: a perfect rainfall or demand forecast has value only if reservoir releases use this forecast according to an optimal or near-optimal rule.

The paper is organized as follows. The IR framework is introduced in the next section. The choice of a decision criterion is examined within the context of optimal decision-making under imperfect information, and an example in reservoir control is given. Then the case of nonoptimal decision making under imperfect information is investigated, using a flood forecasting example. It is noted that availability of imperfect information includes, as a special case, the availability of perfect information.

2. GENERAL FRAMEWORK: THE INFORMATION RESPONSE (IR) SYSTEM

The approach to decision theory proposed in this chapter may be viewed as an extension of the preposterior analysis of statistical decision theory (15, 1), that allows consideration of nonoptimal decision rules. The fundamental concept here is an Information Response (IR) system that quantitatively links data, information, decision rules, and the resulting operation of a complex water resource project. Quantitative measures of performance and efficiency stem from measures first developed in Sniedovich and Davis (17) for evaluation flood forecasting response systems and then generalized in Davis et al., (2).

The design operation of a water resource project often may be improved by the acquisition of more data or by the better use of the data currently available. There are instances, however, where the acquisition and use of additional data is minimally useful or even counterproductive (13).

The key to analyzing such apparent anomalies is the joint study of the data acquisition process and the manner in which the data are used in the design and operation of a project. Previous studies on the worth of hydrologic data and their stochastic properties (3, 7), have been framed within a Bayesian decision theoretic apporach. This approach assumes that the best use of the data is being made; that is, based on the data and the objectives of the project, optimal design and operation decisions are being made. The value of the additional data is defined to be the incremental change expected in payoff or loss. If the cost of obtaining the data is zero, and if the decision based on these data is optimal, then if follows that the value of the data is

228

always nonnegative and ususally positive. Often, decisions based on hydrologic data are not optimal, bacause of, for example, regulations governing the use of the data, lack of information needed to make an optimal decision, or even lack of interest in obtaining an optimal decision.

2.1 Information Response System Model

The performance of a water resource project is measured by, for example, increase of a benefit. Variables that affect the performance of the project may be either controllable or uncontrollable. Controllable or decision variables determine the operation of an existing project. Uncontrollable variables or states of nature determine the actual performance of the project once the design and operation have been chosen. Data are taken to provide information about the values of the states of nature. The decision-maker responds to this information by choosing values for the decision variables. This choice, and the values of the state variables determine the effectiveness of the project. The overall process of obtaining information and making the decisions in response to this information is viewed as an IR system.

The systems approach to operating a project is based on the following four elements of and IR system: (1) data collection, (2) information compilation, (3) decision and (4) project operation itself. The first two elements form the information subsystem and the latter two form the response subsystem. This general scheme is depicted in Figure 1.

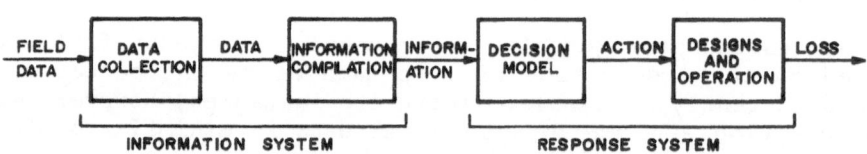

Figure 1. Information Response System

In the information subsystem, the field measurements are collected and transformed to information which is provided to the response subsystem. In the response subsystem, the decision rule is used to respond to this information. This decision rule assigns values to the decision variables which determine the operation of the project. In conjunction with the values of the states of nature, the decision variables determine the loss or benefit to be expected.

To formalize the IR process, let I, a random variable, be the information generated by the information sybsystem, say, areal precipitation. The probability distribution function of I is $H_I(i|\theta)$ where θ is the state of nature. The imperfect knowledge of the state of nature is described by the probability distribution $p(\theta)$. In the response subsystem, the response to the information is obtained by the use of a function termed the decision rule: $d(I)$. For a given project, an associated loss $\ell(d(I),\theta)$ is incurred if the decision (response) $d(I)$ is made and the state of nature takes on the value θ. Specification of the distribution H_I is necessary to account for imperfection of the information provided by the information subsystem. The decision rule d is a function of the information I so that a mechanism be available for adapting the response to new information. Evaluation of loss function $\ell(d(I),\theta)$ often involves the intermediate step of determining the outcome $z\varepsilon Z$, as it will be explained in section 3.

Given a decision rule d, the expected loss or risk for the project is:

$$R(I,d) = R[H_I(\cdot),d] = \iint \ell(d(i),\theta)dH_I(i|\theta) \, p(\theta) \, d\theta \qquad (1)$$

An optimal decision rule d* minimizes the risk and is termed the Bayes decision rule. The Bayes risk $R(I,d*)$ corresponding to the use of an optimal decision rule is a function of the information I. How perfect information may reduce this risk is shown next.

2.2 Case of Perfect Information

Information is considered to be perfect, I*, if there is a unique value of the state of nature corresponding to each realization of the information I. Formally, information is defined as being perfect if there exists a function $\phi(i)$, such that if $H_I(i \; \theta) \neq 0$, then

$$\theta = \phi(i).$$

The risk with perfect information $R(I*,d)$ is calculated in the usual manner from Eq. (1). If the function ϕ has an inverse, the calculation of risk with perfect information may be simplified:

$$R(I*,d) = \int \ell(d(\phi^{-1}(\theta)),\theta) \, p(\theta) \, d\theta \qquad (2)$$

In this case the risk is a simply functional of the decision rule; by using an optimal decision rule, the risk may be reduced.

With perfect information and an optimal decision rule the risk is

$$R(I^*,d^*) = \min_{d(I)} \iint \ell(d(i),\theta)dH_{I^*}(i|\theta)\ p(\theta)\ d\theta \qquad (3)$$

Since perfect information implies knowledge of θ, the optimal response to perfect information is one that minimizes the loss given θ. Thus the risk may also be calculated as

$$R(I^*,d^*) = \int \min_{a\epsilon A} \ell(a,\theta)\ p(\theta)\ d\theta \qquad (4)$$

where A is the set of actions or decisions $\{a\}$ given θ.

From Eqs. (3) and (4), the optimal decision rule for perfect information may be determined as follows:

$$d^*(i^*) = b, \text{ where } \ell(b,\phi(i^*)) = \min_{a} \ell(a,\phi(i^*)) \qquad (5)$$

and i^* is a realization of I^*.

2.3 Measures of Effectiveness

In determining the effectiveness of the IR system, two conditions for each subsystem are considered. For the output of the information subsystem, we consider either perfect information or the actual information (usually not perfect); while for the behavior of the response subsystem, we consider either the optimal response or the actual response (usually nonoptimal). These conditions determine four modes of operation of the IR system as shown in Table 1. As expected loss, or risk, R, is associated with each mode. As a reference value for comparing various risks, the risk incurred with no response R^0 is taken. The term "no response" may indicate either zero response or the response that would be made without benefit of information from the information subsystem.

Evaluation of the IR system consists of a performance vector and an efficiency vector. The components of the performance vector are defined as four "values", described as follows:

(1) The potential value (PV) of the IR system is the difference between the risk with no response and the risk with perfect information--optimal response:

$$PV = R^0 - R(I^*,d^*).$$

Table 1. Expected Risks in the Four Modes
of Operation of an IR System

| | | RESPONSE SUBSYSTEM | |
		OPTIMAL RESPONSE d*	NONOPTIMAL RESPONSE d
INFORMATION SUBSYSTEM	PERFECT INFORMATION I*	$R(I^*,d^*)$	$R(I^*,d)$
	IMPERFECT INFORMATION I	$R(I,d^*)$	$R(I,d)$

(2) The realizable value (RV) of the IR system is the difference between the risk with no response and the risk with perfect information--actual response:

$$RV = R^0 - R(I^*,d).$$

(3) The optimal value (OV) of the IR system is the difference between the risk with no response and the risk with actual information--optimal response:

$$OV = R^0 - R(I,d^*).$$

(4) The actual value (AV) of the IR system is the difference between the risk with no response and the risk with actual information--actual response:

$$AV = R^0 - R(I,d).$$

In Bayesian terminology (Raiffa and Schlaifer, 1961), the difference between the potential value, PV, and the optimal value, OV, is the expected opportunity loss:

$$XOL = PV - OV = R(I,d^*) - R(I^*,d^*). \tag{7}$$

When the response is not optimal, the expected actual opportunity loss can be defined as

$$XAOL = PV - AV = R(I,d) - R(I^*,d^*). \tag{7}$$

a comparison of Eqs. (6) and (7) shows that XAOL $\geq$ XOL.

The efficiency vector has three components: the information efficience IE, the response efficiency RE, and the overall efficiency OE:

$$IE = OV/PV,$$

$$RE = AV/OV, \tag{8}$$

$$OE = AV/PV = IE \cdot RE$$

Intuitively, the information efficiency, IE, is a measure of how well the information subsystem meets the needs of the response subsystem, and the response efficiency, RE, is a measure of how well the response subsystem is utilizing the information subsystem.

When the actual information I converges to the perfect information I*, for example, as a result of increasing sample size, then AV $\rightarrow$ RV and OV $\rightarrow$ PV; consequently, XOL $\rightarrow$ 0. Also IE $\rightarrow$ 1 and RE $\rightarrow$ RV/PV which means that the information efficiency IE attains its maximum, while the response efficiency RE is bounded by RV/PV. Since AV $\rightarrow$ RV, XAOL does not approach zero and may actually increase. Any further improvement in the response efficiency RE and reduction in the XAOL can be achieved only through an improvement in the response. If such an improvement is enforced in the proper direction so that d $\rightarrow$ d*, then RV $\rightarrow$ PV; consequently XAOL $\rightarrow$ 0 and OE $\rightarrow$ 1.

The efficiency vector is a good point to start a comparison between system operation schemes. the performance vector is necessary for intersystem comparison and benefit-cost analyses of system modifications.

3. OPTIMAL DECISION-MAKING WITH IMPERFECT INFORMATION

3.1 Choice of a Decision Criterion

If the information sybsystem provides imperfect information I and the response subsystem follows an optimal decision rule d*, then the system operates within the context of statistical decision theory. Essentially, such formulation requires that the possible decisions, that is, the options available to the decision-maker, be specified. It requires also that the outcome of each decision be quantified for each possible situation or state of nature. This quantification is summarized in a loss table or by a loss function ℓ as defined in section 2. In addition to stating the problem in this format, the decision-maker must select a criterion for comparing or choosing among alternative decisions. It is this selection of decision criterion which is examined closely in this section. The manner in which the criterion effects a decision is discussed within the format of statistical decision theory.

Statistical Decision Theory. Let the basic elements of a decision problem be denoted:

D = decision space, Z = outcome space, and g(·) = probability distributions mapped on the outcome space. The outcome space Z consists of elements z, each of which represents the outcome of a particular decision dεD for a given state of nature θεΘ. choosing among probability distributions g(·) over Z, which may consist of choosing the parameters of a pdf or selecting the type of pdf, is not a trivial matter; several criteria for making such choices have been proposed (Raiffa and Schlaifer, 1961). Presently, two of these criteria, Bayes and minimax, are contrasted after Weber et al., (1980). Also, the important distinction between absolute and expected minimax is pointed out and the proposition is put forth that most decisions are made on the basis of a mix of criteria.

The elements of Z must be well-defined, but they may represent complex outcomes or occurrences. Decision theory assumes that the decision-maker has preferences among the elements of Z. Preferences among monetary outcomes are determined naturally; preferences among multi-dimensional or vector-valued outcomes are generally more difficult to estimate. Several loss functions can be defined: one for economic aspects, another for environmental impact, a third for risk, and so on. The problems involved in establishing consistent preferences among multi-dimenstional outcomes have been considered extensively in the context of utility theory (9, 11,). In the loss function $\ell(d,\theta)$, the outcome is implicitly expressed as as single number for each combination of decision and state of nature, but this is an over-simplification done for illustrative purposes only.

Bayes and Minimax Criteria. Bayes decisions have already been introduced in the context of the IR system after Eq. (1). Formally, if G is any distribution of θ, then for any dεD the expected loss or risk R(G,d) is given by

$$R(G,d) = \int \ell(d,\theta) \, dG(\theta) \qquad\qquad (9)$$

The Bayes risk is then defined as: $R^*(G) = \inf_d R(G,D)$. A decision d* is a Bayes decision against the distribution G if and only if $R(G,d^*) = R^*(G)$. Thus a Bayes decision against G minimizes the risk $R(G,d)$.

Clearly, minimizing risk is not the only possible criterion in choosing a decision, although it is frequently the criterion used when the prior probability distribution G on θ can be specified. Such a distribution frequently is subjective and, in practice, it may be difficult to choose an appropriate prior distribution. The most frequently used criterion which does not require specification of a prior distribution is the minimax criterion. The alternative d** is a minimax decision if $\sup R(G,d^{**}) = \inf_d \sup R(G,d)$, where sup is taken with respect to Gεϝ. Minimax decisions can be made under quite general conditions (DeGroot, 1970; Ferguson, 1967), but in some cases they do not exist. Further, for many problems the minimax decision consists of a mixed or probabilistic choice of alternatives, that is, d** is a probability distribution of alternatives. This type of result may be inconvenient or unappealing in practice. In terms of a reservoir example, a mixed strategy would consist of choosing a release policy for the satisfaction of several objectives, say conservation and flood control using a specified random mechanism. On the other hand, a Bayes decision need never be mixed; it can always consist of one alternative.

The minimax criterion is conservative or pessimistic; essentially, the worst possible distribution of the states of nature is assumed as a prior distribution. If the worst possible or least favorable prior distribution from the point of view of the decision-maker is assumed, a minimax decision is a Bayes decision for that least favorable prior distribution. A distribution is least favorable if it results in the largest minimax risk, that is, γεϝ is a least favorable distribution if:

$$\min_{d\in D} R(\gamma,d) \geq \min_{d\in D} R(G,d) \qquad \text{for all } G\varepsilon\varGamma \qquad (10)$$

The minimax decision criterion originated in the context of games of strategy (18). If two rational decision-makers playing a zero-sum game randomly choose alternatives according to specified probability distributions, and if expected value is assumed to indicate the value of the possible loss associated with a particular probability distribution, then minimax is the appropriate choice criterion. In this case, each decision maker can be assumed to choose the alternative associated with the distribution which is most favorable for his opponent, and thus the alternative each decision-maker chooses is a Bayes decision for the distribution which is least favorable for him. This would be applicable to two unfriendly contries sharing a water body and trying to protect their own interest--say, minimize damage due to poor water quality.

In the context of games of strategy, assuming a least favorable prior distribution is easily justified. For decision made in other contexts, this assumption implies a very conservative approach in which the decision-maker assumes the worst and then minimizes expected loss. Although nature cannot generally be assumed to be perverse with respect to the preferences of the decision-maker, there are some problems for which this conservative approach seems reasonable, for example, if human lives or irreversible damage are involved. In this case, the decision-maker may choose to avoid any possibility of a catastrophic occurrence without consideration of the associated probabilities.

For any problem, an expected minimax decision exists if and only if a least favorable distribution can be specified. It might seem that a least favorable distribution exists for all or at least most problems; actually this is not the case, especially if $dG(\theta)$ is continuous.

It can be shown that a least favorable distribution exists if $R(\theta,d)$ is bounded for all d, even though θ is unbounded. In practice, this is probably most likely to occur when the bound on $R(\theta,d)$ is constant for all d. If θ is bounded and if $R(G,d)$ is bounded for all d, then the least favorable prior distribution is a point distribution which assigns probability one to the value for θ for which $\min_{d \in D} R(\theta,d)$ is maximized.

As an example of a least favorable distribution, assume that the loss $\ell(\alpha,\beta)$ is proportional to the minimum expected squared error (variance), where α and β are the parameters of a beta distribution. Then

$$\ell(\alpha,\beta) = \frac{k\alpha\beta}{(\alpha+\beta+1)(\alpha+\beta)^2} \tag{11}$$

If β is fixed, the least favorable distribution is obtained by maximizing $\ell(\alpha,\beta)$ with respect to α. A little algebra yields that

$$\frac{\partial \ell(\alpha,\beta)}{\partial \alpha} = 0,$$

is attained if

$$\alpha = 1/2\{-(1 + \beta) \pm [(1 + \beta)(1 + 9\beta)]^{1/2}\} \qquad (12)$$

For example, if $\beta = 3/5$, the least favorable distribution $\gamma(\alpha,\beta)$ is obtained if $\alpha = 2/5$. In this case, a minimax decision would be Bayes using $\beta(2/5,3/5)$ as the distribution. Note that this least favorable distribution does not necessarily correspond with a subjectively reasonable view of the probabilities of the corresponding states of nature.

A distinction must be made between an expected minimax criterion and an absolute minimax criterion, particularly since both are frequently referred to as minimax. In contrast with the (expected) minimax criterion defined above, the absolute minimax criterion minimizes the maximum possible loss, not the maximum possible expected loss. Using the absolute minimax criterion, a decision-maker determines the maximum possible loss for each of his alternative strategies and chooses the strategy for which this loss is minimum.

If the expected minimax strategy is pure, that is, consists of a specific alternative and not a distribution of alternatives, then there is a saddle point corresponding to the strategy for which

$$\sup_{G\varepsilon\Gamma} \inf_{d\varepsilon D} R(G,d) = \inf_{d\varepsilon D} \sup_{G\varepsilon\Gamma} R(G,d) \qquad (13)$$

The decision d^{**} is a minimax decision if $\sup_{d\varepsilon D} R(\theta,d^{**}) = \inf_{d\varepsilon D} \sup_{\theta\varepsilon\theta} R(\theta,d)$ and is pure if the saddle point condition holds. In this case, the two types of minimax decsions, expected and absolute, are equivalent. This equivalence always occurs when there is a saddle point, since for a saddle point the payoff value is both the maximum of its row and the minimum of its column. Note that is is conventional to write losses without the negative sign. Thus minimizing a loss is equivalent to maximizing its positive value.

However, if there is no saddle point, the two minimax criteria are not equivalent. In complex cases, especially when the possible losses vary greatly, the differences between the two minimax criteria can be substantial.

Choice of a Decision Criterion. The essential differences amoung the absolute and expected minimax and Bayes criteria result from the different assumptions concerning the prior distribution of the states of nature. The absolute minimax criterion is determined without consideration of the prior distribution of the states of nature, although of course the loss depends on the state of nature that actually occurs. This criterion is sometimes considered particularly appropriate for unique decisions, that is, for decisions which are not repeated, such as response to an extreme hydrologic event.

The expected minimax criterion is optimal, in the sense of expected value, against the least favorable prior distribution. Thus, this criterion is appropriate when nature is assumed to be perverse. There is some doubt that the expected minimax criterion can be justified in many practical cases. The theory of games, in which the expected minimax criterion originated, assumes that two rational decision makers oppose each other in a zero-sum situation, each knowing the payoff matrix. The zero-sum characteristic of games of strategy is usually not present in other decisions made under uncertainty. In many cases, the decision maker wishes to choose the optimal decision in the presence of uncertainty about the state of nature. However, there is no reason to assume that nature is somehow perverse and chooses its state in order to minimize the decision-maker's expected or average payoff.

The Bayes criterion is optimal, by definition, if the prior probability distribution accurately reflects the decision-maker's view of the probabilities of occurrence of the states of nature, and the decision is made so that maximizing expected value is reasonable. However, if either of these assumptions does not hold, another criterion may be preferable to Bayes.

In practice, a decision maker may compensate for an unrealistic loss function, for example, by adjustments in his prior distribution, or vice versa; or he may use a decision criterion that adjusts for inaccuracies in either or both his loss function and his prior distribution. Such a consideration of risk aversion is discussed, for example, in Davis et al., (3). In many cases, a decision-maker does not formulate precisely a prior distribution or a loss function in terms of utilities. Instead, he attempts to protect himself against certain outcomes which he views as extremely distasteful, without considering their exact utility or the exact probability of their occurrence for alternative decisions. Thus, in practice, decisions frequently are not based on the absolute minimax, expected minimax or Bayes criterion, but represent a combination of these criteria which the decision-maker feels, intuitively, is appropriate for his particular problem. If the nature of this mix can be identified, more accurate models of a decision-maker's behavior can be formulated. These models, in turn, lead to a better definition of the decision rule d(I) which is a necessary component of the IR system.

3.2 Example: Reservoir Control Under Uncertainty

The Decision-Making Situation. Consider a connected system of surface and ground reservoirs utilized for the dual objectives of flood control and water supply. The optimal operating policy of the reservoir system has to be decided for one month only. Since at least one element of the system is a random variable, the problem is to make an optimal decision under uncertain information. Let:

Q = inflow into the reservoir

W = end storage = demand

X = release or pumpage from the reservoir (decision variable)

S = starting storage

V = maximum usable storage

θ = parameter vector of pdf $f(\cdot\ \theta)$ of random variable $(\cdot)$

Let the components of the objective function be:

(a) A quadratic flood damage function for the case when the amount of water X taken out of the reservoir is insufficient:

$$C(X) = 2(Q-X-V+S)^2 \quad \text{for} \quad Q > X + V - S \qquad (14)$$

(b) A linear shortage penalty function for the case when too much water is released:

$$K(X) = 10(W-S-Q+X) \quad \text{for} \quad Q < X + W - S \qquad (15)$$

In system "response", the decision rule consists of minimizing (over X) the expected value of the sum of Eqs. (14) and (15):

$$g(X) = E[Z(X)] = E[C(X) + K(X)] \qquad (16)$$

This expectation is to be taken upon all random variables present in the model.

Natural Uncertainty in Q and Economic Uncertainty in W. Let us minimize Eq. (16) under uncertainty in Q, W or both. Other uncertainties may be considered in an analogous manner (Duckstein and Bogardi, 1978). Letting $S = 4$, $V = 9$, the goal function Z in Eq. (15) becomes

$$Z(X,Q,W) = \begin{array}{ll} 2(Q-X-5)^2 & \text{if } Q > X-5 \\ +10[W+X-Q-4] & \text{if } Q < W + X-4 \end{array} \qquad (17)$$

The realistic situation would be that Q and W are jointly distributed f(Q,W) with a negative correlation coefficient--when inflow in high, demand is likely to be low, and vice versa. In this case, the objective function would be written:

$$g(X) = 10 \int_0^{12} \int_0^{W+X-4} (W+X-4-Q)f(Q,W)dQdW$$

$$+ 2\int_0^{12} \int_{X-5}^{12} (Q-X-5)^2 f(Q,W)dQdW \qquad (18)$$

However, one need not look at such a computationally complicated case to illustrate the influence of randomness: Table 2 shows the optimum decision X* for four combinations of deterministic and

Table 2. Illustration of Hydrologic and Economic Uncertainty (Natural and Model)

Input Q	Demand W	X*	Z*
6	6	0	0
uniform in (0,12)	6	2.34	13.45
6	uniform in (0,12)	0.717	3.23
uniform* in (0,12)	uniform* in (0,12)	0.460	8.62
Poisson with mean 6	6	2.0	5.79

*Q and W are assumed to be independent

random Q and W, respectively, calculated as particular cases of Eq. (18). For example, if Q = 6 and W is uniform in the interval (0,12), Eq. (16) is written:

$$g(X) = \frac{10}{12} \int_{10-X}^{12} (W+X-10)dW + 2(1-X)^2 \qquad (19)$$

The optimum release X* = 0.717 is readily found from solving the equation g'(X) = 0. Each row of Table 2 represents a different hydrologic and economic model choice, with, however, a constant expected value of Q and W: E(Q) = E(W) = 6. Analyses such as the one illustrated in Table 2 are necessary to study the sensitivity of decisions to uncertainties.

A Bayesian Analysis of Input Uncertainties. An analysis of input uncertainties consists of a decision-making phase and a decision evaluation phase. They are described in the following paragraphs.

(1) Decision-making phase. Referring to Eqs. (14) and (15), the preceding section has suggested an approach to estimate the pdf describing natural uncertainty in the input Q. Similar methods could be used to investigate the natural and sample uncertainty in V, S and W. For sake of simplicity, a complete Bayesain decision analysis (including natural and sample uncertainty) will now be presented considering uncertainty on Q only. In Eq. (17), let W = 6, which is the expected value used in Table 2. Then, with V - S = 5,

$$Z(X,Q) = \begin{array}{ll} 2(Q-X-5)^2 = C(Q,X) & \text{if } Q > X - 5 \\ \\ 10(X+2-Q) = K(Q,X) & \text{if } Q < X + 2 \end{array} \qquad (20)$$

Assume that, from a combination of historical data and prior knowledge gained from hydrometeorological studies the distribution of the discretized random inflow Q (natural uncertainty) can be fitted to a Poisson pdf with mean $\hat{\lambda}$ = 6. In actual examples, the Poisson distribution does not appear to be a common choice for modeling input volume pdf. It is clear that the validity of the present analysis would in no way be affected if any other pdf for Q were chosen. The main reason for choosing a Poisson model is the case of computation provided by a well-known one-parameter pdf especially when sample uncertainty is accounted for. Consider the case when $\hat{\lambda}$ has been computed from a sample size $\hat{t}$ = 2 (case A), $\hat{t}$ = 5 (case B), or $\hat{t}$ = 8 (case C). The likehood of a sample q is thus:

$$L(q\,\lambda) = \frac{\lambda^q e^{-\lambda}}{q!} \quad , \qquad q = 0,1,2,\ldots \tag{21}$$

From Eq. (16), the goal function to be minimized is

$$g(X|\lambda) = E[C(Q,X) + K(Q,X)]$$

$$= \sum_{q=X-5}^{\infty} C(q,X)\cdot L(q\,\lambda) + \sum_{q=0}^{X+2} K(q,X)L(q\,\lambda) \tag{22}$$

The uncertainty in λ (parameter uncertainty) is described by the prior pdf $f'(\lambda)$, which, in the case of a Poisson pdf, is a normalized gamma-1 with parameters r and t, defined in Raiffa and Schlaifer (15) as:

$$f'(\lambda) = \frac{e^{-t\lambda}(t\lambda)^{r-1}t}{(r-1)!} \quad \begin{array}{l} \lambda > 0 \\ r,t > 0 \end{array} \tag{23}$$

with mean $E(\lambda|r,t) = \dfrac{r}{t}$ and variance $V(\lambda|r,t) = \dfrac{r}{t^2}$.

Hence, the estimated parameters are, for the three cases A, B, C considered:

	A	B	C
$\hat{t}$	2	5	8
$\hat{r} = \hat{\lambda}\hat{t}$	12	30	48
$s = \sqrt{\hat{r}/\hat{t}^2}$	1.73	1.095	.866

Our decision rule states that the optimal release from the reservoir minimizes the expected value of $g(X,\lambda)$ given by Eq. (22), to yield:

$$R(X^*) = \min_{R} E_\lambda[g(X,\lambda)]$$

$$= \min_{X} \int_{0}^{\infty} g(X,\lambda)\cdot f'(\lambda)d\lambda \tag{24}$$

(2) Decision evaluation phase. To evaluate the decision, the expected opportunity loss (XOL) is calculated. As given in Eq. (6) and explained further with hydrologic examples in Duckstein and Davis (1976), the XOL represents a measure of the expected value of perfect information. If the true value λ_T of λ

were known, the loss would be $g(X,\lambda_T)$ and the optimum decision X_T. Since the decision X* has been taken, an opportunity loss OL has been incurred, namely,

$$OL(X^*,\lambda_T) = g(X^*,\lambda_T) - g(X_T,\lambda_T)$$

$$= g(X^*,\lambda_T) - \min_X g(X,\lambda_T) \qquad (25)$$

Under sample or parameter uncertainty, λ_T is unknown by definition, but the expectation of Eq. (25) can be taken using the prior pdf $f'(\lambda_T)$ of λ_T. This yields, as in Eq. (6), the XOL:

$$XOL[f'(\lambda)] = \int_0^\infty [g(X^*,\lambda_T) - \min_X g(X,\lambda_T)]\cdot f(\lambda_T)d\lambda_T \qquad (26)$$

When one more observation q is available, the prior distribution $f'(\lambda)$ is updated using Bayes Rule, to become the posterior distribution $f''(\lambda)$, which is of the same family of gamma-1 as the prior; such a family is labeled a conjugate family (4):

$$f''(\lambda|q) = \frac{e^{-\lambda(t+1)}\lambda^{r-1+q}(t+1)^{r+q}}{(r-1+q)!} \qquad (27)$$

The predictive distribution of 0 becomes the negative binomial distribution:

$$L(q) = \frac{t^r(r-1+q)!}{(r-1)!q!(t+1)^{r+q}} \qquad (28)$$

The expected opportunity loss, given one more observation q is

$$XOL[f''(\lambda|q)] = \int_0^\infty [g(X^*,\lambda) - \min_R g(X,\lambda)]f''(\lambda\ q)d\lambda \qquad (29)$$

and the expected opportunity loss is, using the predictive or Bayesian pdf L(q):

$$XXOL = \sum_{q=0}^\infty XOL[f''(\lambda|q)]\cdot L(q)$$

$$= \sum_{q=0}^\infty \int_0^\infty [g(X^*,\lambda) - \min_R g(X,\lambda)f''(\lambda|q]d\lambda\cdot L(q) \qquad (30)$$

Finally, the expected value of sample information is (3):

$$EVSI = XOL[f(\lambda)] - XXOL \qquad (31)$$

The results of the analysis are as follows:

Case	A	B	C
Record Length t	2	5	8
Minimum Bayes Risk R(X*)	10.26	7.617	6.972
Optimal Decision X*	2	2	2
XOL	3.807	1.590	.755
XXOL	2.458	1.392	.706
EVSI	1.349	0.198	0.049

As more data become available (cases B and C), the risk or expected cost of the decision decreases, as well as the expected opportunity loss. The lower limit of the risk is equal to 5.79 which is the value of the expected cost calculated with respect to natural uncertainty only. The expected value of sample information, which is relatively small even when $\hat{t} = 2$, decreases rapidly with record length which means that an additional data point would hardly ever cause a change in the optimal decision $X* = 2$.

Under what conditions would the optimum release X* be different from 2? Changes in the mean input should be considered for this purpose. Since $r = \lambda t$, it is convenient to vary r and t as integers; specifically, let the record length be t = 2 for $9 \leq r \leq 15$; then, t = 5 for $27 \leq r \leq 33$. The corresponding values of λ, X*, minimum Bayes Risk $\overline{R*}$ and EVSI are

t 2

r	9	10	11	12	13	14	15
λ	4.5	5	5.5	6	6.5	7	7.5
R(X*)	7.296	8.152	9.343	10.26	11.35	12.35	13.32
X*	1	1	2	2	3	3	4
EVSI	1.147	1.120	1.380	1.343	1.434	1.320	1.003

t = 5

r	27	28	29	30	31	32	33
λ	5.4	5.6	5.8	6	6.2	6.4	6.6
R(X*)	6.944	7.062	7.283	7.617	8.066	8.530	8.661
X*	1	2	2	2	2	3	3
EVSI	0.477	0.272	0.177	0.198	0.332	0.476	0.288

The optimum decision value X* and risk R(X*) both increase as the average input $\hat{\lambda}$ is increased, as it may be expected. For t = 5 years of record, both R(X*) and EVSI are smaller than for t = 2 years of record. Finally, the EVSI starts decreasing as $\hat{\lambda}$ becomes larger than the demand W = 6: If the water supply is abundant it becomes less useful to study its uncertainty. But note that the EVSI is always positive, as long as a Bayes criterion, which is an optimal decision rule by definition, is being used.

4. NONOPTIMAL DECISION-MAKING WITH IMPERFECT INFORMATION

4.1 Statistical Example

If we no longer use an optimal decision rule such as the Bayes rule in the above reservoir system example, it will be shown that the worth of sample information EVSI may actually be negative! The performance and efficiency vectors will be calculated for an IR system, using the peak flow Q for the 6 1/4 year flood with a data record of size two and four, respectively. The underlying hydrology is greatly simplified in order to keep the calculations simple.

Procedures for determining the T year flood must take both natural and sample uncertainty into consideration if an unbiased estimate is to be obtained. In this example, the underlying distribution of the random variable X is assumed to be normal with a known variance σ^2 taken equal to be unity to simplify the presentation, but with an unknown mean μ. Information about the value of μ might be available from regional studies; in this example, prior to sampling, no information is available and all values are considered equally likely.

If μ were known, the 6 1/4 year flood flow Q could be determined:

$$Q = \mu + \sigma = \mu + 1 \qquad (32)$$

However, μ must be estimated. Let $\hat{\mu}$ be the estimator and $\sigma_{\hat{\mu}}^2$ be the variance of the estimate. The sample of peak annual flows leading to the estimate $\hat{\mu}$ could have come from distributions having many different values of the parameter μ. Averaging all these distributions gives the expected probability distribution of the peak annual flows which is approximately normal with mean $\hat{\mu}$ and variance $(\sigma^2 + \sigma_{\hat{\mu}}^2)$. The 6 1/4 year flood may now be estimated:

$$Q = \hat{\mu} + (2 + \sigma_{\hat{\mu}}^2)^{\frac{1}{2}} \tag{33}$$

The factor $[2 + \sigma_{\hat{\mu}}^2)/2]^{\frac{1}{2}}$ is called the expected probability correction and is applied to the second term in the right hand side of Eq. (32).

The information available consists of the estimate of the population mean and the number of samples, i.e., $I = (\hat{\mu}, n)$. Two different estimates of the mean are used, the sample mean

$$\bar{X}_n = \frac{1}{n} (X_1 + X_2 + \ldots + X_n), \tag{34}$$

and the sample mid-range

$$M_n = \frac{1}{2} (X_{max} + X_{min}). \tag{35}$$

The variance of $\bar{X}_n$ is $1/n$. The variance of M_n for $n = 2$ is $1/2$, for $n = 4$ is 1.092 1.4, and for large n is $\pi^2/24 \log_e n$ (10). When using the sample mean to estimate μ the expected probability correction is $(1 + \frac{1}{n})^{1/2}$.

Four decision rules of a general form

$$Q = d(I) = d(\hat{\mu}, n) = \hat{\mu} + k \qquad k > 0 \tag{36}$$

where k may depend on n, were examined as shown in Table 3. For samples of size two, the first three decision rules give the same value for Q, while rule 4 gives an estimate of Q considerably higher than the first three rules. Decision rule 1 uses the sample mean and the expected probability correction. Decision rule 2 uses the sample mean plus a constant multipled by the standard deviation which yields the expected probability estimate of Q for a sample of size two, but floods of longer return periods

Table 3.   Decision Rules for Estimating the
6 1/4 Year Flood, $\sigma^2 = 1$

Decision Rule	General Form	For n = 2	For n = 4
1.	$\bar{X}_n + (1 + \frac{1}{n})^{\frac{1}{2}}$	$\bar{X}_2 + 1.225$	$\bar{X}_4 + 1.118$
2.	$\bar{X}_n + 1.225$	$\bar{X}_2 + 1.225$	$\bar{X}_4 + 1.225$
3.	$M_n + 1.225$	$M_2 + 1.225$	$M_4 + 1.225$
4.	$\bar{X}_n + 2.200$	$\bar{X}_2 + 2.200$	$\bar{X}_4 + 2.200$

$\bar{X}_n$ = sample mean

M_n = sample midrange

σ = population variance (=1)

n = size of sample

for larger samples.   Rule 3 substitutes the sample midrange for
the sample mean in rule 2.   Decision rule 4 is rule 2 with a
higher value for k.   Thus, all decision rules give different
values of Q when used with a sample of size four.

Let the loss be the square of the difference between the mean
of the expected probability estimate of 6 1/4 year flood, Eq. (33)
and the value given by the decision rule:

$$\ell(d(I),\mu) = [d(I) - \mu - (1 + \frac{1}{n})^{1/2}]^2 \tag{37}$$

If μ is known the expected loss may be calculated as

$$E(\ell) = \int \ell(d(i),\mu)dH_I(i|\mu) \tag{38}$$

For this example the expected loss is equal to the variance of the
estimate of the mean plus the square of the bias of the estimate
of Q:

$$E(\ell(I,d)) = \sigma_{\hat{\mu}}^2 + [E(d(I)) - \mu - (1 + \frac{1}{n})^{1/2}]^2 \tag{39}$$

For the decision rules used in this example $E[d(I)] = \mu + k$;
substituting in Eq. (38) gives

$$E(\ell(I,d) = \sigma_{\hat{\mu}}^2 + [k - (1 + \frac{1}{n})^{1/2}]^2 \tag{40}$$

since the expected loss is independent of the parameter μ, the risk is equal to the expected loss, i.e., $R(I,d) = E[L(I,d)]$

The optimal decision rule which minimizes the expected loss (and thus the risk) is

$$d*(I) = \hat{\mu} + (1 + \frac{1}{n})^{1/2} \tag{41}$$

From Eq. (39) the risk is

$$R(I,d*) = \sigma_{\hat{\mu}}^2$$

Perfect information is either the sample mean or midrange from a sample of infinite size; both are unbiased, consistent estimators of the parameter μ. From the definition of perfect information

$$\mu = \Phi(\bar{x}_\infty) = \Phi(M_\infty) = \Phi(\hat{\mu}*)$$

Also $\sigma_{\hat{\mu}}^2 = 0$. From Eqs. (2) and (40) the risk with perfect information is

$$R(I*,d) = [k - (1 + \frac{1}{n})^{1/2}]^2 \tag{42}$$

Since $k = (1 + \frac{1}{n})^{1/2}$ in the optimal decision rule, the risk with perfect information is using an optimal decision rule is zero. For this example, it is postulated that the risk associated with no response, R^0, is $2\sigma^2 = 2$.

The risks for an IR system using each of these four decision rules, as well as the system performance and efficiency vectors, are shown in Table 4 for samples of size 2 and in Table 5 for samples of size 4.

Comparison of the two tables shows that, for the same decision rule, the actual value of the IR system is greater with four years of data than with two years of data. This difference represents the value of the additional two years of information. Two factors contribute to the difference: the reduction in variance in the estimate of the mean and the change in the bias of the decision rule.

Decision rules 2, 3 and 4 have a bias increasing with sample size. For decision rules 2 and 3, the decrease in the variance of

Table 4. Risks, Values and Efficiencies for an IR System
         Estimating the 6 1/4 Year Flood From Two
         Years of Data, $\sigma = 1$

Decision Rule	1	2	3	4
$\hat{\mu}$	$\bar{X}_2$	$\bar{X}_2$	M_2	$\bar{X}_2$
$k\sigma$	1.225	1.225	1.225	2.200
Risks:				
$R(I^*,d^*)$	0.	0.	0.	0.
$R(I,d^*)$	.5	.5	.5	.5
$R(I^*,d)$	0.	.050	.050	1.440
$R(I,d)$	.5	.5	.5	1.451
Values:				
Realizable, RV	2.	1.950	1.950	.560
Optimal, OV	1.5	1.5	1.5	1.5
Actual, AV	1.5	1.5	1.5	.549
Efficiencies:				
Response, RE	1.	1.	1.	.366
Overall, OE	.75	.75	.75	.275

For all decision rules, $R^0 = 2$, potential value PV = 2 and
Information efficiency IE = .75.

the estimate is larger than the increase in the square of the
bias; thus, the marginal value of increasing sample size is
positive. However, for heavily biased decision rule 4, with
sample size larger than five, the increase in the square of the
bias is faster than the reduction in the variance of the estimate.
Figure 2 shows how the risk decreases and then increases as the
sample size gets larger.

     The use of decision rule 4 in this IR system produces the
paradoxical result that additional data have a negative worth. In
this example, it is easily seen that the paradox is caused by the
changing relative contributions of the information and response
subsystems to the overall IR system risk. The information subsystem provides an expected loss due to the variance of the
estimate of the mean while the response subsystem provides an
expected loss due to bias in the decision rule. In most IR
systems, the risks from the subsystems are not separable.
Examination of the efficiencies of the subsystems and the overall

Table 5.   Risks, Values and Efficiencies for an IR System
           Estimating the 6 1/4 Year Flood From Four Years
           of Data, $\sigma = 1$

Decision Rule	1	2	3	4
$\hat{\mu}$	$\overline{X}_4$	$\overline{X}_4$	M_4	$\overline{X}_4$
$k\sigma$	1.118	1.225	1.225	2.200
Risks:				
$R(I^*,d^*)$	0.	0.	0.	0.
$R(I,d^*)$	.250	.250	.273	.250
$R(I^*,d)$	0.	.051	.051	1.440
$R(I,d)$	.250	.261	.284	1.421
Values:				
Realizable, RV	2.	1.949	1.949	.560
Optimal, OV	1.75	1.75	1.727	1.75
Actual, AV	1.75	1.739	1.716	.579
Efficiencies:				
Information, IF	.875	.875	.863	.875
Response, RE	1.	.994	.994	.331
Overall, OE	.875	.870	.858	.290
Value of increasing sample size from two to four	.250	.239	.216	.030

For all decision rules, $R^0 = 2$ and PV = 2

system can be used to determine the contribution of each subsystem
to risk reduction. The overall efficiency OE, the information
efficiency IE and the response efficiency RE are plotted in
Figure 3 for the example IR system using decision rule 4. Note
that the information efficiency IE increases with sample size,
while the response efficiency RE decreases with sample size.
Their product, the overall efficiency OE, first increases and then
decreases as would be expected from Figure 2.

Decision rules 1, 2 and 3 are all optimal for a sample of
size two. For larger samples only rule 1 is optimal. All
efficiencies are higher when using rule 1. Realizable and actual
values are also higher. For samples of size four, comparison of

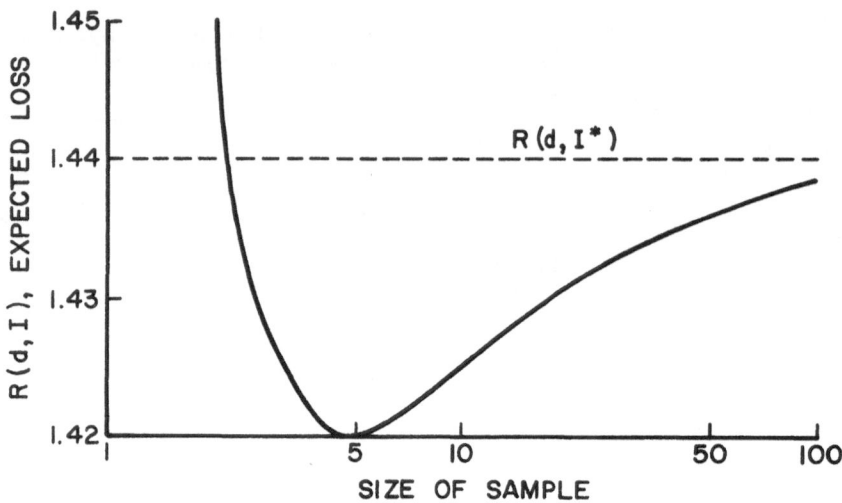

Figure 2.　Risk Versus Sample for Decision Rule 4

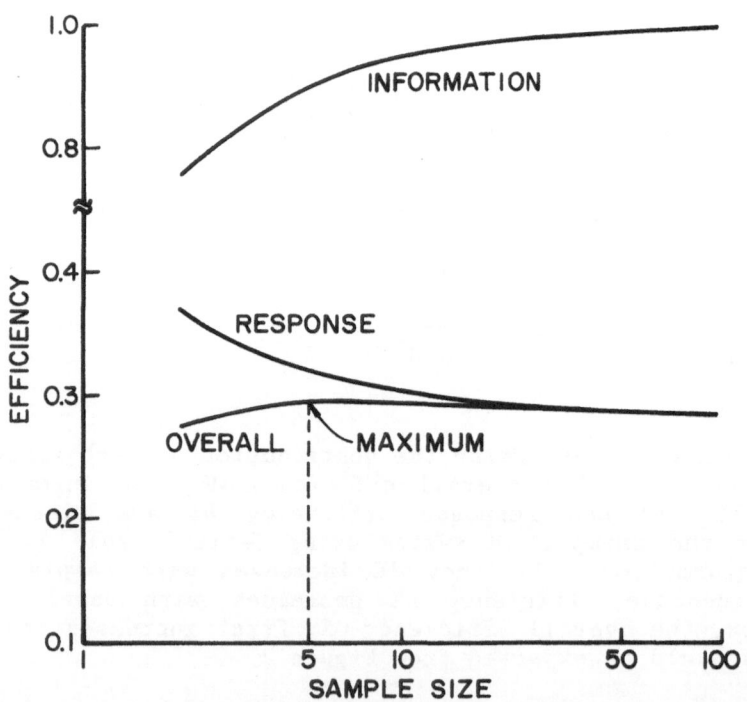

Figure 3.　Information Efficiency IE, Response Efficiency RE
and Overall Efficiency OE Versus Sample Size for
Decision Rule 4

decision rule 2, which uses the sample mean to estimate population mean, shows that potential and realizable values are the same for both rules; however, the optimal and actual values are higher for the decision rule utilizing the sample mean.

Comparison of the actual value of the IR system with each decision rule for the samples of size two and four shows that the value of the additional two samples is positive and maximum when optimal decision rule is being used. Anomalously, there are other IR systems where additional data are of greater value to the nonoptimal decision rule. In the study of the worth of additional data, one finds many such apparent anomalies and paradoxes, several of which are analyzed in Davis et al., (1979).

4.2 Value of Flood Forecasting

A model of the flood forecasting-response system has been developed in Sniedovich and Davis (17) and Krzysztofowicz et al., (12) for the purpose of assessing the value of data for flood forecasting and the value of the forecasts to the decision-maker who responds to the forecast by taking an action to decrease potential flood loss. In this model, which is similar to the IR system model, losses are the sum of the cost of taking mitigating action and the expected damage due to inundation. Three types of decision rules are used to model the flood plain dweller's response to the flood forecasts: (1) the "optimal" response which minimizes expected flood losses; (2) the "pure" response which is initiated as rapidly as possible when the forecast indicates inundation, but only at that time; and (3) the "human factors" response which is obtained from a learning model based on man-machine considerations such as the floodplain dweller's previous experience with floods.

Sniedovich and Davis (17) developed a comparison between two flood forecast-response systems A and B in which system A produces more accurate forecasts than system B. The more accurate forecasts produced by system A have shorter lead time than those produced by system B, thus the cost of responding is higher in sytem A. The tradeoff point between lead time and accuracy depends essentially on the decision rule. The performance and efficiency vectors have been calculated using both optimal and pure response for each system. Under optimal decision rules, system B, which is characterized by a longer lead time, less accurate forecast and lower response cost, has a higher actual value AV than System A. However, if the pure response decision rule is used, then system A, which is characterized by more accurate forecasts, shorter lead time and higher cost of response, has the higher value Av.

The model developed in Krzysztofowicz et al., (12) evaluates flood forecast-response systems, accounting for the sequential and stochastic nature of flood forecasts as well as the cost and time required to respond. In the work, evaluations were made of the forecast-response systems for Milton, Pennsylvania, on the West Branch of the Susquehanna River, and for Columbus, Mississippi, on the Tombigbee River. The objective of these systems was the reduction of economic flood losses to residential, commercial and industrial structures. Actions taken to mitigate potential flood damage are costly and take considerable time to complete. The Milton and Columbus evaluations have shown that timeliness of the forecast information and the optimality of the decision rule determining the floodplain dweller's response to the forecast are most important determinants of the perfromance of the flood forecast-response system.

Milton has a relatively short forecast lead time as compared with the time required to complete mitigating action, while the lead time for Columbus is close to the time needed to complete mitigating actions. If the pure response is used, the actual value AV of the system for Milton is negative; but if the human factors model response is used, the value AV is positive, though the efficiencies are low. For Columbus, the system AV is positive for both modes of response; the pure response gives the higher value, and the system has moderately high efficiencies. It is believed that the short lead time provided by Milton's forecasts does not allow the mitigation actions called for by the pure response to be effective. In this case, the human factors model response which is based on both the floodplain dweller's past experience with floods, and the current forecast, makes better use of the forecast information available to Milton residents bacause mitigating activities are started earlier. However, for Columbus, use of the pure response, a decision rule which does not call for mitigating action until the forecasts indicate inundation, is superior to the human factors response. This is because a long lead time allows completion of mitigating actions started at a later time, when there is more certainty that such actions will be needed.

In a sensitivity anaylsis, the same forecasts were delivered to the floodplain dweller two hours earlier. The performance for both communities using both response strategies improved. Again, there were significant differences between the two communities. In Milton, the two hour increase in forecast time raised the values AV and OV of the forecast-response system almost equally. Since the efficiencies were low to start with, this increase in system values led to a larger increase in response efficiency RE than in information efficiency IE. In Columbus, the two hour increase in lead time produced a negligible increase in either OV or IE but a considerable increase in AV and RE.

An explanation for the considerable difference in the effect of the earlier forecasts on the vlues AV and OV of a flood forecast-response system may be as follows. When it takes a large amount of time to complete mitigating actions, an optimal response "anticipates" future forecasts, while the pure response rule does not. If an earlier forecast confirms what had been anticipated, little has been gained by that earlier forecast. For those flood-plain dwellers who do not anticipate, but only respond when the forecast indicates thay may be flooded, the earlier forecast is of value as it provides more time to take mitigating action. The larger increase in RE as compared to the increase in IE indicates that the value of an earlier forecast rests more in improving the ability of the floodplain dweller to use the available information than in providing improved information.

5. DISCUSSION AND CONCLUSIONS

5.1 Discussion

The preposterior analysis of statistical decision theory has been extended to allow for consideration of nonoptimal decision rules which are likely to be applied in the operation of complex water resources systems. For this purpose, an IR system model and accociated measures of performance and effectiveness have been constructed. With these measures of performance and effective-ness, it is, for example, possible to assess the value of a monitoring network, and to evaluate quantitatively the effect of possible changes in the IR system.

When the data and information from a network are shared by a variety of different users, there are many IR systems having common information subsystems but different response subsystems. The systems may even be using the same decision criterion and the same decision rule. Care must be taken in analyzing IR systems having a shared information component to insure that each individual loss function be considered. If the loss functions or decision criteria are different, a common decision rule, for example a common reservoir release rule, is often nonoptimal.

As a first step, a decision criterion (Bayes, miminax...) should be chosen. Then the three basic elements of the risk (Eq. (1)) can be defined, namely: (a) the loss function, (b) the decision rule, and (c) the probability distributions describing the stochastic properties of the information. An optimal decision rule minimizes the risk. If any one of these elements changes, the IR system may no longer be optimal. Decision rules and criteria used at present may have been obtained, explicitly or implicitly, using loss functions and statistical concepts of the past. When the loss function and the statistical concepts are brought "up to date" the decision rule must be adjusted in order to maintain optimality. Reduction in the error of estimate of a statistic used in the decision rule does not guarantee that a previously optimal decision rule will remain optimal. Further, in most instances, reduction in the error of estimate will reduce the risk of the IR system only if the optimality of the decision rule is maintained.

To determine the worth of data, the information required to calculate risk must be known. Most often the decision rules, criteria and the forms of the requisite probability distributions are known. The loss functions are often not known to the decision-maker, especially if there are many users of the information. Davis et al., (2) discuss this problem in some detail. In any case, the hydrologist need not wait for the date to improve the form of the decision rule: techniques of statistical decision theory (15) as applied, for example, to reservoir operation in section 3, can be used for this purpose.

Returning to the problem of choosing a decision criterion, difficulties are involved in determining states of nature, prior probability distributions and loss functions. These problems result in decisions which, strictly speaking, satisfy neither Bayes, expected minimax or any other specified criterion. Such decisions are based on a mix of criteria--perhaps with the objective of trying to prevent disastrous outcomes and "average out" other possible outcomes. The outcomes and associated loss functions are not precisely formulated and the prior probability distribution is intuitive and imprecise. But, even under these conditions, the decision theory approach can help in organizing the problem, identifying the difficulties and suggesting a rational choice. It is important to identify the mix of criteria used in such practical problems in order to understand more completely the nature of the decisions made.

One of the persistent criticisms of decision theory has been with respect ot its applicability in practical problems. Usually the difficulty is that the decision-maker is unable to specify a loss function, to quantify the probabilities assigned to the uncertain states of nature, to fix a decision criterion to specify a decision rule, and then to make the decision accordingly. Recent studies concerning multiobjective decisions and forecasting techniques are of help in construction loss functions. Also, progress which has recently been made in the evaluation of subjective probabilities should resolve some of the previous difficulties (16).

After a decision criterion has been chosen and the elements necessary for the calculation of IR system risk are known, the measures of system performance and efficiency may be obtained in a straight forward manner. The efficiency measures indicate whether improvement to the information subsystem or response subsystem would be the most beneficial. These performance measures provide a quantitative measure of the benefits to be expected from a change in the IR system and may be used in a benefit-cost analysis, or, better, a multiobjective analysis of the propsed change. Also, these measures can indicate whether or not better use of presently available date and information can be more effective than additional or improved date.

The optimal or Bayes decision rule anchors the IR system and provides the basis for the performance and efficiency calculations. An IR system with an optimal decision rule alwyas has a 100% response efficiency RE. In this case, more data and/or improvements in the information system are needed only if the information efficiency IE is below 100%. More information which is related to the uncertain state of nature, will, by definition, increase the information efficiency IE and thus the overall efficiency OE. On the other hand, RE value of an IR system with a nonoptimal decision rule is less than 100%. In this case, additional data and information will result in an increase of the information efficiency IE; however, the response efficiency RE may either increase or decrease. The ratio of the relative increase or decrease of response efficiency RE to information efficiency IE,

$$\frac{\Delta RE/RE}{\Delta IE/IE} \tag{43}$$

provides a simple indication of the value of additional data for nonoptimal decision-making. If the ratio is less than $-(1+\Delta IE/IE)^{-1}$, it can be shown that the overall efficiency OE decreases, which indicates that the value of additional data is negative.

5.2 Summary and Conclusions

A general framework for use of decision theory in the operation of complex water resource systems, such as a network of reservoirs, has been provided under the form of an Information-Response system. Measure of performance of the two subsystems and the overall system have been provided.

Since decisions are made on the basis of a criterion, this chapter has outlined the differences among the Bayes, expected minimax, and absolute minimax criteria. A summary of these differences are summarized in Table 6. The main contention is in practice, decision-makers frequently use a mix of these and

Table 6.   Comparison of Absolute Minimax, Expected Minimax
and Bayes Decision Criteria

	Requires Specification of All Possible States of Nature	Requires Specification of Loss Function	Requires pdf of States of Nature	Requires Specification of Least Favorable Distribution	Always Provides Unique (Nonprobabilistic) Decision	Protects Against Catastrophic through Unlikely Losses
Absolute Minimax	no	no	no	no	yes	yes
Expected Minimax	yes	yes	yes	yes	no	no
Bayes	yes	yes	yes	no	yes	no

perhaps other criteria, especially if large losses are possible.
In any case, the relative effect of uncertainties should be
weighed; it would be useless to have a detailed physical model
when economic factors can only be estimated with a large error.

The material presented above leads to the following
concluding remarks:

(1) The worth of data and information is always nonnegative
for optimal decision making, but can be negative for nonoptimal
decision making. Sensitivity analyses investigating the value of
data and information can lead to inconsistent results in the
absence of optimal decision rules. A numerical example presenting
such an inconsistent feature is explained in terms of the
Information Response system and its measures of performance and
efficiency.

(2) An expected value or Bayes criterion is optimal provided
the loss function corresponds to the decision-maker's preferences

and the prior probabilities of the states of nature can be quantified. Such a Bayesian approach enables both natural and sample uncertainties to be accounted for in decision making. More importantly, if provides a measure of the economic worth of perfect information and sample information.

(3) Care must be taken when using an optimal decision rule such as a Bayes criterion with new and improved data, to ensure that the rule remains optimal. If the method for calculating statistics is changed, an optimal decision rule should be adjusted accordingly to maintain optimality.

(4) A minimax criterion if frequently appropriate when possible payoffs or losses are large and difficult to quantify, or when decisions cannot be made repeatedly. Minimax provides a viewpoint from which to replace or modulate Bayes decisions. Furthermore, it is important to distinguish between absolute and expected minimax decisions. Absolute minimax protects against the worst possible outcome while expected minimax protects against the least favorable distribution of the states of nature.

(5) Expected minimax and Bayes decisions differ with respect to the prior distribution assumed, which is a Response Subsystem characteristic. Minimax decisions often involve mixed strategies, and hence are difficult to apply, while Bayes decisions correspond to pure strategies which give uniquely defined decisions.

(6) Real decisions appear to represent a mix of absolute minimax, expected minimax, and expected value criteria. The type of decision to be made and the availability of information condition this mix. Implications for the the Information subsytem are substantial. Also, failure to recognize the existence of diverse criteria may result in a breakdown in communications between opposing groups.

(7) Additional data may simultaneously reduce the error of estimate of a statistic used in the decision rule and increase the risk of the Information Response system. Using the measures of performance and efficiency defined for the IR model, additional data always improve information efficiency IE but may increase or decrease response efficiency RE. The relative increase or decrease of these efficiencies determines the worth of additional data for nonoptimal decision making.

From the practical standpoint, the IR system model places operations research and Bayesian decision theory in proper perspective; it points out that improved information does not necessarily mean improved performance of a system. Also, once a decision criterion has been chosen, say a 50-year flood, then the introduction of more precise estimation methods of the 50-year flood into the design may lead to a worse decision than keeping the original procedure. Finally, the IR model provides a rational way to design and operate the monitoring or surveillance network of a complex water resource system, such as a regional flood warning system.

ACKNOWLEDGEMENTS

This research was supported in part by National Science Foundation Grants #INT 78-12184, "Decision-Making in Natural Resources Management," and #ENG 78-09365, "Bayesian Methodology for Rainflood Forecasting and Reservoir Control." Furthermore, the substantial contributions of Jean Weber, Roman Krzysztofowicz and Donald Davis to the material presented in this chapter are gratefully acknowledged.

REFERENCES

1. Benjamin, J. and C. Cornell (1970). Probability, Statistics and for Civil Engineers, McGraw-Hill, New York.

2. Davis, D., L. Duckstein and R. Krzysztofowicz (1979). "The worth of hydrologic data for nonoptimal decision making," Water Resources Research, Vol. 15, No. 6, pp. 1733-1742.

3. Davis, D., C. Kisiel and L. Duckstein (1972). "Bayesian decision theory applied to design in hydrology," Water Resources Research, Vol. 8, No. 1, pp. 3341.

4. DeGroot, M. (1970). Optimal Statistical Decision, McGraw-Hill, New York.

5. Duckstein, L. and I. Bogardi (1978). "Uncertainties in lake management," Proceedings, Int'l Symp. on Risk and Reliability in Water Res., University of Waterloo, Ontario, Canada, June 26-28, pp. 638-661.

6. Duckstein, L. and D. Davis (1976). "Applications of statistical decision theory," Chap. 11, Stochastic Approaches to Water Resources, Vol. I, H. W. Shen (ed.), Water Res. Publs., Ft. Collins, Colorado, pp. 11-1 - 11-52.

7. Duckstein, L., R. Krzysztofowicz and D. Davis (1978). "To build or not to build: a Bayesian analysis," J. Hydrological Sciences, Vol. 5, No. 1, pp. 55-68.

8. Ferguson, T. (1967). Mathematical Statistics, Academic Press, New York.

9. Keeney, R. and H. Raiffa (1976). Decisions withMultiple Objectives: Preferences and Value Tradeoffs, John Wiley & Sons, New York, 569 p.

10. Kendall, M. and A. Stuart (1973). The Advanced Theory of Statistics, Vol. 1, 3rd edition, Hafner Pub. Co., New York.

11. Krzysztofowicz, R. (1978). "Preference criterion and group utility model for reservoir control under uncertainty," Natural Resource Systems Technical Report #30, Department of Hydrology and Water Resources, University of Arizona, Tucson, Arizona 85721.

12. Krzysztofowicz, R., D. Davis, L. Duckstein and M. Fogel (1978). "Stoachastic model of a flood forecast-response process," Proceedings, Int'l. Symp. on Risk and Reliability in Water Resources, University of Waterloo, Ontario, Canada, June 26-28, pp. 697-712.

13. Moss, M., D. Lettenmaier and E. Wood (1978). "On the design of hydrologic data networks," EOS, Vol. 59, No. 8, pp. 772-775.

14. Musy, A. and L. Duckstein (1976). "Bayesian approach to tile drain design," J. of Irrigation and Drainage Div., ASCE, Vol. 102, No. IR3, pp. 317-334.

15. Raiffa, H. and R. Schlaifer (1961). Applied Statistical Decision Theory, Harvard University, Boston.

16. Smith, M. and W. Ferrell (1981). "Base rate effects on subjective probability calibration: a model and experiment," Working paper #81-16, Department of Systems and Industrial Engineering, University of Arizona, Tucson, Arizona 85721.

17. Sniedovich, M. and D. Davis (1977). "Evaluation of flood forecasting-response systems," J. of the Water Resources Planning and Management Div., ASCE, Vol. 103, No. WR1, pp. 83-97.

18. von Neumann, J. and O. Morgenstern (1947). Theory of Games and Economic Behavior, Second Edition, Princeton University Press, Princeton, N.J.

19. Weber, J., D. Davis and L. Duckstein (1980). "Bayes Versus minimax criteria: a source of conflict in decision making," Working paper #80-4, Department of Systems and Industrial Engineering, University of Arizona, Tucson, Arizona 85721.

15. SOCIETAL CONSTRAINTS IN DECISION MAKING

Norman Wengert
Colorado State University

This chapter examines theories and concepts of decision making and outlines how the context within which decisions are made affect the result. It stresses significant nonquantifiable factors which may receive little or no careful appraisal. Treatises on engineering and economics usually recognize the importance of human values, environmental concerns, community preferences and expectations, and similar socio-political and administrative factors. Systems models seek to provide for such inputs. Yet difficulties in identifying these factors and giving them appropriate weights often result in their being minimized or ignored--with possible subsequent consequences for the operation of the primary water system. This paper seeks to analyze inherent limitations of so-called "rational-comprehensive models" and outlines the more pragmatic "incremental" approach to decision making.

1. THE CONTEXT OF DECISION CONTRAINTS

This chapter considers first, decision making generally, and second, examples of decision constraints in the operation of complex water systems. Primary concern is with a water system essentially in place. Changes, improvements, renovations, modifications may be necessary as new technology and better knowledge become available, or as those served by the system change their needs, desires, and expectations with respect to the system (3).

The systems we are considering are complex. A simple water system might be a village well, or a small stream or brook. It might be a primitive Persian wheel by which water is diverted from a well or stream to irrigate a few hectares. From a navigational perspective, a simple system would be an estuary or fiord permitting easy access by fishing boats or ferries. Perhaps somewhat more complex might be a water wheel used to run a grist mill or a saw mill or perhaps a small turbine and generator for electric power.

Complexity enters when from the supply side a large water shed or river basin is involved, perhaps including ground and surface water interrelationships. The characteristics and breadth of the data used may also affect the complexity, e.g., elaborate climatological and hydrological data, based on careful studies of historic trends and projections of such trends to forecast the probable future water situation in the system. Reservoirs, control structures, channel improvements and regulation, interbasin transfers, all may add to the complexity.

On the demand or use side complexity may be introduced by technological developments, as well as by extensive distribution networks--ditches, aqueducts, pipes, pumps--designed to convey the water to where it might be used in elaborate irrigation systems, or to provide urban water with special treatment facilities including provisions for possible reuse to supplement existing supply or to expand benefits and increase flexibility of the system. Clearly a part of the complexity on the demand or use side is associated with the functions that the delivered water is to serve. In New York City (Manhattan), for example, the fire protection system simply pumps raw water from the rivers that surround the island.

In some of the Western States of the United States, a region of frequent drought and short water supplies, use priorities have been established by law with "domestic" use given first priority, "agricultural" use second priority, and "industrial" use third. In a simpler day one hundred years ago "domestic" tended to be understood as primarily "household" use, "agricultural" as irrigation and livestock use, and "industrial" as factory use in the production process. But as the economy changed, as population in the West increased and became highly urban, these simple definitions for water allocation broke down. Confronted with elaborate urban water distribution networks serving many different types of users, some American courts dealing with legal issues of allocation have tended to equate domestic with urban including interrelated commercial and industrial uses. But the allocation issue is by no means finally settled (1, 2)!

It is important, too, to recognize the "spillover" or secondary effects from complex water system operations. So long as the water system is simple, the possibilities of drastic spillover consequences are small. Thus the ancient English Common Law could permit use of water in the stream so long as it did not diminish downstream flows. This principle is, of course, a logical impossibility since taking even a cup of water upstream will reduce by that much the flow downstream. But since the volume of flow was usually so large, the law could tolerate this kind of diminishment fiction. In modern times in the Eastern United States this doctrine has been modified so as to permit "reasonable use," without stressing the extent to which diminishment might occur.

Complexities of waste water management, of flood control, and of how water is generally handled reflect directly the more intricate relationships between water and society, the economy, and the quality of life--all these add to the complexity of water system operations. And as these changes occur, the nature of the system also changes, both within system boundaries and in terms of the linkages to other systems. It is appropriate in this connection to refer to the conclusion of British medical historian, Frederick F. Cartwright (5), as illustrating complex changes in relationships. He argues that the decline and fall of the Roman Empire was not due simply to moral decay or even to barbarian invasions but to the neglect and deterioration of the aqueducts and pure water delivery system!

Finally, a focus of this book includes the word system, not in its general or loose sense, but in the special and specific sense associated with systems analysis as developed in engineering and economics over the past thirty years or so (Hall and Dracup, 1970). This chapter outlines how decision making may be accomplished with full cognizance of this theme.

2. THEORIES AND CONCEPTS OF THE DECISION MAKING PROCESS

This discussion of decision making is based on American experience. The generalizations about decision making theories and practice in the U.S. probably have their analogues in other countries.

Models of decision making in business and in government tend to represent what has come to be called "the rational-comprehensive mode," which emphasized a progression of logical steps to the best decision. Thus the first step stressed recognition and logical diagnosis of the problem in all of its facets and permutations. The second step in turn required the identification of all possible alternatives for dealing with or solving the defined problem. The third logical step involved in investigation and appraisal of each alternative, with the fourth step requiring a systematic comparison of the consequences of each alternative or combination of alternatives. The fifth step involved selection of the best and most desirable alternative to solve the problem and realize stated goals and objectives (8).

By definition, since this theory of decision making is comprehensive as well as rational, the analytical and informational demands even for relatively minor decisions are likely to be overwhelming (22). In most cases, no matter how dedicated we might be, we cannot identify all possible alternatives. We just do not know enough, we cannot marshall the needed facts, particularly concerning the future; we do not control important independent variables; and most often we do not have the time to do all the investigations that the rational-comprehensive approach would seem to demand (14).

In his classic work, Inside Bureaucracy (10), Anthony Downs lists the inherent limitations of the rational-comprehensive theory of decision making. His conclusions are paraphrased here:

1. Decision-makers lack the time such an approach seems to demand.

2. Decision-makers (like all of us) have restricted spans of attention and are only able to comprehend and integrate a limited amount of information and bring together only a limited number of subjects--even with the help of computers.

3. Especially at the top of an organization decision-makers are forced to confront so many subjects simultaneously that they usually can focus attention on only a small part of any one particular problem without neglecting all the rest.

4. As a result, top decision-makers receive only a small fraction of the information relevant to a particular problem and usually only to those aspects that are of particular interest to them; in the process of preparing executive summaries key data may be misinterpreted or distorted or omitted. What to include and what not to include requires sophisticated and informed staff judgements and may unconsciously or deliberately skew the resulting decision. (It is popular gossip that President Eisenhower demanded one page summaries of all decision or action documents!)

5. The cost--in money and time--of attempting to secure all relevant information on most decisions could be tremendous, even if one assumes the capacity to correlate and integrate such information.

6. With respect to many important decisions, there will always be information which it is impossible to get because:

    a. Decisions by definition deal with the future, and there are no unclouded crystal balls.

    b. Independent or exogenous variables cannot, by definition, be taken into account except on a probality basis.

    c. Where human behavior is involved nonrational factors, personal preferences, individual values and perceptions, personal beliefs, needs, goals and objectives may be completely unpredictable.

7. Finally, few important decisions can avoid judgements on a wide range of factors. Even with the best information, judgments are not automatic; nor are they necessarily self evident emerging inexorably from the data.

Until World War II, most American administrators, teachers of public administration, and management consultants subscribed to general aspects of the rational-comprehensive decision model. The behavioral implications of the 1930's Hawthorne Experiments (Roethlisberger) were only beginning to have a disturbing impact. It remained for Charles Lindblom in a 1959 article in the Public Administration Review (22), entitled "The Science of Muddling Through" to shake the rational-comprehensive belief systems to their very roots (23). Lindblom's critique had been preceded by that of Nobel Laureate Herbert Simon in a 1947 book entitled Administrative Behavior (27). Simon showed rather convincingly that what had been considered "principles" of administration were really only "Proverbs," normative rules of behavior, one rule often conflicting in practice with another. The result, for decision theory, was a new pragmatic relativism to which the term "incrementalism" was applied. Among the axioms of the incremental approach are:

1. In any particular decision, only a small number of alternatives can be considered.

2. Those alternative analyzed, in the absence of a crisis or climactic situation, tend to be those which differ only incrementally or marginally from existing policy.

3. Only a few--often the more obvious policy consequences--are considered in any depth.

4. Decision making tends to involve continuous ends-means and means-ends adjustments based on experience and practical reality.

5. The idea of a single "right" decision, particularly at the general policy level, is rejected, the stance of the administrator being one of flexibility and readiness to adjust. This has lead to the word "satisficing" to denote such behavior, rather than optimizing or maximizing.

6. Incremental decisions may be wrong and may generate undesirable consequences, but since they tend to be small decisions, changes in policy direction is easy and losses may be reduced.

Not all who are concerned with decision theory, however, have been satisfied with incrementalism and the incremental approach which has come to dominate American political science especially as it sought to explain governmental decision making (16). The pluralistic and pragmatic, incremental theories were regarded as lacking recognition of those elements which many felt distinguished human society. There is today in the United States no single, unified theory or set of concepts on decision making in government or in the private sector. A great variety of ideas and approaches characterize the subject of decision making (18).

In a very recent book critizing the Lindblom approach, professor Robert Mowitz (25), asserted

"...that we do have the knowledge and skill to design and to develop governmental decision systems...and that such systems are necessary if we are to deal with the increasing complexity and interdependence of the future."

Mowitz qualified his more optimistic view, stating that "...the difficulties and obstacles in the way of establishing such decision systems must also be carefully weighed."

The criticisms of rational-comprehensive models raised two important sets of problems: those which rested on the lack of internal logic and on unreasonable information requirements, and those which stressed behavioral or social-phychological realities of interpersonal relationships in decision making. The last of these problems remains most difficult to deal with even though volumes of behavioral research have been undertaken.

At about the same time that the simple doctrines of rational-comprehensive decision making were being challenged, a new and more sophisticated rationality was gaining popularity. This was not in response to the breakdown of old doctrines, but as a result of improved technology for dealing with old and for formulating new problems, utilizing developing techniques of analysis and improved data and information. There were several manifestations of this new rationalism. One, called "operations research," grew out of World War II and represented the application of mathematics and mathematical theories to the solution of public and private decisional problems (21). Another, arising out of a growing concern with public investment efficiency, became what we now know as "benefit-cost analysis." The third, and the one to which Professor Mowitz's comments were specifically addressed, represented a kind of combination of the other two and ultimately became known as "program budgeting" or by its acronym "PPBS" (26). It should be noted that the latter became feasible on a large scale only with the development of the computer. All three of these manifestations of the new rationalism in decision making, were highly dependent on data, i.e., on quantitative information. The evolutionary connection of the fourth development to the other three is not so clear, but the necessary relationships are obvious, and the mutual reinforcement evident. This fourth manifestation of new rationalism in decision making has been "systems analysis." It grew out of solving of engineering problems and application of mathematics and computers to such problems. In one major respect, however, systems analysis is different from the other three techniques; this is in the deliberate delineation of the analytical system--the focus for organizing and analyzing data interrelationships (17, 18, 19, 24).

Over ten years ago in a pioneering work on the subject, Professors Hall and Dracup (17), defined the "concept of a system" as follows:

"A system may be defined as a set of objects which interact in a regular interdependent manner. Systems engineering is concerned with making decisions with respect to those aspects of the system which are subject to some degree of control in order to attain certain objectives. Since it is virtually impossible to isolate all interacting objects, the principal interacting elements are termed the system and those interacting objects not included are considered through the specification of the interactions with the environment in the form of inputs to and outputs from the system. All systems are implicitly dynamic, a fact all too often forgotten in the preoccupation with equilibrium models of real systems...."

Hall and Dracup then suggest three critical elements of system design;

a.   determination of system boundaries,

b.   the imputs to and outputs from the system,

c.   and the interrelationships between the elements of the system, the inputs and outputs, and any external interactions between output and input i.e., feedback (11, 12, 13).

Quantification was the key to all four of these manifestations. And for many, a new concern arose over the difficulties and even impossibilities of quantifying important aspects of the human situation. To a large extent in the U.S., such doubts about quantification were embodied in the statutory language of the National Evnironmental Policy Act of 1969 (N E P A) which stated explicitly (42 U.S.C. 4332):

> That a systematic interdisciplinary approach applying natural and social sciences was to be used in environmental decision making and that presently unquantified envrionmental amenities and values should be given consideration in decisions along with economic and technical considerations. (Paraphrased; not a direct quotation).

3. DECISIONAL CONSTRAINTS

Against the background of these briefly sketched theories, concepts, methodological values, and belief systems, it is perhaps feasible to comment on characteristics of decisional constraints.

First, I suggest that decisions are constrained by how they are perceived and how the area of decision making is defined. Second, decisions are constrained by the technology available for collecting and organizing data and by the knowledge and skills of the analysts. Third, at the same time, as the quantity of data increases, structuring and simplification becomes desirable, if not necessary. The boundaries of the analytical or decisional system must be set and the causal relationships identified. Fourth, the very process of boundary setting, given our limitations of knowledge, has certain arbitrary aspects to it and may exclude variables which may later turn out to be most significant. Fifth, no matter how hard we may try, we cannot avoid the fact that the world, including that concerning which we are making decisions, is seen through our own eyes. Our own values, personalities, and predilections influence even the most detached decisions at various and sundry points (1, 2).

Thus the importance of considering decisional constraints requires identification of key factors which affect the system being analyzed and which may prevent it from functioning in the way in which the model implies. Decisional constraints are, in a way, like friction in mechanics. They represent those factors or forces which may prevent a system from operating at optimal efficiency. In some cases, constraints on decision making in the operation of complex water resources systems can be identified. Some are related to a particular system or situation. Others are of a more general character, having a pervasive effect on system decisions. In either case, examination of particular situations is needed to provide a basis for analysis and comment. Neither time nor space permits such case studies. Instead, several broad categories will be identified as the source of decisional constraints often overlooked or ignored.

First, consider human behavioral influences on water system decisions. Like in many engineering systems or even some natural systems, there is considerable evidence of how easy it is to overlook the human or bahavioral influences. For example, Americans drink much water e.g., at meals, between meals, on going to bed, on arising. Now the quantity of water involved may not be great, but the habit has direct effect on water quality. Odd tastes, odors, or other objectionable features will not be tolerated. Hence water treatment becomes very expensive. As another example, the region west of the 100th meridian, the area of less than twenty inches of rain per year, is arid and semiarid. Yet it is into this large region that millions of people are migrating each decade, and with them they bring humid area water values and perceptions. Green landscaped lawns, large broadleafed trees, stable stream flow, and other evidences of the humid areas from which they have come are insisted upon (3, 6).

A second set of constraints may be categorized as political or governmental (29). In each country these will tend to be special and unique. In the United States, for example, the fact of our federal system with fifty states and about 3,000 counties has a far-reaching impact on how we deal with water problems. Jurisdictional controversies are frequent, and for special historic reasons, the provision of water and sewer services is regarded as a function of local governments--cities and counties. In fact, the jurisdictional situation with respect to water functions is so complex that some problems just go unresolved. Not irrelevant in this context, has been the development at state levels of independent systems of water law limited only by some of the broad protections of the National Constitution and its Bill of Rights including the very special protection given property rights (30).

A third category of constraints might be classified as social or societal. Here we are dealing with group behavior, customs, traditions, as well as with community norms and values. Thus in Salt Lake City, Utah, it has for many years been customary to flush the streets with huge quantities of water. In many parts of the U.S., moreover, use of water is considered almost a "natural right;" and attempts to curtail uses runs into deep resistance. Or, as in New York City, the use of Hudson River water has for long been considered unattractive, even in drought situations. Moreover, New York City reservoirs are fenced in to prevent public access, while in many other parts of the U.S. water supply reservoirs provide boating, swimming and other recreation. In those areas the esthetic senses of the ultimate consumers are not violated by such prior uses. After all, cities like St. Louis and Cincinnati have for years been drinking sewage of Chicago and Pittsburgh.

A fourth category of constraints might be labelled environmental. These are relatively new, and their character has by no means been fully devloped. But clearly water systems in the U.S. can no longer be operated without careful attention to environ mental concerns; these values may be expressed in statutes, or in public protests and demonstrations (20).

Finally, I want to refer to fiscal or economic constraints. It is almost axiomatic that the demands on government for services generally at any particular time exceed the capacity of government to pay for such services. Even where service is provided on a cost basis, fiscal or economic constraints can prevent the realization of the most rational, the most efficient, the most beneficial water plan. Conversely, where those affected possess the requisite political power, fiscal or economic constraints may not limit the planning, development, and operation of economically unsound water plans (9).

For each category listed above, concrete examples could illustrate decisional constraints of various kinds. Perhaps, however, sufficient illustrations have been indicated to suggest the importance of such constraints and to indicate the need for more thorough analysis of them in the operation of complex water systems. To put it differently, when factors of these kinds are ignored, the system may appear less complex, but its operation is likely to be more difficult.

REFERENCES

1. Anderson, R. L. and Norman Wengert, "Water systems '79," Proceedings of the ASCE Water Resources Planning and Management Division Specialty Conference, New York, New York: ASCE, February 25-28, 1979.

2. Anderson, R. L., Wengert, N. I. and Heil, R. D. The Physical and Economic Effects on the Local Agricultural Economy of Water Transfer from Irrigation Companies to Cities in the Northern Denver Metropolitan Areas, Fort Collins, Colorado: Environmental Resources Center, Colorado State University, 1976.

3. Biswas, A. K. (Editor). Systems Approach to Water Management, New York, New York: McGraw-Hill, Inc., 1976.

4. Burchell, R. W. and Listokin, D., The Environmental Impact Handbook, New Brunswick, New Jersey: Center for Urban Policy Research, Rutgers University, 1975.

5. Cartwright, F. F., Disease and History, New York, New York, Crowell, 1922.

6. Collins, B. E. and Guetzkow, H. A Social Psychology of Group Processes for Decision Making, New York, New York: John Wiley & Sons, Inc., 1964.

7. Conn, P. H., Conflict and Decision Making, New York, New York: Harper & Row, Publishers, 1971.

8. Diesing, P., Reason in Society, Five Types of Decisions and Their Social Conditions, Champaign-Urbana, Illinois: University of Illinois Press, 1962.

9. Dorfman, R. and Jacoby, H. D., "A model of public decisions illustrated by a water pollution policy problem," in The Analysis and Evaluation of Public Expenditures: The PPB System; A Compendium of Papers Submitted to the Subcommittee on the Economy in Government of the Joint Economic Committee, Congress of the United States, Volumes 1, The Appropriate Functions of Government in an Enterprise System, Volume 2, Institutional Factors Affecting Efficient Public Expenditure Policy. Volume 3, Some Problems of Analysis in Evaluating Public Expenditure Alternatives, Washington, D.C.: U.S. Government Print Office, Three Volumes, Paged Continuously; Volume 1, pp. 226-276, 1969.

10. Downs, A., Inside Bureaucracy. Boston, Massachusetts: Little, Brown and Company, 1967.

11. Easton, D., The Political System, New York, New York: Alfred Knopf, 1953.

12. Easton, D., A Framework for Political Analysis, Englewood Cliffs, New Jersey: Prentice-Hall, Inc., 1965a.

13. Easton, D., A Systems Analysis of Political Life, New York, New York: John Wiley & Sons, Inc., 1965b.

14. Friedrich, C. J. (Editor), "Rational Decision: Nomos VII, Yearbook of the American Society for Political and Legal Philosophy, New York, New York: Atherton Press, 1967.

15. Grigg, N. S. et al., Metropolitan Water Intelligence Sytems, Fort Collins, Colorado: Department of Civil Engineering, Colorado State University, 1973.

16. Gore, W. J., Administrative Decision Making: A Heuristic Model, New York, New York: John Wiley & Sons, Inc., 1964.

17. Hall, W. A. and Dracup, J. A., Water Resources Systems Engineering, New York, New York: McGraw-Hill Book Company, 1970.

18. Hoos, I. R., Systems Analysis in Public Policy: A Critique, Berkley, California: University of California Press, 1972.

19. Hufschmidt, M. M. and Fiering, M. B., Simulation Techniques for Design of Water Resource Systems, Cambridge, Masachusetts, Harvard University Press, 1966.

20. Jain, R. K., Urban, L. V., and Stacey, G. S., Environmental Impact Analysis: A New Dimension in Decision Making, second edition, New York: Van Nostrand Reinhold Company, 1981.

21. Kisiel, C. C. and Duckstein, L., "Operations research study of water resources, Part I and Part II," in Water Resources Bulletin, 6:737-745 and 857-867, 1970.

22. Lindblom, C. E., "The science of 'muddling through'," in Public Administration Review 19:79-88, 1959.

23. Lindblom, C. E., The Intelligence of Democracy, Decision Making through Mutual Adjustment, New York, New York: The Free Press, 1965.

24. McKean, R. N., Efficiency in Government Through System Analysis, with Emphasis on Water Resources Development, New York, New York: John Wiley & Sons, Inc., 1958.

25. Mowitz, R. J., The Design of Public Decision Systems, Baltimore, Maryland: University Park Press, 1980.

26. Novick, D. (Editor), Progran Budgeting, Cambridge, Massachusetts: Harvard University Press, 1965.

27. Simon, H. A., Administrative Behavior: A Study of Decision Making Processes in Administrative Organization, third edition, New York, New York: The Free Press, 1945, 1947, 1957, 1976.

28. Walker, G. M., Jr. and Wengert, N., Urban Water Policies and Decision Making in the Detroit Metropolitan Region, Ann Arbor, Michigan: University of Michigan, 1970.

29. Wengert, N., "Politican and social accommodation: the political process and environmental preservation," in Natural Resources Journal 11:437-446, 1971.

30. Wengert, N., The Political Allocation of Benefits and Burdens, Berkeley, California: Institute of Governmental Studies, University of California, 1976.

31. Wengert, N., "Reflections of information needs and the integration of social science perspectives in resource planning," in Proceedings, Social Scientists Conference: Social aspects of Comprehensive Planning. Three Volumes, Fort Belvoir, Virginia: U.S. Army Corps of Engineers, Institute for Water Resources, 1977.

PART IV: INSTITUTIONAL AND ECONOMIC PERSPECTIVES

PART III. PERSONAL AND ECONOMIC FACTORS

16. WATER MANAGEMENT INSTITUTIONS

Norman Wengert
Colorado State University, Fort Collins, Colorado

To understand how institutions impact the operation of complex water systems, this chapter first reviews definitions of the term "institution" in the literature and in common usage. In U.S. Government reports the term is used frequently as a synonym for "organization." But not all organizations are institutions and not all institutions are organizations! The definition used here draws on concepts of sociology and stresses the process of institutionalization, suggesting that an institution is the structured outcome of a process by which values (norms) are articulated, arranged, communicated, and applied. There is continuity over time, and influence or control on the behavior of persons involved with the institution but who may not have participated in the process of institutionalization. In the analysis of complex water system operations, it is necessary to consider how cultural norms, societal rules, values people pursue, individual group, community and organization bahavior, as well as structures and organizations that determine and support regularized social interaction have become institutionalized and continue dynamically to change and evolve.

1. THE CONCEPT OF AN INSTITUTION

The term institution (including variants such as institutional arrangements, institutional factors, or institutional constraints) occurs frequently in American literature on water management. Two diverging definitions dominate. Among practical administrators and in government jargon, the term is frequently used as a synonym for organization. Among some social scientists, however, the term institution has a broader and more significant meaning, drawing on concepts of sociology and political science, and stressing the dynamic process of institutionalization rather than simply the end product desginated "the institution," which (it should be stressed) need not be an organization (1).

In an investigation on the state-of-the-art of institutions for urban water management (25) several conclusions were drawn relevant to this chapter:

(1) Many writers used the term institution without defining it or being clear as to its meaning.

(2) While inferences from particular usages could sometimes be derived, they were often confusing and inconsistent, failing to explain the term.

(3) Often the word served as a kind of "black box," a means for seeming to include human and social factors without really analyzing or explaining their relationships to the system being analyzed.

(4) Least common in the literature dealing with water was the sociological usage which recognized how institutions evolve and are structured, through social processes, over time; they have continuity, are rooted in practice, and reflect widely held values or norms influencing or limiting human behavior.

(5) A common meaning especially among American federal govenrment agencies was to use the word institution as a simple synonym for organization.

In reviewing the literature on institutions in a more general context, Professor Harold E. Smith (30) concluded:

"Eight common elements found in the seventy definitions analyzed in this study are as follows:

(1) cultural norms,

(2) interrelated parts or structure,

(3) persistence and stability,

(4) functions,

(5) sanctions,

(6) cognitive elements,

(7)   regularized social interaction,

(8)   material culture traits."

From my research, drawing on concepts of sociology, I suggest that stress be placed on the process of institutionalization. An institution is the structured outcome of a process by which values, or norms, are articulated, arranged, communicated and applied. It has continuity over time, and influences or controls the behavior of persons involved with the institution but who may not have participated in the process of institutionalization (30).

The question arises, does the process of institutionalization make a difference in considering the operation of complex water systems? My answer is unequivocally "yes," because it is through institutions that society operates. Institutions are the means by which societal goals are implemented and actions directed and constrained.

Many aspects in the operation of complex water systems illustrate the process of institutionalization. But only three institutional developments which have shaped the course of complex water system operations in the United States are cited here. First, has been the evolution and development of American water law (9); Second, intricate relationships and patterns of responsibility and action have developed from the federal system of American government in which thousands of governmental units share responsibility for operating complex water systems--a situation which increases the complexities manifold; Third, a system of public involvement and citizen participation in decisions affecting the enviroment has evolved over the past ten to twenty years (29). These three developments are used as themes to explain, in the following pages, the role of institutions in the operation of complex water systems (2).

2. ORGANIZATIONS AS INSTITUTIONS

To summarize a pervious clarification not all organizations are institutions, nor are all institutions organizations (18). To illustrate, a formal declaration by appropriate authorities might create an Erice Symphony Orchestra, giving it legal and administrative existence. But this act of creation would not automatically make of the Erice Symphony Orchestra an institution. In contrast, without knowing anything about how it was created, we can all agree that La Scala is an institution of world renown. Wherein lies the difference? When (and how) does an organization become an institution? Simple organizational changes are often superficial and do not result in institutionalization unless related to fundamental changes in attitudes, perceptions, and expectations of the affected citizens. Putting a policeman's uniform on a man does not make him a policeman. Thus to suggest that organizational tinkering will create institutions to solve urban water problems may lead to misdiagnoses and false solutions to the problems involved.

To illustrate the above point, numerous studies such as those of the United States National Water Commission (23) have pointed to the fact that in many metropolitan areas of the United States, water functions are highly fragmented, responsibility being decentralized among many governments and agencies. There is convincing evidence that this fragmentation may increase costs; result in poor service, waste, and bad technology; and may contribute to other problems such as unwise land use. Frequently it is proposed that this fragmentation of organizations dealing with water functions should be eliminated through consolidation and reorganization (33, 34).

But the resulting organizational changes would not automatically create "new institutions," since the present fragmented system reflects deep seated values and historical traditions. In short, fragmentation, characterizing the decentralization of political power, has become institutionalized. Fragmentation rests on a web of values and norms which cannot be changed by simply passing a law or issuing an administrative order. The institutions are not simply the organizational pattern which exists at a particular time, but rather the set of socio-political values and norms which support and reinforce the exaggerated localism which characterizes urban water administration, in the United States as well as land use planning and control and numerous other aspects of American government. Unless values and perceptions are changed, reshuffling of functions and agencies may accomplish very little.

It might be more useful in approaching reorganization, as Kaynor has suggested (11), to stress the process of institutionalization and the inputs to that process, regarding the institutions as end products of complex interactive processes over time. In Kaynor's view the way in which government activities are institutionalized to plan or manage public services such as water and to control and influence citizen behavior with respect thereto should be the focus of analysis in reorganization decisions. The term institution is thus a way to designate the network of interdependencies or the cluster of relationships among law, agency operations, public finance, political values and influence in the context of the general culture. The following sections describe these aspects of institutional processes.

3. THE SYSTEM OF WATER LAW

In the United States to about the hundredth meridian, the quantity of water available has been so plentiful that questions of allocation have seldom arisen. The eastern states adopted English common law principles which specified that water could be used by owners of land along the banks of a stream or the shores of a lake. Riparian water law included the doctrine of non-degradation and nondiminution in flow. What this meant was that those who used the water were not to significantly degrade it or lessen its flow.

Rainfall patterns in the eastern United States in most years are so regular, amounting to about three inches each month, that the use of water appears not to diminish its quantity. As a matter of logic, of course, anyone who takes a bucket of water from a stream reduces the quantity of water in that stream. But the law overlooks this technicality, since the useful supply is not diminished; thus the modern riparian concept is stated in terms of reasonable use. With respect to degrading the quality of water through pollution, the English common law allowed for nuisance suits whereby an injured party downstream could seek damages in a nuisance action against the industry, the city, or the individual who caused pollution or diminution.

Western water law developed quite differently. The amount of rainfall in the states west of the one hundred meridian averages less than 20 inches per year, irregularly distributed by season, and often concentrated in a few heavy storms. Many months are without any appreciable precipitation. Thus in the west, problems of water allocation cause critical social, political and economic issues (9).

For curious historic reasons the Congress of the United States, which had initial jurisdiction, failed to establish policies, laws, or institutions for allocating water in the western regions. As a result, a unique set of indigenous institutions developed, based essentially on the rule of taking and maintaining possession. Water uses of the time were primarily for mining and agriculture. Water in the west became subject to property rights, the priority of these rights described by the slogan, "first in time, first in right." The first settler to get to a lake or stream established a legal right to use the quantity of water he claimed, as against all other users. His claim, however, was regarded as limited to that amount of water which he, himself, could reasonably use; and in some states the rule required that the water had to be diverted from its natural source to indicate intent to use.

To the "first in time" rule was added the "use it or lose it" rule, which provided a basis for challenging waste and attempts to claim greater quantities of water than seemed fair or reasonable.

In most of the seventeen western states, the courts became the instrumentalities for settling disputes over water rights among conflicting claimants. In some states special water courts were established. In other states an administrative system was also set up for recording the ownership rights and determining rights. In addition, ground water rights have been related to surface water rights. This is true particularly where the inter-connection is geologically or hydrologically provable.

It must be emphasized that since water rights in the west have been regarded as property, they gain the protection of the fifth and fourteenth amendments of the U.S. Constitution, which say that property shall not be taken by government without just compensation and in accordance with due process of law. Ownership has, thus, an unique tenacity! The law has not been clarified concerning the extent to which use must be efficient in terms of either purpose or the social utility of that purpose.

4. AMERICAN FEDERALISM AND COMPLEX WATER SYSTEMS

4.1 The Role of the Federal Government

Consistent with general American doctrines of Constitutional limitations the role of the federal government in operation of water systems has been indirect. It is based on legal authority and obligations in which water was not a primary factor. Federal water development and associated institutions were by products of other responsibilities. The oldest federal water responsibility related to water was navigation. But the authority for navigation development was military, having to do with the need to protect westward moving settlers from Indian hostilities, and in this connection to supply frontier forts with munitions, men, food, and supplies. Because of its basis in military necessity, responsibility for river and harbor improvement was considered appropriate for the Army Corps of Engineers, a function which it exercises still today (6, 7).

The second major water responsibility assumed by the federal government, flood control, was based also on a military concern. This was supplemented by a growing interest in commerce and in the general welfare--both now regarded as proper federal responsibilities. Flood control planning was begun first on the lower Mississippi River and ultimately (after 1936) on all major rivers (6). The federal jurisdiction over interstate commerce was the basis for the authority concerning flood control matters; this seems fairly clear, especially since stream navigability is related to entire river systems. But equally important for federal intervention in river development has been the "spending power," which many Constitutional lawyers regard as virtually unlimited.

The flood control role of the federal government was fully articulated in the Flood Control Act of 1936. But before that a federal role in irrigation development in the seventeen western states had been authorized by the Reclamation Act of 1902. The constitutional authority for this act stemmed from the responsibility granted the federal government to manage federal property--in this case public lands in the western states. The commerce and spending powers provided another basis.

The Tennessee Valley Authority, created in 1933 is another well known organization of the federal government. Its authority came from these same constitutional provisions, as well as from a national defense obligation to produce nitrates at the original Muscle Shoals plant authorized in World War I.

Initially at reclamation projects and then later at Army Corps of Engineers and TVA projects, provision was made for the generation of electric hydro power. While the constitutional bases for electric power generation were more complex than in the case of the other federal water activities, a major constitutional justification rested on the inference that the federal government had the right to manage its property efficiently, which allowed generation and sale of electric power as a kind of by product of other activities. TVA's thermal power production, initiated on a large scale during World War II, rests on both national defense, property management and spending power authority.

This brief review of the traditional federal water programs has suggested that though based on the constitution and on statutes, the implementation of these programs has been dependent on concommitant institutional development involving complex interactions among people and groups and their perceptions of public problems and interests. In all of the federal programs, complex patterns of intergovernmental relations developed. Initially these were between the responsible federal agencies and local governments, and associations of local people. (6, 7). In most cases, state governments played minor roles.

Particularly noteworthy has been the encouragement of associations of irrigation farmers to allocate and manage water produced by the Bureau of Reclamation in the Department of the Interior. A similar approach was used by TVA for marketing electric power; rural cooperatives were used to serve rural areas and city owned systems for urban areas. This TVA policy, ultimately written into statute law--known as the "preference clause"--gives cooperatives and cities first claim on power generated at all federal facitities no matter which agency is responsible.

The Rural Electric Cooperative Associations have blanketed the country, often receiving low interest funding from the Rural Electrification Administration (7 USC 901), a bank-type agency in the U.S. Department of Agriculture, so that today virtually all of rural America is electrified. The affairs of the hundreds of distribution cooperatives are managed by the farmer members and their elected boards of directors. In many cases, especially where federal public power has not been available, the REA cooperatives purchase power from private companies or have established their own generating facilities, often as joint ventures of a consortium of cooperatives and municipal systems.

4.2 Institutions for River Development Planning

Partly because of the pervasive influence of the American federal system, partly for political reasons including the need for support, and partly because it seemed to make sense, river development planning became more and more comprehensive (Schad, 1964). At the same time, responsible federal agencies sought in various ways to involve states and local governments. Some collaboration was voluntary, but ultimately statutory encouragement was given by the Congress authorizing the federal water agencies to focus on comprehensive regional, river basin planning (26, 32).

The degree of intergovernmental collaboration has varied considerably, but the federal government was the dominant partner (16). A cynical critic might question whether the pattern of collaboration ever became institutionalized to the point where local, state and federal planners shared comprehensive planning objectives and responsibilities. About all that can be asserted is that the idea of collaboration has become institutionalized. This has taken shape in a negative sense, however, so that today no federal agency would launch a river development project without discussion with state and local interests (7).

4.3 Water Supply and Wastewater Management

When a flood had occurred, when a stream needed dredging, when water storage or water control works were desired, Americans--particularly their local governments--turned naturally to the U.S. Army Corps of Engineers, an agency whose functions in river development had been institutionalized for many decades. In contrast urban water supply and development of wastewater disposal systems have been institutionalized as local municipal functions. The evolution of city government as water supplier was often not a deliberate decision but a response to the default of private systems. In Detroit, for example, when the private system failed in the 1830's, the city assumed responsibility for the system (14, 19).

The institutions for urban water supply continue the local mode (8) although in some of the larger metropolitan areas-- New York, Philadelphia, Detroit, Chicago, Milwaukee, Denver, Los Angeles, Seattle--metropolitan systems evolved after World War I to serve the entire or a major part of a metropolitan region. There seems little doubt that from the view of engineering and economic efficiencies of scale in planning and in operation, substantial improvements in system integration would be desirable. But the institutional strength of extreme localism has often precluded such collaboration. Selfish power interests and hostilities among jurisdictions have also played a part in the over-fragmentation of urban water functions (19).

The concern for purity of urban water in the early days was largely aesthetic, stressing smell and appearance. Once bacteria were discovered, a technical basis was provided for concern with water purity (20).

Development of wastewater management in most cities was far less systematic than was water supply (Kneese and Bower, 1968). This reflected the fact that sewage had negative value to the community and the primary concern was simply to get rid of it by discharge into the nearest stream. Industries often had their own outfalls, and there was relatively little health control over sewage systems (5).

As urban sewage systems developed (17) there was the tendency to combine sewage and storm water mamagement in one system, but many of these systems were under-designed. The result was mal- function during storm periods. Increasingly in newer suburbs and subdivisions storm water management has been separated from the sanitary sewer systems, but much remains to be done in this regard. In the eastern United States until World War II where rainfall was plentiful, cities dumped sewage into rivers or lakes without any treatment. Some of the more advanced cities provided for primary treatment, but secondary treatment became the norm only after World War II. A few cities on their own had developed secondary treatment and Milwaukee processed its sewage by the activated sludge process into a fertilizer material called Milorganite, marketed through much of the country.

Until after World War II, water supply and wastewater management on farms remained primitive. Because many farms were still without electrification, water pumping was either by gasoline motor or by hand. In some areas windmills were used to pump water for stock tanks. Rural sewage systems were virtually nonexistent. In a few cases septic tanks had been installed, but the basic rural system for managing human waste was the privy or outhouse.

State governments have exerted almost no influence on urban water supply beyond regulation of water purity as a health function. There has been no involvement otherwise.

The river basin planning process, referred to above, paid scant attention to municipal water supply or waste water disposal unless these functions were directly related to flood control and navigation. Where storage reservoirs had an impact upon urban water supply or sewage disposal, engineering plans took such factors into account, but largely on a passive basis. This neglect of water supply and sewage wastewater management in river basin planning made little sense, since often the waters from which cities drew their supplies and into which they dumped their sewage were the same waters which were subject to river basin planning.

But several forces were at work to change this local concern into an intergovernmental concern, involving local, state and federal governments--not through the institutions of river basin planning but through building a largely independent set of institutions and relationships for joint collaboration. Ultimately the federal government, because of its fiscal strength and political support, became the dominant partner, but in a manner very different from that associated with river basin development.

Although there were exceptions, prior to World War II, most urban and industrial waste disposal systems used only primary treatment methods. The oxygenation capacity of receiving streams and lakes were considered sufficient to handle the BOD loads. Many smaller communities and even large cities e.g., New York, on the ocean dumped raw sewage into the receiving waters. But population densities were generally low, (i.e., total U.S. population in 1940 was just over 150 million, with an average overall density of 50 per square mile) and most industrial wastes were still biodegradable.

Only after World War II, with its accompanying large-scale industrial expansion and rapid urban growth, did perceptions develop concerning the serious impairment of the ecology and aesthetic characteristics of natural waters (3). Thus in 1948 the

U.S. Congress took the first steps toward creating a national water pollution control program and strengthening local and state institutions for dealing with the subject. They began cautiously in the 1950's by stressing research and close collaboration, and relied on persuasion and limited financial support. Major revisions in the 1960's and again in the 1970's increased substantially the federal regulatory role and also provided billions as grants for local treatment works. The expanded federal programs sought to push the state governments into a larger regulatory role, in setting standards, and in forcing compliance. In general, this has been achieved. At the local government level federal funds became available for building secondary treatment plants, extending interceptor sewer lines, and generally improving local systems.

4.4 Drinking Water

As indicated, water supply has been even more of a local function than wastewater disposal. Several major exceptions might be noted--in all of which the local municipality continued to play a major role. One exception has been water supply for Los angeles which has included access to waters stored by Hoover Dam, built by the federal government and more recently, waters diverted from Northern California to Southern California under the State of California Water Development Plan. And as indicated, a number of other cities have improved their water supply by being able to draw on water impounded by Federal dams.

But the legislative history of 1974 the Safe Drinking Water Act which placed the federal government in a dominant position with respect to certain aspects of water supply--reflects not processes of institution building but rather public panic, precipitated by the perceived problem of trihalomethanes. For decades, chlorination was used to disinfect water for drinking and sewage before discharging it into a receiving stream or lake. Then in the early 1970's, it was discovered on the Lower Mississippi River that chlorine reacted with certain synthetic chemicals already in the river to create carcinogenic compounds. When this was learned, Congress reacted almost immediately with the Safe Drinking Water Act. But the act although precipitated by a perceived crises, was passed within an established framework of thinking, which was the idea of safe drinking water which was firmly institutionalized. It has not necessarily resulted, however, in further organizational institutions. This is one of the subtle points regarding the idea of institutions.

5. CITIZEN PARTICIPATION AND PUBLIC INVOLVEMENT

Most public programs in the United States, including those involving complex water systems, have required some form of citizen participation. Before the 1960's, formal citizen participation was primarily of two types:

(1) local, state and federal legislators were selected and they established programs and policies, and

(2) administrative hearings specified by statute or ordinance (21).

In varying degrees the public and various types of groups had access to both legislators and administrators. But prior to the 1970's, there was little uniformity in approach and little system in the search for public input (29).

Two events, both federal legislative enactments, have served to institutionalize the process and the substance of citizen participation. The first was President Johnson's "War on Poverty," enacted in the mid-1960's. It required "maximum feasible participation" of the poor in planning and implementing programs which were to benefit or affect them. The second was the National Environmental Policy Act of 1969, also called NEPA.

Two provisions of NEPA were of great importance for citizen involvement, especially as interpreted by the federal courts. One was that all federal agencies must consider and evaluate possible alternative courses of action. The other was assessment concerning how proposed actions were likely to impact the environment. As interpreted by the courts, and reinforced by a broadening doctrine of who had "standing to sue," individual citizens and particularly citizen groups were given a basis for challenging federal agency actions, including the question of whether the views of the public were given adequate attention. As a result, public participation, and citizen involvement have been institutionalized in the processes of the federal government. It should be stressed that this was not a necessary result of the legislative enactments. The Congress did not command the type of participation now so familiar. Rather the complex and and subtle processes of institutionalization were at work as suggested in the definition given at the beginning of this chapter.

The way in which we in the United States deal with the operation of complex water systems, how we make policy and program decisions in this substantive field may fall far short of the "rational comprehensive decision making mode" which I discussed in chapter 14. But it does recognize the pluralist and pragmatic forces which interact in the American system. It suggests how individual values and institutional conditions constrain policy and program decisions. It also suggests some of the characteristics of a very open decisional system. To give a sense of how that system operates was the purpose of this chapter.

REFERENCES

1. American Society of Civil Engineers. "Legal, institutional and social aspects of irrigation and drainage and water resources planning and mamagement," Proceedings of Specialty Conference, ASCS Irrigation and Drainage Division and ASCE Water Resources Planning and Management Division, New York, 1979.

2. Anderson, R. L., Wengert, N. I. and Heil, R. D., The Physical and Economic Effects on the Local Agricultural Economy of Water Transfer from Irrigation Companies to Cities in the Northern Denver Metropolitan Areas, Fort Collins, Colorado, Environmental Resources Center, Colorado State University, 1976.

3. Dahl, R. A., Pluralist Democracy in the United States, Conflict and Concent, Chicago, Illinois, Rand McNally and Company, 1967.

4. Fox, I. K., "Review and interpretations of experiences in water resources planning," in Organization and Methodology for River Basin Planning, C. E. Kindsvater (editor), Atlanta, Georgia, Water Resources Center, Georgia Institute of Technology, pp. 61-87, 1974.

5. Grigg, N. S. et al., Metropolitan Water Intelligence Systems, Department of Civil Engineering, Colorado State University, Fort Collins, Colorado, 1973.

6. Holmes, B. H., A History of Federal Water Resources Programs, 1800-1960, Washington, D. C., Economic Research Service, U.S. Department of Agriculture, Miscellaneous Publication No. 1233, 1972.

7. Holmes, G. H., History of Federal Water Resources Programs and Policies, 1961-1970, Washington, D.C., Economics, Statistics, and Cooperative Service, U.S. Department of Agriculture, Miscellaneous Publication No. 1379, 1979.

8. Holtz, D. and Sebastian, S. (Editors), Municipal Water Systems, Bloomington, Indiana, Indiana University Press, 1978.

9. Hutchins, W. et al., Water Rights in the Ninteen Western States, U.S. Dept. of Agriculture, 3 Vols., 1971, 1974.

10. Jones, C. O. and Thomas, R. D. (Editors), Public Policy Making in a Federal System, Beverly Hills, California, Sage Publications, 1976.

11. Kaynor, E. R. and Howards, I., "Limits on the institutional frame of reference in water resource decision making," in Water Resources Bulletin 7:1117-1127, 1971.

12. Kaynor, E. R. and Howards, I., Attitudes, Values, and Perceptions in Water Resource Decision Making within a Metropolitan Area, Amherst, Massachusetts, Water Resources Research Center, University of Massachusetts, 1973.

13. Kneese, A. V. and Bower, B. T., Managing Water Quality: Economics, Technology, Institutions, Baltimore, Maryland, The Johns Hopkins University Press for Resources for the Future, 1968.

14. Koelzer, V. A. and Bigler, A. B., Policies and Organization for Urban Water Management, Fort Collins, Colorado, Water Resources Publications, 1975.

15 Langton, S. (Editor), Citizen Participation in America, Lexington, Massachusetts, Lexington Books, D. C. Heath and Company, 1978.

16. Leven, Charles, An Analytical Framework for Regional Development Policy, Cambridge, Massachusetts, MIT press, 1970.

17. Major, D. C. and Lenton, R. L., Applied Water Resource Systems Planning, Englewood Cliffs, New Jersey, Prentice-Hall, Inc., 1979.

18. March, J. G., Handbook of Organizations, Chicago, Illinois, Rand McNally and Company, 1965.

19. McPherson, M. B., Prospects for Metropolitan Water Management New York, New York, American Society of Civil Engineers, 1970.

20. Minton, G. R., Williams, R. and Murdock, T., "Institutional Analysis Criteria for Water Supply Planning," in Water Resources Bulletin, ff. 486, June 1980,

21. Pierce, J. C. and Doerksen, H. R., Water Politics and Public Involvement, Ann Arbor, Michigan, Ann Arbor Science Publishers, Inc., 1976.

22. Schad, T. M., "Legislative history of federal river basin planning organizations," in Organization and Methodology for River Basin Planning, C. E. Kindsvater (Editor), Atlanta, Georgia, Water Resources Center, Georgia Institute of Technology, pp. 41-60, 1964.

23. U.S. National Water Commission, <u>Water Policies for the Future</u>, Washington, D.C., U.S. Government Printing Office, 1973.

24. Walker, G. M., Jr. and Wengert, N., <u>Urban Water Policies and Decision Making in the Detroit Metropolitan Region</u>, Ann Arbor, Michigan, University of Michigan, 1970.

25. Wengert, N., "What do we mean by 'Metropolitan Water Management Institutions'?" in Water Resources Bulletin, Vol. 9, #3, pp. 512-521, June 1973.

26. Wengert, N., "A critical review of the river basin as a focus for resources planning, development and management," in Proceedings, American Water Resources Association Symposium on Unified River Basin Management, Gatlinburg, Tennessee, (in press) May 4-7, 1980.

27. Wengert, N., "The politics of water resources development as exemplified by TVA," in <u>The Economic Impact of TVA</u>, John R. Moore (Editor), Knoxville, Tennessee, The University of Tennessee Press, pp. 57-79, 1967.

28. Wengert, N., "Political and social accommodation: the political process and environmental preservation," in Natural Resources Journal 11:437-446, 1971.

29. Wengert, N., <u>The Political Allocation of Benefits and Burdens</u>, Berkeley, California, Institute of Governmental Studies, University of California, 1976.

30. Wengert, N., <u>Urban-Metropolitan Institutions for Water Planning, Development and management</u>, Fort Collins, Colorado, Environmental Resources Center, Colorado State University, 1972.

31. Wengert, N., "Citizen participation: practice in search of a theory," in Natural Resources Journal 16:23-40, 1976.

32. White, G. F., "A perspective of river basin development," <u>Law and Contemporary Problems</u> 22(2):156-187, 1957.

33. White, G. F., <u>Strategies of American Water Management</u>, Ann Arbor, Michigan, Ann Arbor Paperbacks, The University of Michigan Press, 1971.

34. Widstrand, C., <u>Water and Society, Conflicts in Development</u>, Oxford, England, Pergamon Press, 1978.

17. ORGANIZATIONAL RELATIONSHIPS

Luis Veiga da Cunha
Laboratorio Nacional de Engenharia Civil,
Lisbon, Portugal

As long as water is plentiful, and there are no serious problems of pollution, water mangement most often is divided among various administrative departments. This is true almost regardless of country. But as competition for water increases, and as other problems are accentuated the activities of various administrative units may begin to conflict or overlap, with consequent loss of efficiency. It then becomes necessary to plan and coordinate water use by setting up institutional frameworks that can ensure an overall approach to water management.

At this point water resource management requires coordinated actions of various authorities, e.g., legislative, executive, advisory. These authorities may be classified as:

1. those responsible for management,

2. those responsible for economic, social, and land use planning, and

3. those having jurisdiction over water related matters.

The associated activities may include agriculture, industry, power production, human settlements, transports, fishing and tourism.

Ancillary domains affected by water management activities include the environment and quality of life; health and sanitary conditions; education; culture and training and research organizations; recreational water uses; public works; and finances.

The authorities responsible for water activities should be part of a management framework, which may have a hierarchal relationship from national level, to regional and local levels, which insures coordination between their functions.

The functions of these authorities can be analyzed according to:

1. their categories of authorities, and

2. their territorial jurisdiction.

These ideas are reflected in the organization chart of Figure 1. This diagram shows a general water management framework. This diagram is the result of a comparative analysis of the water management institutional frameworks in various countries (1).

This chapter takes into consideration previous work on the subject, particularly the work reported by OECD countries (2, 3, 4, 6, 7, 8, 9).

It is difficult to set forth a universal model for the water resources institutional framework, because the organization, composition, and functions of water management institutions depends on a set of contextual factors. The most important of these factors include:

1. regional and physical characteristics such as climate, physiography, area, population density and distribution by urban and rural areas, and economic and social development,

2. system of water rights and water administration,

3. amount of public participation in the decision making process,

4. type and efficiency of state organizations, which may or may not justify the setting up of management authorities with administrative and financial autonomy,

5. forms of political and administrative organization, and their relationships to the central government.

Along with the influences of these factors, there are also certain basic rules needed for efficient operation of a water management institutional framework. One of these basic rules is that coordinated action is essential between the authorities responsible for water management and those responsible for economic, social and land use planning, and those with jurisdiction over water related matters. In fact, the various government departments normally have limited perspectives of water problems; thus it is essential to reconcile these different attitudes by means of coordinative authorities capable of formulating an overall water management policy. These authorities must consequently have coordinative functions and include representatives of the departments more directly involved with water problems. At the top of the water management framework there must be an interdepartmental coordinating body. For obvious reasons, it is also fundamental that responsibility for the executive functions of water management policy should be concentrated in a coherent set of authorities.

When a country intends to implement a new water management framework along the above lines, the executive framework may either coordinate the action of authorities and bodies that have been traditionally concerned with the conservation and development of water resources, or else, in a more radical way, the existing system may be entirely replaced by new systems that hold full executive authority concerning water. Note that attempts at securing coordination through centralization are not always successful, since functional coordination is a process often independent of structural centralization.

More specific rules concerning water institutions are outlined in a United Nations publication (10). The key points quoted as follows:

1. "there is certainly no single way to organize or administer a river basin program,

2. the plan of organization must in each case be fitted into the general governmental structure and into the cultural patterns and political traditions of the countries and regions which are involved,

3. there may be alternative kinds of feasible organizational schemes and structural devices for different countries in view of the contrasting kinds and combinations of water functions exercised at varying stages of administrative development: construction and maintenance, planning and operations, budgeting and spending, funding and vending, scientific and technical, service functions and regulatory functions,

4. there has been considerable hesitancy to arrive at organizational models in the realm of water administration,

5. most often, in fact, the prime organizational problem is coordination as process rather than organization as structure,

6.   in search for effectual coordination of policies and operations, the structural choices are quite varied and the organizational alternatives cover a broad range of possibliities. A common trend seems to be the creation of an increasing number of agencies with coordinating functions at higher and higher organizational levels and, as part of this trend an overall ministry of resources is sometimes established."

In analyzing the relationship between the various levels of intervention, the following additional points are made:

1.   "decentralization of central government activities is merely part of the general process by which modern organizations, both public and private become deconcentrated and disencumbered,

2.   the simulation of local initiative and the encouragement of of active popular participation at the subnational level are being widely recognized as value impulses to effective planning and successful development,

3.   responsibility for administrative activity must be shared if it is to be effectively carried out. Such sharing may take the form of constant delegation of functions to subordinate administrative branches to autonomous bodies or corporations; to private or cooperative institutions, to local authorities and to individuals at various levels of the administrative structure. Sharing of responsibility among civil servants and participation of citizens tend to bring out the higher loyalties and the best administrative abilities of the nation."

Concerning the institutional framework of water resources management the United Nations Conference (11) stated that institutional arrangements adopted by each country should ensure that the development and management of water resources take place in the context of national planning and that there is real coordination among all bodies responsible for the investigation, development and management of water resources. The problem of creating an adequate institutional infrastructure should be kept constantly under review and consideration should be given to the establishing of efficient water authorities to provide for proper coordination.

To attain these goals the United Nations Water Conference recommends that the various member states adhere to the points enumerated below:

1. "Adapt the institutional framework for efficient planning and use of water resources and the use of advanced technologies where appropriate. Institutional organization for water management should be reformed whenever appropriate so as to secure adequate coodination of central and local administrative authorities. Coordination should include the allocation of resources with complementary programs.

2. Promote interest in water management among users of water; users should be given adequate representation and participation in management.

3. Consider, where necessary, the desirability of establishing suitable organizations to deal with rural water supply as distinct from urban water supply, in view of the differences between the two in technologies, priorities, etc.

4. Consider as a matter of urgency and importance the establishment and strengthening of river basin authorities, with a view to achieving a more efficient, integrated planning and development of the river basins concerned for all water uses when warranted by administrative and financial advantages.

5. Secure proper linkage between the administrative coordinating agency and the decision-maker."

The organization chart of Figure 1, was in accordance with the basic principles set forth. It is intended as a general institutional model of water management. According to this model the water management structure should include three types of bodies and institutions:

1. decision-making and coordinative authorities, which define policies, are responsible for water planning, and coordinate the action of different departments concerned, and make the decisions;

2. executive authorities and bodies, which carry out action and provide support to the decision-making and coordinative authorities;

3. advisory bodies, which collaborate with the decision-making and coordinative authorities, and enable parties with interest in water matters to voice their opinions.

298

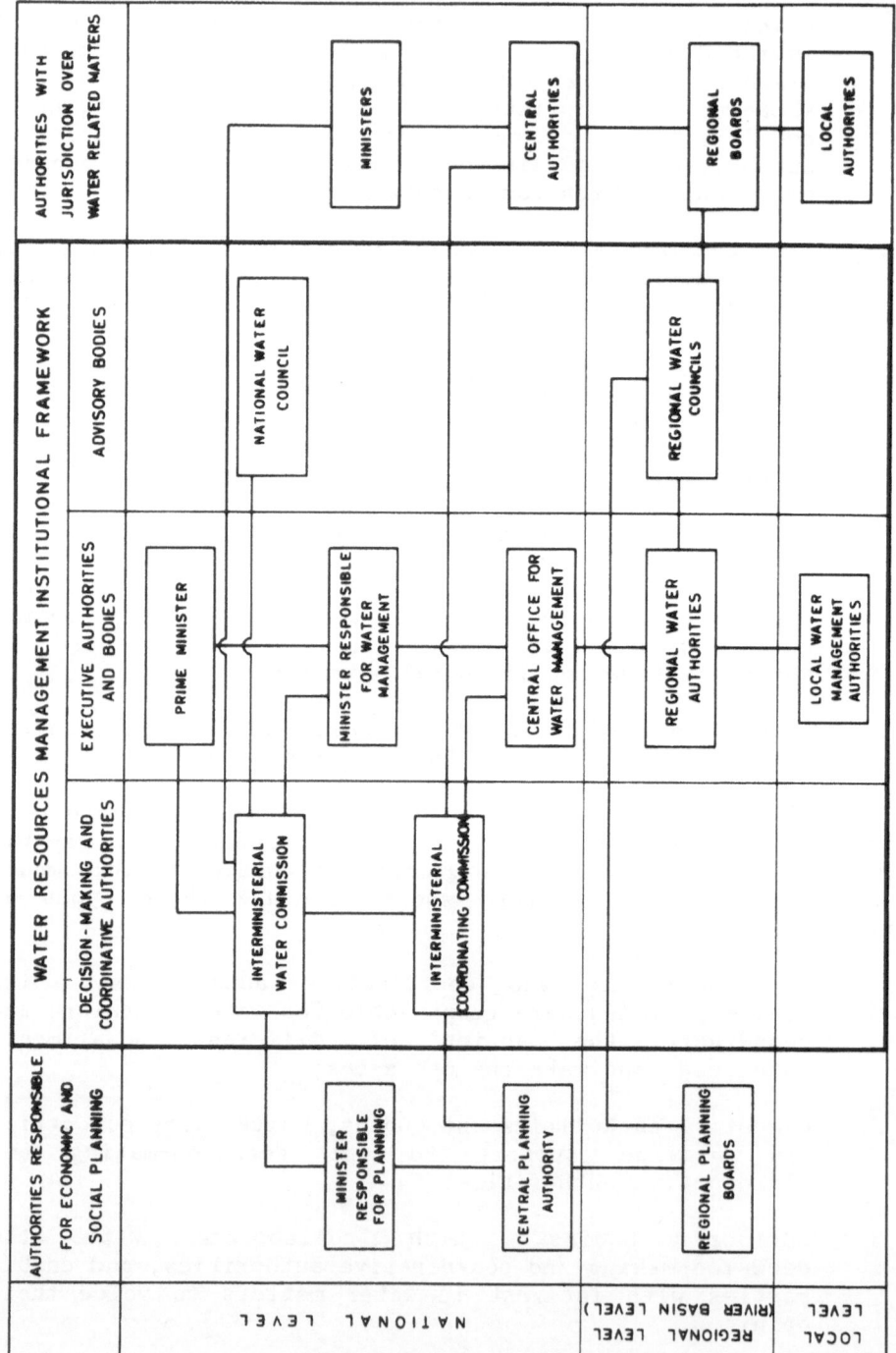

Figure 1. Outline for a General Water Resources Institutional Framework

The set of authorities and bodies of each type is represented at national, regional and local levels, thus ensuring coverage of the territory by decentralized management units, with hierarchical relationship within the first and second types. Figure 1 is an organizational chart which shows in its three central columns the authorities and bodies that make up the proposed water management institutional framework. These authorities and bodies are represented at national, regional and local levels, as indicated in the major rows. Details of the composition and general functions of the authorities and bodies included in the proposed framework are outlined below.

National Level Authorities The water management institutional framework depends on the minister responsible for water management. An interministerial commission, called "Inter-ministerial Water Commission" would be responsible for the formulation of the broad lines of water policy and for the definition of the guidelines for the execution of this policy. This Commission would consist of the minister in charge of water management, who would chair as a delegate of the Prime Minister, the minister in charge of economic and social planning, the ministers with jurisdiction over water-using activities such as agriculture, energy, industry, fishing, tourism and inland navigation, and the ministers who do not have jurisdiction over water related matters but have nevertheless a decisive influence over the use of water resources, particularly the environment, health, education, culture, public works, and finances.

The Interministerial Commission would in fact correspond to a cabinet council for water resources problems. Its implementation would take place only in the countries where the existence of limited cabinet councils is current practice. Otherwise the competences envisaged for the Interministerial Water Commission would fall on the cabinet council itself.

There would be also an "Interministerial Coordinating Commission" consisting of qualified representatives of the Director of the Central Office for Water Management, described later on in this chapter, presided over by the minister that chairs the Interministerial Water Commission or by his represen-tative. His charge would be to promote the studies needed for preparing the decisions of the Interministerial Water Commission as well as to take decisions for this Commission. This coordi-nating Commission would actually act as an extension of the Inter-ministerial Water Commission but would meet more often in order to carry on the tasks of the Interministerial Commission, following the guidance lines set by the latter.

The executive function of the water management policy would be carried out by a minister of a specific Ministry for Water

Problems or a minister of a sectorial ministry, appointed for coordinative or traditional reasons, to accumulate the executive responsibilities of water resources management with other responsibilities of its own. This Ministry would depend upon the Central Office for Water Management which would be a central executive institution. But as the regional bodies, henceforth designated as "Regional Water Authorities," are being set up, the functions of this central institution would become merely coordinating in nature.

Regional Level or River Basin Level

It is widely accepted that the regions naturally defined for water resources management are river basins. The water management executive authorities are placed at this level. In the organization chart, shown in Figure 1, they are called Regional Water Authorities. Attached to these authorities are advisory bodies, which appear as "Regional Water Councils." Their composition is parallel to that of the National Water Council. They include representatives of the Regional Planning Boards and of the departments with jurisdiction over water use activities. It should be noted that the areas of jurisdiction of Regional Water Authorities may not coincide with the planning regions, or with the areas of influence of the other regional boards. Moreover, in the Regional Water Councils there must be representatives of all boards covered by the areas of concern to the Council.

The functions and capacities of the Regional Water Councils could, in some countries, evolve with time as the structure for water management is implemented. It is thought that they might start by having only consultive functions for the Regional Water Authorities and with time ascend to increasing decision-making powers.

One of these would be the capacity to approve the Regional Water Resources Plan and the activity program of the respective authority. The Regional Water Councils would thus have very wide deliberative and coordinating powers within the jurisdiction area of the Regional Water Authorities. In fact they would become true "water parliaments," the name by which such bodies are known in some countries.

Each Regional Water Authority would be in charge of preparing the proposals of the Regional Water Resources Plan and the proposals for its action programs.

The Regional Water Authorities would also hold administrative functions through the licensing of water use and laying down of the conditions for this use. The financing of programs in their area of jurisdiction could be through the levy of water use charges and pollutions charges.

The regional water authorities could also have responsibilities concerning the operation of hydrological networks and the collecting, processing and retrieval of hydrologic data, and design, construction and operation of works of regional interest. The key bodies of the proposed structure would be the regional water authorities who would superintend the water resources within large river basins. In the countries that wish to set up an entirely new water management institutional framework the establishment of these regional water authorities would correspond, in a first phase, to the deconcentration of power decentralization.

The consultive functions confered to the regional water councils as envisaged in Figure 1, correspond to a first stage in which deconcentration is the aim. In a subsequent phase in which decentralization is the aim the councils should be placed in the column to the left of the central column in Figure 1, as deliberative and coordinating bodies. It should be obligatory for them at this stage to include elected representatives of regional water use interests, as it is done in certain countries.

In France, for example, each regional water council is composed of an equal number of representatives appointed by the government, the municipalities and by the water users associations. The administrative councils of the regional water authorities include eight state level representatives, four of the local municipalities and four of the different types of users; the president is appointed by the prime imnister. In England the executive boards of the regional water authorities include a majority of local representatives and also representatives of the state. The president is appointed by the government.

It would be convenient if the area jurisdictions of the regional water authorities would coincide with the administrative regions and with the planning regions. In the frequent cases in which this coincidence is not possible adequate coordination forms should be implemented between the regional water authorities and the two types of regions above referred to. To mention again the case of France, this country has technical water commissions within the scope of the regional planning framework to establish the connection with the planning bodies of the regional water authorities.

Local Level The local level of water resources management is the project level. Typical kinds of projects are hydroelectric, irrigation, water supply, and municipal sewage. The management of these systems is generally up to public or private organizations. For local management of projects, regional water authorities may have local delegations, but in most cases power will be a delegated to these local agencies. In local management of water resources it is important to accomodate public participation, which is one of the conditions for the success in implementating a water resources management policy.

As a final remark it must be emphasized that the model presented is not intended to offer the universal pattern of a water resources management framework. Such frameworks must be estblished taking into account the relative importance of the different contextual factors prevailing in each country. The model is general and flexible enough, however, for a large number of situations. Adjustments may be achieved through development of the authorities described. The model may be helpful to those who must establish or reform a water management framework in a given country. It provides a way to check whether the important aspects have been taked into account. The model better fits unitary states than federal states, since in the latter case two criteria of regional subdivision have to be considered, one based on states, the other on river basins, which as a rule are not easy to bring into agreement.

An important question is that of international river basins, in which the basic unit of water management should be controlled by more than one country. In this case also the model stands valid. But a new dimension should be added to ensure the cross connections between the different national water management frameworks established according to the proposed principles.

Finally it has to be mentioned that the simultaneous implementation of the whole framework is not feasible. On the contrary a progressive process should be adopted, whose description, although of interest, is beyond the scope of this chapter.

REFERENCES

1. Cunha, L. V. et al. Management and Law for Water Resources. Water Resources Publicatons, Fort Collins, Colorado, 1977.

2. Water Management in Canada. OECD, Paris, 1976.

3. Water Management in the United States. OECD, Paris, 1976.

4. Study of Economic and Policy Insturments for Water Management in Finland. OECD, Paris, 1976.

5. Studies on Economic and Policy Instruments for Water Management in France. OECD, Paris, 1978.

6. Water Management in Japan. OECD, Paris, 1976.

7. Study on Economic and Policy Instruments for Water Management in the Netherlands. OECD, Paris, 1976.

8. Policy Instruments in Water Management in the Federal Republic of Germany. OECD, Paris, 1976.

9. Study of Economic and Policy Instruments for Water Management in the United Kingdom. OECD, Paris, 1976.

10. National Systems of Water Administration. United Nations, New York, 1974.

11. Report of the United Nations Water Conference. United Nations, New York, 1977.

18. LEGAL ASPECTS OF WATER POLLUTION CONTROL

Giovanni Torregrossa
Counciller of State, Ministry of Public Works, Rome

The legal aspects of water pollution control have roots in a wider political and cultural movement concerning environmental conservation. One of the early political stands on the topic was Kennedy's "New Frontiers" speech in 1961, which addressed the need to fight environmental degradation. This was followed in 1962 by Rachael Carson's book <u>Silent</u> <u>Spring</u>, which molded the public conscious and precipitated a plethora of laws. These laws started in 1965 and extended through the late 1970's. Some of the tangible political actions during this period included Johnsons's "Water for Peace" conference in 1967, and Nixon's 1970 State of the Union speech, which laid the basis for an environmental policy encompassing 23 draft bills and 14 administrative measures. The central points were water pollution control, air pollution control, solid waste management, parks, wildlife, and the more elusive idea, "quality of life."

On the subject of water pollution control, the speech contains the important statement,

"the water courses belong to all of us and neither a municipal council nor an industry may be allowed to discharge their waste matter into these courses, beyond the limits imposed by the waste matter being diluted without polluting."

Another important declaration in the same speech concerns the need for a policy to facilitate elimination of wastewater and the idea of water reuse. It encompassed also the area of solid waste management.

The idea of reuse is important to the theme of this book, which stresses the fact that we are dealing with complex water systems. This theme captures the idea that we are coming full circle back to the goal that we should not squander our natural resources and that recycle and reclamation will help alleviate our pollution problems.

The political tone in Nixon's initiative is not an isolated example since it can be seen in many other places. Other examples include Article 2 of the North Atlantic Treaty and the 1969 NATO Commission Institute on the problems of modern society. Daniel Moynihan's comment on the NATO initiative is significant,

"The role of NATO in finding a solution to environmental problems is a determinant since we are at the dawning of an era which our decendants will consider decisive for the servival of the human race."

The political and cultural drive in the United States was reflected immediately in Europe. These are seen in the initiatives of the European Council in Strasbourg which proclaimed 1970 the European Year for Nature Conservation and in the initiatives of the Italian Senate with its well-known Franceschini Commission investigation, and Senator Medici's conference on water.

The political stimulus contained in the above initiatives went through a fundamental articulation in the 1972 United Nations Stockholm Conference on the "Problems of the environmental conditions in which man lives." This was reinforced by the European proposal, still under dicussion, on evaluating the "environmental impact."

The cultural movement which went along with the ecological challenge of the years 1960-1970 involved all sectors of society. Its most important tendency was to formulate a new environmental ethic. In it societal norms of behavior were aimed to protect and preserve the natural and human environments. In this ethical concept it is important to include things that may affect the biological relationship between man and nature. The idea relates to the set of physical, chemical and biological agents, plus social factors, which are liable to have a direct or an indirect effect on living creatures and human acitvities. The effects may be immediate or long-term.

While the environmental movement is recent, it is not unique in history. One of the articles in the Pennsylvania Constitution has similarity to the current environmental movement. In the words of this provision, everybody has a right to "pure air, clean water, maintenance of the landscape, and historical and aesthetic values of the environment" (Art. I-27).

The Italian Constitution, however, has no article defining the concept "environment," and thus it has not created a specific discipline to regulate those activities of society which cause negative environmental effects. In spite of this, there have been many attempts in the context of Italian law to develop a legislative notion on environment. In this respect, although some of the articles of the Italian Constitution are designed to protect specific legislative rights, they also express more general requirements and could be applied more widely. Examples include Article 32 on guarding the right to health, and Article 9 on guarding the landscape. An inquiry into this subject may seem superflous in comparison to a similar one concerning the legislative apects of water pollution control. As a matter of fact, the existence of special ordinary laws, such as No. 319 dated 10th May 1976, and No. 650 dated 24 December 1977, better known as Merli laws 1 and 2, seem to circumscribe the legislative aspects of both pollution and pollution control. Under the terms of these laws waste is acceptable if it occurs within the limits specified. Correlatively, wastewater treatment may be legally necessary to reduce water pollution to these limits. However important this outlook may be, i.e., to define the limits within which industries may operate, it cannot be considered exhaustive. The reasons are discussed in the paragraphs following.

First, even wastewater discharges are reduced to acceptable limits by treatment, this activity must be commensurate the other laws protecting natural resources and ecological values. An example is the eutrophication in the Adriatic which is causing algae blooms.

Second, it must be borne in mind that the current Merli laws are but a beginning in the creation of a legal framework for environmental control. These laws mandate effluent standards in 1984 based upon knowledge of the quality characteristics of various water bodies. But this is an exercise which requires experience and empirical adjustment of the tables which proscribe the expected standards. When these tables are developed finally, we must look still further ahead to the new needs in environmental protection, created by a constantly evolving society. An example of the expanded vision needed is a new law on soil conservation which is being examined by Parliament. The idea of on water reclamation and reuse, as outlined by the already quoted Medici Conference, is another example.

Third, the most important legislative point in the Merli laws is the following question. Can effluent discharges, within the limits fixed by the standards tables, be considered legitimate without reservation, or it it advisable to further evaluate their sufficiency on the basis of provisions and principles coming from other legal sources, mainly consititutional ones?

Because of these more encompassing considerations, e.g., the notion of the "environment," and its relationship to pollution control, the search for a comprehensive legal foundation begins to take shape. In fact, if relationships between certain constitutional laws gave the all clear to the creation of a constitutional notion of "environment," then this fact would prevent any ordinary law--such as the Merli law--from being able to legitimize environmentally harmful acitvities.

The above concerns are more than academic. They cannot be avoided because of two events which, in the future, may modify current law. These are: (1) the current European trend toward defining "environmental impacts" which, as in the United States, should make public sector activities subject to a compatibility check with the environment they are destined to influence, and (2) the new bill recently approved by the Italian Government which recognizes the legitimacy citizen groups wishing to protect their common interests. The latter allows citizens to take legal action to protect their common interests which because of their diffuse nature, do not belong exclusively to any single person. The case of "Italia Nostra" aims to control nature preservation, and is an example of this. These laws also address other kinds of concerns which already have been tried out in America, such as consumer protection.

In a recent study I showed how the constitutionally based laws protecting the rights to health and to the landscape do not mean that there is a more general legislative right which would permit any kind of activity, whether productive or not. While the arguments to establish this assertion are beyond the scope of this chapter, it can be said is that there is extreme uncertainty as to the contents and limits of the above notion. This makes it unacceptable since it could hinder severely the development of any productive activity. Its subjective character--at least while it is being worked out--prevents it from becoming a legislative limit to productive activities. Since the latter are constitutionally protected also, they need clear guidelines. This does not mean, however, that wastewater treatment is subject only to the limits of the "Merli law," and that the constitutional laws do not supply constraints conditioning the very legitimacy of the ordinary laws which fix pollution standards. Thus, it is out of the question for the right to health or landscape conservation, protected by the constitution, to become established as ordinary law. By allowing water pollution to go beyond acceptable limits this would, in effect, violate all constitutional percepts.

Seen this way, as outlined in the foregoing arguments, the problem is to define, from the constitutional point of view, the "acceptable limits" of water pollution control. The answers to this problem are on two very different and contrasting levels. On the one hand, the right to health is considered to be so highly protected by the Constitution that it prevails over any other kind of constitutional protection, including work or economic activity. On the other hand, it is thought that productive activity, a possible source of pollution involving the community as a whole, may call for the individual to sacrifice the right to health, but with compensation for damages.

Both hypotheses, which were outlined during the Bay of Naples pollution control controversy, are clearly untenable. The first relates to the conflict between a polluting industrial activity and the individual's right to health. It ignores the constitutional protection given as the freedom of economic activity (Constitution, Art. 41) and the closely related right to work (Art. 4). These could be seriously compromised if absolute and unconditional protection were given the right to health. The second solution, would end up depriving man of his right to physical integrity. No law can legitimately allow this, nor can it reduce health protection to monetary compensation for damages. The solution must be found within the criteria defining the limits for survival of constitutionally protected rights.

The need for several subjective rights to coexist and the consequent need to harmonize the various interests in potential conflict makes it necessary to fix operational limits. These are rights which apparently should be allowed unlimited protection. The right to health is an example. It is a right recognized historically, as one bounded by limits such that it is compatible with other individual requirements equally worthy of protection. A directive to defining compatibility limits between production and health goals can be seen in Art. 844 of the Civil law. This article provides for regulating noxious inputs such as smoke, heat, etc. from one estate to another. It rules that they become illegal once they exceed the limits of normal tolerance.

Since this directive reflects a profound need, there is nothing to prevent it from being called a general principal which is constitutionally important. Furthermore, Principle No. 6 of the 1972 United Nations Conference in Stockholm is in perfect accord with this notion. It is concerned with avoiding "serious and irreversible damage to the ecological systems." Here the term "irreversible" is equated with "intolerable;" hence irreversible damage is considered damage which is not normally tolerable.

By applying the above principle we can answer the intitial question of this inquiry and define the legitimate limit for the ordinary legislator concerning the right to health, and make it compatible with the protection of other constitutionally guaranteed rights. Hence it is possible to overcome the contrast in doctrine between those believing that respect for laws on pollution also encompasses protection from damage to the physical and mental health of an individual, and those who respect such ordinary standards with the view that measured inputs and pollution are to be tolerated. The latter viewpoint should be understood as qualified with a warning that respecting ordinary standards means that normal tolerance is assumed. Moreover, this does not prevent an adjudication from determining whether the limitations on the health right exclude compatibility between the other constitutionally guaranteed rights and therfore from determining whether any single inviolable right has been offended. In such case the ordinary standard must be considered constitutionally illegal.

The adoption of the aforesaid criterion allows contrasting interests to coexist legally. Also productive activity cannot be considered illegal even when producing psycho-physical states of unease or intolerance, if these are normally tolerable. Such externalities must be considered an inevitable result of a productive organization in an industrial context; inevitably such activity is going to cause some harm to natural resources. In other words, on the one hand, this criterion blocks environmental extremism and allows the socio-economic development of a nation. On the other hand, it prevents the wanton destruction of the natural environment in the name of economic productivity. It requires that technological development must be well conceived.

The survival of the human race and maintaining the integrity of the environment for future generations remain as inseperable limits for any kind of development for some, or the right to work for others, to be constitutionally guaranteed if there is irreversible damage to our natural resources in fulfilling the basic needs of the human race. Among these needs we must include also the aesthetic and cultural needs. Because of Art. 9 of the Italian Constitution, culture is no longer the poor relative in the scale of rights to be protected. It has its own place within the order of the great moral and spiritual values inspiring the constitutional view of comunitary life. In fact it has risen to the dignity of a fundamental principle. In other words, this constitutional provision gives "culture" constitutional importance. In placing "culture" amongst the fundamental principles, not only does it have the political meaning of classifying the Italian Republic as a "Cultured State," but also the legislative one of showing the kind of guarantee given it in comparison to the other constitutionally guaranteed rights.

The legislative criteria limiting pollution, along with its corollary defining the legitimacy of pollution control, must also be compared to the protection given the right of "culture" described by the legislative notion "landscape." Thus, the principle to be applied can be only that of "normal tolerance" between production requirements and the requirement not to alter the external appearance of a territory when it is a legislatively denominated "landscape." The criteria by which territory becomes landscape has been discussed in a previous article, i.e., Torregrossa, "An outline of environmental protection" in Riv. Trim. di dir. proc. civ.; 1980, p. 148. All that need be said here is that not every alteration of the environment must be considered illegal. The illegality pertains only those which alter and damage the cultural heritage of the community, as defined by the prevailing norms. Therefore it will be aesthetic-cultural protecion, rather than hygenic motivations, which will prevent say the construction of an industrial plant whose waste matter could cause alterations in the flora and fauna of a nearby lake, changing it from its integration with the surrounding area.

19. ENVIRONMENTAL IMPACTS OF WATER PROJECTS

Asit K. Biswas
Biswas & Associates, Oxford, England

1. INTRODUCTION

Water has always been considered one of the essential elements to sustain animal and plant life. This is indicated by the writings of early Greek philosophers like Plato and Aristotle who believed water to be one of the five "elements" which constituted every item on earth. Some two thousand years later, Leonardo di Vinci considered water as the driver of nature.

The role of water in overall development of countries became an increasingly important issue during the decade of the 1970's due to several events, some natural and others man-made. First, severe drought in many parts of the world during the early seventies contributed to a major food crisis. During the World Food Conference, convened in 1974 to propose solutions to such a global crisis, it became evident that proper water control is essential not only for further agricultural expansion but also for increasing the overall yield. Second, steadily increasing prices of fossil fuels, especially oil, focussed national attention on development of hydroelectric power. Third, the Lima declaration recommended that by the year 2000, 25 percent of global industrial production should be in developing countries. This means more water will be required to sustain industrial development. Fourth, the UN Conferences on Human Settlements (4) and on Water (6) emphasized the plight of developing countries who do not have access to safe drinking water. At the recommendation of the Water Conference (3), the decade 1981-90 has recently been declared to be International Water Supply and Sanitation Decade by the General Assembly of the United Nations. Finally, pollution of inland and coastal water bodies and the oceans became an increasing national

and international concern, partly through the work of the United
Nations Environment program, which was created in the early
seventies. All these events, individually and cumulatively,
contributed to the realization of the urgent necessity of
rationally managing both the quantity and quality of available
water resources.

2. MAN'S IMPACTS ON WATER RESOURCES

Viewed broadly, man's impacts on water resources can be
classified under the following three categories:

1. environmental impacts due to construction of water
   development projects, i. e., dams and reservoirs,

2. impacts on water quality due to human activities,
   especially discharge of waste materials, and

3. impacts on "mining" of groundwater, where rate of
   withdrawal substantially exceeds the rate of
   replenishment.

Since environmental impacts of large scale water development
projects is the main topic of this chapter, the other two
categories of impacts will be discussed only briefly.

Increasing population and standard of living and inadequate
environmental protection measures have contributed to deterioration of quality of surface and groundwater as well as of the
oceans. So far as inland water bodies and aquifers are concerned,
water pollution unfortunately has become a universal phenomenon.
Even in a developed country like the United States, which
undoubtedly has spent more funds on water quality control, the
1980 annual report of the Council on Environmental Quality (8)
stated that there has been no change nationally in water quality
of rivers and streams during the period 1975 to 1979, if six key
pollution indicators are considered: fecal coliform bacteria,
dissolved oxygen and total phosphorus, cadmium, lead and mercury.
For four of the six indicators; fecal coliform bacteria, total
phosphorus, lead and mercury, violations of EPA criteria for
swimming or preservation of aquatic life remain frequent and
widespread, according to the report. Similarly the 1978 State
Water Quality Reports submitted to the U.S. Congress under the
Clean Water Act indicated that 35 of the 37 states reporting had
problems of pollution by toxic substances (8). In addition,
pollution of rivers and streams by pesticides, hydrocarbon
products, solvents, and acid residues continues to remain a
serious problem. Even in the United States, the lack of reliable
data makes if impossible to estimate the magnitude of the problem

or even whether conditions are improving or deteriorating. The situation is much worse in other countries.

Extensive use of groundwater in many countries of the world, primarily for irrigation purposes, has already contributed to development of many problems. Over exploitation of groundwater results not only in the lowering of the water table but also could contribute to decreased pressure in aquifers, changes in rate and direction of flow, salt water intrusion and land subsidence. Continued over-exploitation could lower the water table to such an extent that it no longer remains economic to pump the water for irrigation. Thus, agricultural developments in such instances are only temporary phenomena, since production declines significantly once the water availability is reduced. This situation has already occurred in many parts of the world. For example, recent data from groundwater monitoring in the province of Tamil Nadu in India indicate that at least 37 observation wells had more than a net fall of more than 6 meters during the period of only six years, between January 1973 and January 1979. One recorded a net fall of as high as 16.40 meters, which means an average lowering of 22.77 cms per month, a rather high figure viewed from any direction (10).

3. ENVIRONMENTAL IMPACTS OF LARGE DAMS

All development projects have social, economic and environmental impacts, some of which are positive and others adverse. The main reason for implementing any development project is that the sum total of the positive benefits outweigh the adverse ones. Large water development projects are no exception.

The social and environmental consequences of water development are many, and the resulting effects often extend much further than the planning area itself. The interaction of diverse forces are often so complex that environmentalists are hard pressed to predict overall effects with any degree of certainty. For example, the present knowledge of ecosystems of man-made lakes leaves much to be desired, and unless planning precedes construction by five to ten years, several unpredictable and unforeseen situations could occur. At the current state-of-the-art, environmentalists often find it impossible to convince engineers, economists and politicians that certain developments are unwise, or of the necessity to spend scarce resources on appropriate remedial measures because of lack of hard facts or solid scientific evidence. In addition, water development projects traditionally have been dominated by engineers and economists, and consequently social and environmental condsiderations have often been sadly neglected during the planning process. In certain cases, social and environmental scientists have been brought in only after large-scale damages have already occurred. Hence, even though much lip service is given to interdisciplinary teamwork, it is still an exception rather than the rule. Since the environmental impacts of large water development projects are many, these will be discussed under three broad categories: physical, biological and human.

4. PHYSICAL SUBSYSTEM

Water development projects invariably change river and ecosystem regimes, and thus the real question is not whether such developments will affect the environment, but how much change is acceptable to society as a whole, and what countermeasures should be taken to keep the adverse changes to a minimum, at a reasonable economic cost. One of the largest dams of the world, the Aswan in Egypt, has been criticized for contributing to major environmental disruptions. The project, completed in 1968, was built primarily for generating hydropower, and to expand perennial irrigation. A detailed objective analysis of the benefits and the costs of the Aswan Dam has yet to be made, but some of these effects can now be perceived.

First is the question of silt. Before the dam was constructed, large amounts of silt used to be either deposited in the Nile Valley or carried all the way to the delta and the sea. These sediments are now being trapped in the reservoir, Lake Nasser, created by the dam. Before the dam was built, suspended matter in the River Nile, passing the Aswan, ranged between 100 to 150 million tons per year. Observations made during the first few years after the completion of the dam indicate that the reservoir is losing about 60 million cubic meters of storage per year due to siltation. At this rate, the dead storage capacity of 30 km^3 will be filled in about 500 years. As a result of the siltation in the reservoir, clean water is flowing downstream, causing erosion to the river bed and banks. One possibility now being considered is to construct a number of barrages to reduce the velocity and force of the clear water.

Another effect of the siltation in the reservoir is erosion of the Nile Delta, some 100 km away. Prior to the construction of the dam, the Delta used to be built up during the flood season, due to the silt carried by the river. This situation in the Delta compensated for the erosion that resulted from the winter waves of the preceding year. Without enough siltation, erosion of the Delta has become a major problem and studies are now being carried out to find a suitable solution.

Loss of silt has further affected the productive capacity of the Nile Valley which used to get regular deposits of sediments every year. Currently studies are being undertaken to assess the actual nutritive value of the silt, and the trace elements present therein, so that this loss can be compensated by using chemical fertilizers.

Lack of sediments downstream of the dam has contributed to the significant reduction of planktons and organic carbons. It has, in turn, reduced the sardine, scrombroid and crustacean population of the area. Loss of sardine along the Eastern Mediterranean has created economic problems for the fishermen who used to depend on the catch for their livelihood. This, however, is more that counterbalanced by the development of new fisheries in the Lake Nasser.

Lack of silt has also created another environmental problem. Prior to the construction of the dam, there was a thriving small-scale industry making brick from the silt dredged from the canals. In the absence of such silts, many such industries have now resorted to using the topsoil near the canals to make bricks, thus contributing further to the loss of productive soil in the country. On the positive side, however, lack of silt has reduced the cost of dredging of channels.

Besides this situation, other environmental problems created by the Aswan Dam that could be included within the physical sub-system, are change of terrestrial system to aquatic system, hydro-meteorological effects, and changes in soil and water quality. The High Dam created a vast reservoir, having a shoreline length of 9,250 km, surfact area of 6,216 km^2 and volume of 156.6 km^3 at 180 meters elevation. It changed 500 km of the River Nile from a riverine to lacustrine system. Though much of the land inundated was thinly populated, it contained areas rich in historical monuments, formost of which was the Abu Simbel temple. Thus, the temples of Abu Simbel and Philae (near Aswan) had to be dismantled and moved to higher location. The huge man-made reservoir also changed the microclimate of the area. It was calculated that the raising of the water level by 20 km, from 160 m to 180 m, more than doubled the lake surface from 2,950 km^2 to 6,118 km^2, which increased the total annual evaopration from 6 km^3 to 10 km^3 (9).

The construction of the High Dam and the canal system for irrigation has tended to increase the water table in many parts of Egypt. Such developments and the tendency to over irrigate is contributing to an increase in soil salinisation problem, requiring expensive and extensive construction of drainage systems. With the disappearance of the annual Nile floods, the

goundwater level has stabilized at a higher level. The salinity
in the irrigation canals is increasing and some of the reclaimed
lands are already facing a salinisation problem.

The discussion of the above environmental effects of the
Aswan High Dam is not meant to be a total condemnation of the
structure, nor does it imply that it should never have been built.
The benefits of the dam are many, and like evaluation of any other
project, the benefits and costs should be evaluated and compared.
It should be noted that the per capita cultivated land in Egypt
has been reduced from 0.41 acres in 1930 to 0.18 acres in 1975,
and to an even lesser figure at present. Without the Aswan, the
situation would certainly have been much worse. The total
construction cost for the project, including subsidiary projects
and extension of electric power lines, was E£450 million (1).
This was more than paid for within only two years of completions,
as the annual return was E£255 million, e. g., E£140 million from
agricultural production, E£100 from hydroelectric generation, E£10
million from flood protection and E£5 million from improved
navigation. Thus, the real question is not whether the Aswan Dam
should have been built, but rather what steps should have been
taken to reduce the environmental impacts to a minumum.

There are cases where water developemnt projects to increase
irrigated agriculture have also contributed to problems which
eventually reduced the total food production. Amony such problems
are deterioration of soil fertility and eventual loss of good
arable land, due to progressive development of salinity or
alkalinity. For example, at one time Pakistan alone was losing
24,280 hectares of fertile cropland every year, and currently
nearly 10 percent of the total Peruvian agricultural area is
affected by land degration due to salinisation. Among other major
areas affected by salinisation are the Helmud Valley in
Afghanistan, the Punjab and Indus Valleys in the Indian sub -
continent, Mexicali Valley in Northern Mexico and the Euphrates
and Tigris basins in Syria and Iraq (5).

5. BIOLOGICAL SUBSYSTEM

One of the most serious impacts of irrigation developments in
the tropical and semitropical regions is the secondary effect of
spreading water borne diseases, and the consequent suffering of
millions of human beings and animals. Irrigation schemes often
have enhanced or created favorable ecological environments for
parasitic and water borne diseases such as schistosomiasis, dengue
plus dengue haemorrhagic fever, liver fluke infections,
bancroftian filariasis and malaria.

Schistosomiasis is currently endemic in over 70 countries, and affects over 200 million people. Prior to the development of the present extensive irrigation networks, and when agriculture depended primarily on seasonal rainfall, the relationship between snail host, schistosome parasite and human host was somewhat stabilized, and infection rates were low. Snail populations increased during the rainy season, when agriculture was possible, which provided the contact between man and parasites. During dry periods, however, there was a lull in infection. With the stabilization of water resource systems through the development of reservoirs and perennial irrigation schemes, the habitats for snails were vastly extended, and they also had a prolonged breeding phase which substantially increased their population.

The relationship between water developments and increase in schistosomiasis has been demonstrated in several countries of the world. In Egypt, the replacement of simple primitive irrigation with perennial irrigation has caused a high incidence of both S. mansoni and S. heamatobium. Where basin irrigation is still practiced, the incidence is much less. Infection rates in four selected areas witin three years of introduction of perennial irrigation, rose from 10 to 44 percent, 7 to 50 percent, 11 to 64 percent and 2 to 75 percent. The life expectancy of males and females in heavily infected areas is estimated to be 27 to 25 years respectively. In Sudan, with the introduction of perennial irrigation to 900,000 acres under the Gezira Scheme, the incidence of blood fluke rose greatly. It also increased the incidence of flukes in cattle and sheep. In Kenya, the Lake Victoria is hyperendemic for schistosomiasis. S. mansoni infection in school children is up to 100 percent in areas associated with irrigation schemes. In Transvaal, South Africa, the S. mansoni infection rate in European farms was 68.5 percent compared with only 33.5 percent in the reserves, because the former had irrigation (5).

Constant availability of large quantity of water in reservoirs and canals is also conducive to the breeding of mosquitoes, which act as intermediate host for diseases like malaria, bancroftian filariasis, yellow fever or arbovirus encephalitides. Currently, it is estimated that over 200 million people are affected with bancroftian filariasis (UNEP, 1977). Similarly, plant growths around water bodies provide a suitable habitat for the tsetse fly to transmit trypanosomiasis to human beings and domestic animals.

In constrast to the diseases discussed above, water developments tend to reduce the incidence of onchocerciasis. The intermediate host, simulum fly, tends to breed in fast flowing water, which are often drowned by the construction of dams. Thus, the construction of the Volta Dam destroyed the breeding ground of simulum fly that existed upstream. However, adequate measures should be taken to ensure that new breeding places do not develop, especially in the fast flowing waters near spillways.

6. HUMAN SUBSYSTEM

The impacts of water developments on human subsystem could be direct or indirect, stemming from direct effects on physical and biological subsystems. These impacts can either be beneficial or adverse.

Many of the major water development projects have created other human problems, especially in terms of dispalcement of local inhabitants. Thus, the Volta Dam in Ghana has inundated an area of about 3,275 sq miles, and the resulting lake has a shoreline of over 4,000 miles. As a result of the development, some 78,000 people and more than 170,000 domestic animals had to be evacuated from over 700 towns and villages of different sizes. Eventually, 52 new settlements were developed to house 69,149 people from 12,789 families. It was a major social problem since a large number of people coming from small villages (600 of the 700 original villages has less than 100 people, and only one has a population of over 4,000), and having different ethnic back-grounds, languages, traditions, religions, social values and cultures, has to be resettled into only 52 locations. The complex emotional relationships between the different tribes and their lands were not properly understood. The development of a socially cohesive and integrated community, having a viable institution infrastructure became hard to achieve.

Similarly, the Kariba Dam on the Zambesi, between Zambia and Rhodesia, displaced approximately 57,000 Tonga tribesmen, who had to pay a major price for this progress. Technology transfer at that level was a major problem. Since many of the planners were from outside Africa. The resettlement program for the Tonga tribesmen left much to be desired; not only did they suffer great cultural shocks when being thrust into communities as different from their own, as theirs from Great Britain, but also it took two years to clear sufficient land to meet their subsistence needs. The government had to step in to avert famine and very serious hardships and, ironically, this good intentioned step became one of the most destructive parts of the process. The food distri-bution centers also became transmission sites for the dreaded sleeping sickness disease.

Results from water development projects have been similar. Approximately 100,000 people had to be relocated for the Aswan High Dam without sufficient planning, and the World Food Program had to rush in famine relief for the Nubians. Other examples of lakes and populations displaced are the following: Lake Kainji in Nigeria - 42,000; Keban Dam in Turkey - 30,000; and Ubolratana Dam in Thailand -30,000 (5).

Resettlement of population due to water development projects in many developing countries has not been a satisfactory experience. Inadequate planning, insufficient budget, incomplete execution of plans and little appreciation of the problems of technology transfer have all contributed to the failure of plans. The fact that much of the population to be resettled were rural and illiterate, and thus had very little political power, did not help either. The direct beneficiaries of the projects were often the educated elites, who are in power, whereas the direct social costs were borne mostly by the rural poor.

7. CONCLUSION

There is no doubt that the primary effects of the vast majority of water development projects around the world have been beneficial. Equally, however, there is no doubt that many of these development projects have contributed to unanticipated adverse secondary effects. Some of these could have been eliminated and others reduced in magnitude by remedial actions. While very few people would argue with the necessity of implementing further water development projects, especially in developing countries, for reasons discussed earlier, it is imperative that development process be sustainable over a long term (2, 3). Environmentally inappropriate policies often eventually become self defeating. It is important that we learn from our errors from past development projects and the mistakes are not repeated again.

320

REFERENCES

1. Abul-Ata, A. A., "After the Aswan," Mazingira, Vol. 3, No. 3, 28-34, 1977.

2. Biswas, A. K., "Sustainable development," Mazingira, Vol. 4, No. 1, 4-13, 1980.

3. Biswas, A. K., "Environment and water development in the third world," Journal of Water Resources Planning and Management Division, American Society of Civil Engineers, Vol. 106, No. WRI, 319-332, 1980a.

4. Biswas, A. K., United Nations Water Conference: Summary and Main Documents, Pergamon Press, Oxford 217, p., 1978.

5. Biswas, M. R., "Environment and food production," in Food, Climate and Man, Margaret A. Biswas and Asit K Biswas (Editors), John Wiley & Sons, New York, 125-158, 1979.

6. Biswas, M. R., "Habitat in retrospect," International Journal for Environmental Studies, Vol. 11, 276-279, 1977.

7. Biswas, M. R., "United nations water conference: a perspective," Water Supply and Management, Vol. 1, No. 3, 255-272, 1977.

8. Council on Environmental Qaulity, "Environmental quality - 1980," Council on Environmental Quality, Washington, D.C., p. 497, 1980.

9. Hafez, M. and Shenouda, W. K., "The environmental impacts of the Aswan High Dam," in Water Development and Management: Proceeings of the United Nations Water Conference," Part 4, 1777-1785, 1978.

10. Srinivasen, C. A., "Groundwater monitoring in Tamil Nadu," Irrigation and Power, Vol. 36, No. 4, 435-466, 1979.

20. COST SHARING IN MULTI-USER WATER PROJECTS

A. G. Conybeare Williams
South West Water Authority, England

1. INTRODUCTION

The chapter addresses issues related to the allocation of costs in a multi-user water project. Those who have been concerned with the promotion, the construction and the operating of a scheme which benefits or serves more than one user will be familiar with the problems that can arise: difficulties which occur in the allocation of costs as between the beneficiaries of, for example, a water resource project which is designed to serve two or more different users. Before, however, I examine in detail actual cases I propose to review, in brief, some of the main principles that seem generally to be available.

2. PRINCIPLES OF ALLOCATION

A simple method of allocation is, of course, to divide costs pro rata according to the amount of capacity reserved for each user. This method may be refined by dividing costs according to the quantities of water to be supplied to the different users. In this case, both the projected cash flows and the projected quantities of water supplied would be discounted at an appropriate rate. The discounted unit costs derived in his way can be used as a basis for charging the users.

As R. J. White (1) points out, however, there are common situations in which these methods lead to unsatisfactory results. He argues, "A marginal use of water from a project may be economic on the basis of the marginal costs which justify its inclusion in the project, but it may become uneconomic if charged a full share of the capacity costs or the discounted unit costs. In such cases an equitable charge to the marginal user would lie somewhere between the marginal costs and the figure derived from capacity costs or discounted unit costs. The problem which then arises is how to determine an equitable level of charging."

A particular difficulty in allocation of costs can arise where a project is designed for more than one purpose. An example could be the inclusion of both hydroelectric power generation and the provision of water supply. Again the application of capacity costs or discounted unit costs is not appropriate because a cubic meter of water at the top of the reservoir when the level is high has a greater value for power than when the resevoir is drawn down. But for water supply purposes the value of a cubic meter of water is generally independent of the reservoir level at any particular time. Admittedly there can be exceptions to this latter point as when the level in the reservoir is so low that the remaining water is not suitable for quality or chemical reasons for supply.

A way in which these difficulties can, in principle, be overcome is by the "separable costs - remaining benefits" method. This method, as R. J. White demonstrates, distributes costs equitable in a multi-user project. The method provides for:

1. assigning to each user its separable cost, i.e., the added cost of including the user in the project,

2. assigning to each user a share of the residual or remaining costs in proportion to the remaining benefits, i.e., the benefits less the separable costs.

The method can achieve an equitable sharing of the savings from multi-user development amoung the various users.

3. MULTIPURPOSE AS OPPOSED TO MULTI-USER

The inclusion of hydroelectric power generation in a reservoir project raises several issues. Consider a case where a reservoir scheme is to be built and where hydropower facilities could be incorporated. The Electricity Undertaking could bear the direct costs of its own specific installation and pay to the Water Undertaing the cost of the additional works. The integration of the generation project in the reservoir scheme might well be attractive to the Water Undertaking as, by a charging arrangement with the Electricity Board to reflect the value in productivity terms of water used in generation, it would receive a source of income without additional outlay. Equally, the Electricity Board could benefit in that, despite bearing the costs of its own added facilities in the reservoir scheme, it gains by the provision of power more cheaply than otherwise would be the case. The Water Undertaking might, however, seek a weighted capital contribution from the Electricity Board. The community benefit argument does not apply because water is provided (and this is substantially the case in England and Wales) on a different geographical basis from electricity.

One cannot wholly separate the economic arguments from the practicalities of situations. More and more power these days is and will be generated by nuclear means. Apart from this the needs of generating boards are for regular levels of output of electricity, and this situation is more likely to prevail with a direct supply reservoir. River regulating reservoirs are less likely to be of value in this context. But it is interesting to consider how the views of the Electricity Board might change with respect to these "limited" summer supplies from regulating reservoirs as the costs of generation by fossil fuels increase in an inflationary situation.

4. IS THERE A "CORRECT" METHOD?

Concerning the question of whether there is a correct method of cost allocation for the case where two water authorities have interest in the joint resource, David Walker (2), has suggested that one could proceed to devise a formula based upon the following elements:

1. The maximum contribution from either of the two undertakers should not exceed, in present value terms, the cost of either to meet his own separate needs by an alternative scheme.

2. The minimum contribution from either of the two undertakers should equal the savings which would accrue if either were left out of or not provided for by the joint scheme.

3. Between the two limits there is perhaps no "correct" way to allocate costs, albeit lawyers, accountants and engineers have over the years devised various arbitrary rules for differing situations.

The real problem is to arrive at fair estimates of the actual costs involved and this difficulty will bring into question the validity of the applied formula. The "separable costs -remaining benefits" method of the foregoing analysis has been established practice in the United States since it is discussed in Chapter IV of the Geeen Book in 1958 (3).

Before discussing principles of cost sharing in a multi-user project, it is logical to consider the most fundamental situation: i.e., where one Water Undertaking embarks on a long-term scheme of its own. In such a situation the Undertaking places a burden on its present consumers, which is only in part to secure the future supplies to those same consumers. The rest of the cost which the present consumers are called upon to bear is to secure the supplies for consumers who do not even exist at the time of the achievement of the new Scheme. This circumstance which is commonplace, has been summarized by S. W. Hill (4), one of the leading authorities on public finance in the United Kingdom, whose views are given in the following paragraphs.

When new consumers arrive increasing water demand, they usually take from the system at the then prevailing rate of charge. The new consumers are not called upon to "pay back," through increased contributions the investment which their foresighted predecessors made partly for their benefit. The new consumers share costs only from their advent. This is a fundamental principle of public utiltiy finance, which has worked well and is generally acceptable.

When two or more separate undertakings establish a joint scheme they are in effect each staking their claim to a quantitiy of water in exactly the same way as if each had embarked on its own scheme. Pursuing this point Hill (4) has suggested that if any participating Undertaking can be assured:

1.   that the water which it has agreed to pay for will be available when needed, and

2.   it is an economic scheme relative to any alternatives available (considering quantity, cost and time),

there is really no reason why that Undertaking should be concerned about the proportion of the cost that it pays. The Undertaking does, of course, need to be assured that it will not be called upon to pay more than it has bargained for in event the other participating Water Undertakers seek to take more or less water than agreed upon.

5. EXPERIENCE PRIOR TO WATER REORGANIZATION

As might be expected there have been, over the years, many multi-user and some multipurpose schemes in Great Britain. Thus a variety of projects could demonstrate principles of rational and fair cost sharing arrangements.

Prior to 1974, water services were the responsibility of many statutory bodies, identified below:

1. Twenty-nine river basin authorities (5) were charged with the duties of water resource planning and conservation, land drainage and flood alleviation, river water quality and pollution prevention and fisheries.

2. One hundred and fifty-seven water boards (6) undertook the supply of treated water to domestic, industrial and commercial users.

3. Some 1400 local authorities (7), (councils of county boroughs, boroughs, urban and rural districts) undertook the functions of sewerage and sewage disposal.

As a result of legislation (8) all these statutory bodies were abolished and the services and functions mentioned above were vested in ten large regional water authorities. The entire managed portion of the water cycle - from the raindrop to the tap to the drain to the sea - became the responsibility of one multi-purpose authority in each region of England and Wales. This regional approach to water management is absolutely essential from many points of view, but especially in terms of the financial arrangements for multi-user water schemes.

Before the 1974 reorganization, major reservoir or sewage disposal schemes often were promoted by consortiums of under-taking. The more participants, the more complex the financial arrangements.

6. FINANCING A WATER RESOURCES SCHEME — THE REGIONAL CONCEPT

The 1963 Water Resources Act created twenty-nine river authorities in England and Wales and had invested them with the new function of the management and conservation of water as a resource on a river basin concept. The act introduced a system which did not permit withdrawal of water without a license from the relevant river authority. The river authority was empowered to levy a charge based upon volume of water abstracted. Water Supply Undertaking who were licensed by the river authority to abstract water from reservoirs, rivers and aquifers were required to pay charges to the river authority. The basis of charge was volumetric and the rate of charge per 1000 gallons was fixed annually to reflect the financing requirements of the river authoritiy's "water resource account" which had to be balanced, expenditure against income, taking one year with another.

The 1963 Water Resources Act envisaged that new major water resource projects would be undertaken by a river authority on broad strategic hydrological grounds with reference to the general needs and water demands of the area. This contrasts with the earlier practice whereby each Undertaking (and some were small in terms of area, population and supply needs) constructed new reservoirs to meet their own individual requirements. By contrast, the river authority, after 1963, used the water resources account to meet the annual costs of water resource schemes. These expenditures would be met by adjusting the volumetric charges for water taken from all sources by all abstractors according to the terms of the licenses granted to them by the river authority.

7. THE WIMBLEBALL RESERVOIR SCHEME

In 1972 the Devon River Authority, of which I was then the chief executive, promoted the Wimbleball Reservoir Scheme on Exmoor in Somerset. The Scheme was designed to have a storage capacity of well over 4000 million gallons and to provide an estimated daily yield of 20 million gallons. The Wimbleball Scheme was devised to augment the supplies of three water supply boards. Figure 1 is a map showing the water transfers involved. They are summarized as follows:

1. The West Somerset Water Bord required 7.0 million gallons per day and proposed to take this by direct abstraction from the reservoir.

2. The North Devon Water Board required 2.5 million gallons per day and proposed to abstract this downstream of the reservoir at Tiverton.

3. The East Devon Water Board required 5.3 million gallons per day and proposed to abstract this downstream of the reservior at Exeter.

There remained some 5 million gallons a day which the Devon River Authority decided to hold in reserve to meet future needs. The basis for the division of the yield of 20 million gallons per day as between the four statutory bodies was comparatively simple. The difficulties arose when the cost sharing arrangements were explored.

The first difficulty arose when it was considered, following the concept of the Water Resources Act 1963, that the whole of the costs of the Scheme should be "spread" across all the water abstractors in the area of the Devon River Authority through the operation of the charging scheme. The problem arose because, within the area of the Devon River Authority, there were also two

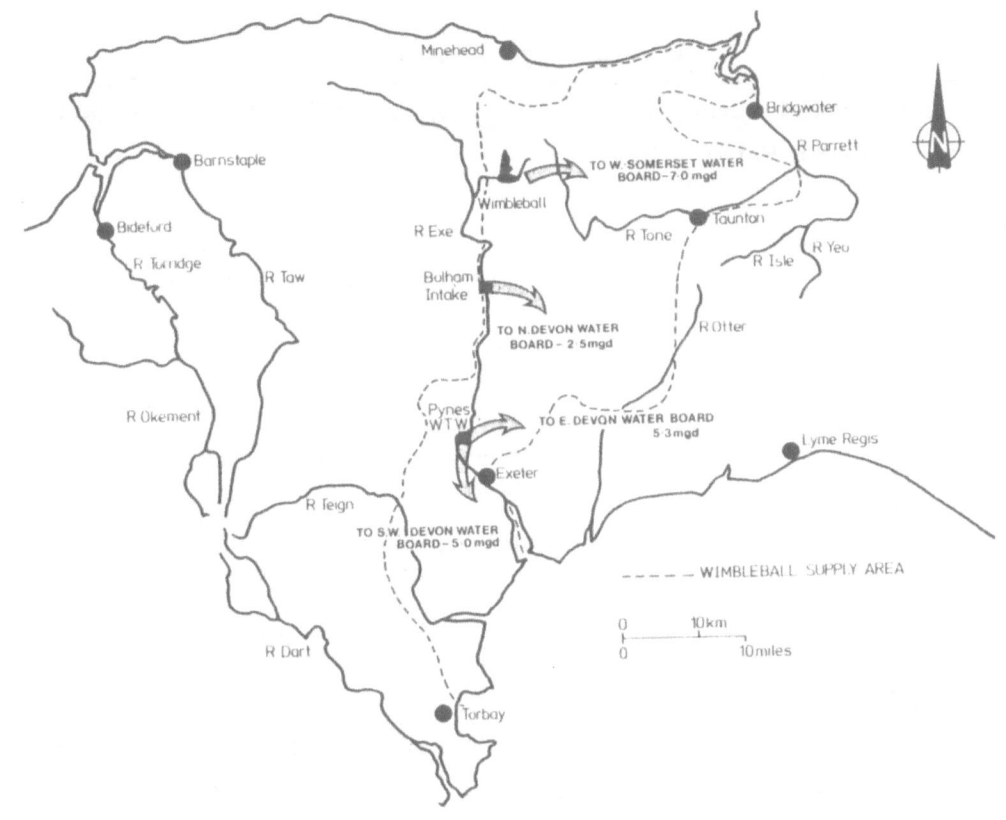

Figure 1.   Wimbleball Reservoir Scheme

other undertakings, i. e., the South West Devon Water Board and the Plymouth City Water Undertaking. Neither of these undertakings were thought to have any interest in the scheme and it was considered not fair that a substantial amount of the costs should fall upon them as they would derive no benefit from the new resource.

The next basis was that the costs of the scheme should be apportioned according to the benefit of "take" from the reservoir. On this basis the apportionment of cost would have been as given in Table 1.

It was argued that there was a defect in this approach and that the equity of an apportionment of cost on this basis is highly suspect since the required storage capacity is not directly proportional to "take" unless all the abstractions come direct from the reservoir or all come from the lower reaches of a regulated river. In the case   under   consideration   the West

Table 1. Cost Apportionment Among Water Boards
According to Take

Statutory Authority	Take (m.g.d)	Percent of Cost
West Somerset Water Board	7.0	35.35
North Devon Water Board	2.5	12.63
East Devon Water Board	5.3	26.77
Devon River Authority	5.0	25.25
	19.8	100.00

Somerset Water Board would be taking direct from the reservoir and the two other Water Boards from downstream so that the storage capacity used for each was disproportionate. For example the amount of water released during dry periods to enable the East Devon Water Board to take 5.3 m.g.d. at Exeter downstream could be of the order of 2 m.g.d. and storage for this might be for say 60 days. West Somerset Water Board, however, would be taking direct from the reservoir and would require, to support a daily abstraction of 7 million gallons, a storage capacity for, perhaps, six months. They should, on this basis, be bearing a much higher percentage of the cost. Table 2 which follows demonstrates the cost sharing on the basis of storage assuming the following factors for "reservation" in the reservoir:

1. Direct Supply   300/1 m.g.d. supply

2. Regulation   150/1 m.g.d. supply

The reservoir storage would be as follows:

Table 2. Cost Sharing on the Basis of Storage

Statutory Authority	Storage Reservation Million Gallons	Percent Reservation
West Somerset	7.0 x 300 = 2100	52.2 %
North Devon	2.5 x 150 = 375	9.3 %
East Devon	5.3 x 150 = 795	19.9 %
Devon River Authority	5.0 x 150 = 750	18.6 %
Total	4020	100.00%

The changes in cost apportionment on the alternative basis would be as follows as shown in Table 3.

In discussing a scheme, e.g., Wimbleball which is a multi-million pound project, one can appreciate the significance of the shift in cost sharing from one undertaker to another of just a few percent. The situation outlined above produced much argument between the four participants, and particularly between the three water supply boards who had to consider the consequential effects of such changes of incidence upon their respective consumers.

Table 3. Differences in Cost Apportionment between "Take" and "Storage" Methods

West Somerset	52.2% - 35.35% = + 16.85%
North Devon	9.3% - 12.63% = - 3.33%
East Devon	19.9% - 26.77% = - 6.87%
Devon River Authority (reserve)	18.6% - 25.30% = - 6.65%

There was another complicating factor under the legal framework which then existed. I have already referred to the licensing system under the Water Resources Act 1963 and to the water resources account to which were credited all charges recovered from licensed water users in proportion to their "permitted abstraction and use of water from the resources of the area." The reservoir costs associated with the 5 m.g.d. reserve of Devon River Authority would become a debit against the water resources account. The water resources account deficit had to be met by annual charges per 1000 gallons of water used, levied by the Devon River Authority upon all licensed abstractors whether public bodies or private individuals. I have already referred to the position of the South West Devon Water Board and the Plymouth City Water Undertaking. Although neither of these Boards had an immediate interest in the Wimbleball scheme, a proportion of the cost of providing the 5 m.g.d. Devon River Authority reserve would, by virtue of the charging scheme, nevertheless fall upon them, and therefore upon their consumers. A proportion would fall upon other licensed abstractors also, such as industry and agriculture. Furthermore, some elements of the reservoir cost for the 5 m.g.d. would be reflected in the higher charges paid by the three participating Water Boards, who had a direct interest in the scheme, for all other water (quite apart from the Wimbleball Reservoir Scheme) which they were licensed to take. Tables 4 and 5 illustrate the point.

When the reservoir costs for the 5 m.g.d. reserve were taken into account the three participating water boards as well as the nonparticiapting abstractors would on the basis of Table 4 share as seen in Table 5.

These complicated financial arrangements took many months of argument and negotiation to settle. After all this and at a late stage, literally on the eve of the promotion of the scheme by the Devon River Authority, the South West Devon Water Board maintained that the 5 m.g.d. unallocated reserve should be licensed to them and that they should abstract this into the Torbay area. As a result of this charge the bases of cost sharing explained above

Table 4. Pre-Wimbleball Share of Payments to River Authorities' Charging Scheme

Abstractor Under Charging Scheme	Percent
West Somerset Water Board	1.88
North Devon Water Board	19.69
East Devon Water Board	27.52
Southwest Devon Water Board	29.85
Plymouth City Water Undertaking	12.52
Industrialists, Agriculture and other Abstractors	8.54
	100.00

Table 5. Cost Sharing for Direct Abstractions and Reserve

	Percentage Share Direct[1]	Percentage Share for Reserve[2]	Total %
West Somerset	35.35	0.48	35.83
North Devon	12.63	4.97	17.60
East Devon	26.77	6.95	33.72
Southwest Devon		7.54	7.54
Plymouth City Water Undertaking		3.16	3.16
Industrialists, Agricultural and other water abstractors		2.15	2.15
	74.75	25.25	100.00

[1] Direct share of cost Wimblegall Reservoir, from Table 1.

[2] Share of unforseen demands from Wimbleball Reservoir (5 m.g.d)

had to be revised and the river authority and the four participating water boards, i.e., West Somerset, North Devon, East Devon and South West Devon, had to negotiate a fresh basis of cost allocation.

8. PROBLEMS OF PROMOTION OF MAJOR SCHEME

Those who have been involved in the promotion of a major public sector scheme will be well aware of the hazards which have to be faced. In England and Wales any such scheme has to be promoted in accordance with strictly defined procedures. Account must be taken of defined criteria and the need for the particular scheme has to be open to the most stringent test at public inquiry before government inspectors, who will report to the Secretary of State. Those who object to the scheme will be entitled to appear at the public inquiry and to object. All reasonable and necessary safeguards are built in to ensure that public money and scarce resources are properly used.

These in themselves are and were demanding issues which had to be met in the promotion of the Wimbleball Reservoir Scheme. When one considers the complex aspects of cost sharing, it will be appreciated that prior to 1974 there were difficult and conflicting aspects for the promoting river authority to reconcile with the water boards, who were to be supplied from the new resource, before one could reasonable expect to mount a successful promotion.

9. PROCEDURES IN THE NEW WATER INDUSTRY

The 1973 Water Act setting up ten regional water authorities has streamlined the organization and management of the water services in England and Wales. As but one example it has greatly simplified cost sharing in schemes such as the Wimbleball Reservoir.

Under the 1973 Water Act the water functions in Cornwall, Devon, part of Somerset and an area in Dorset are vested in the South West Regional Water Authority, one of ten such authorities. The Devon River Authority has been absorbed in the new regioanl authority; the water boards have dissappeared too. The same situation has occurred in that part of England to the east. The West Somerset Water Board, for example, is no longer in existence. Its territory and its functions are now part of the neighboring Wessex Water Authority. Wimbleball Reservoir has been built and has been commissioned by South West Water, in concert with Wessex Water.

The water yielded by the Reservoir is deployed broadly along the lines discussed and debated ten years ago, namely,

1.　some seven million gallons a day is abstracted directly from the reservoir and is pumped to the Maundown Treatment Works in the region of the Wessex Water Authority, and

2.   abstractions totalling 13 million gallons a day are made from the River Exe by the South West Water Authority from Tiverton, Exeter and Torbay.

The cost sharing arrangement is no longer a matter which concerns so many authorities and undertakings. It is now a legal and financial arrangement between the South West Water Authority and Wessex Water Authority as the only two statutory organizations concerned. In brief the basis of the cost sharing agreement is as follows:

1.   South West Water Authority meets all costs of constructing the reservoir and associated works,

2.   South West Water Authority meets all costs of maintaining the reservoir and associated works except the electricity costs of pumping to the Maundown Treatment Works which is owned and operated by the Wessex River Authority,

3.   Wessex Water Authority will pay to the South West Water authority:

   (i)   112/191 of the capital cost of supplying water to outlying villages,

   (ii) 46 percent of all other capital costs,

   (iii) a capital sum is the net total costs of those parts of the total works as related specifically to Wessex Water Authority,

   (iv) all electricity costs of pumping to Maundown Water Treatment Works,

   (v)   46 percent of all running costs net of income except for the costs of fisheries and recreation at the reservoir,

   (vi) 46 percent of the running costs net of income of fisheries and recreation at the reservoir with provision that if income exceeds expenditure South West Water Authority will pay 46 percent of the excess back to the Wessex Water Authority by way of adjustment.

The reservoir has been in commission for just over a year and all the indications suggest that the foregoing cost sharing basis will work satisfactorily.

10. OTHER COST SHARING ARRANGEMENTS

To further illustrate differing cost sharing formulae and applications I refer to two other multi-user projects. Only the general circumstances are outlined.

The first case is an agreement between the Anglian Water Authority and the Severn Trent Water Authority for a supply of water in bulk from Empingham Reservoir, now known as Rutland Water (9). This large reservoir was completed by the Anglian Water Authority following promotion prior to reorganization by a predecessor River Authority. The reservoir has a daily yield of 65 million gallons. The Severn Trent Water Authority receives from the reservoir 4 million gallons per day, which may be taken from the Anglian Water Authority's reservoir system at two points and transferred to the system of Severn Trent. Payment by the Severn Trent Water Authority is on a formula with several factors, e.g., the payment of reservoir standing charges are in the ratio of the daily "take" of 4 million gallons to the total daily yield from the reservoir of 65 million gallons. There are other elements which relate to standing charges for the construction of allied works, pumping costs, treatment costs, rates levied on the Anglian Water Authority, and administrative charges. This basis of payment is for a term of 5 years although individual factors may be reviewed during that period in the light of changed circumstances.

The second case is the Central Scotland Water Board 1976 scheme for apportionment which concerns two sources of supply, Lock Lomond and Loch Turret (10). In this scheme in which the Board was the "sponsoring authority" there were six constituent water authorities, e.g., the regional councils for Border, Central, Fife, Lothian, Strathclyde and Tayside. The financial arrangements provide that the share of the operating expenses to be paid by each constituent regional council will be in the same proportion to the total operating expenses as the quantities of water estimated to be supplied to the individual constituent bear to the total quantity taken by all constituents. There is a condition which has the effect of "abating" part of the payment by each constituent until such time as 80 percent of the total reliable yield is supplied.

The arrangements make provision for the apportionment of loan charges under the scheme - defined as including repayments and interest on monies borrowed by the Central Scotland Water Board. The amount to be apportioned to each regional council for any financial year is to bear the same proportion to the total loan charges for that year as the reservation of water for supply bears to the aggregate of the reservations of all the constituent councils.

There is a special provision for administrative expenses. The basis of the formula is the same as that applicable in regard to operating expenses and loan charges. The factor employed is the rate product of each individual constituent regional council vis-a-vis the total rate product of the whole of the areas of all the constituent regional councils.

The above three cases are only summaries of water resource schemes with cost sharing arrangements. A more detailed and comparative analysis of the principles which have been applied are reviewed in Tables 6, 7, and 8. The cost sharing factors are outlined in considerable detail such that general principles can be identified with specific actions.

Table 6.   Wimbleball Reservoir Scheme - Cost Sharing Analysis

Main Sponsor	Other Participants in Scheme	Title and Outline of Scheme
South West Water Authority	Wessex Water Authority	Wimbleball Reservoir Scheme Cost Sharing Agreement 1978
(As successors to inter alia, the Devon River Authority and the North Devon; East Devon; and South West Devon Water Boards).	(As successors to inter alia, the West Somerset Water Board).	The Wimbleball Reservoir of the South West Water Authority was promoted by a predecessor authority, the Devon River Authority, under the powers of the Devon River Authority (Wimbleball Reservoir) Order 1970.
		The reservoir has a gross storage storage capacity of 4,500 million gallons (20,456 Ml).
		It is on the fringe of the Exmoor National Park.
		Water is discharged from the Reservoir into the River Haddeo, a tributary of the River Exe, to augment the flows of that River. The water is abstracted from the River Exe downstream at Tiverton and Exeter. The yield from the Reservoir taken in this way - from augmentation of the River Exe - is 12.8 m.g.d. (58.1 Ml/d).
		Water is also taken by direct abstraction from the Reservoir by means of a pumping station, within the dam, and pipeline to supply the Taunton Area of the Wessex Water Authority after treatment at their Maundown Water Treatment Works. The yield taken in this way, by direct abstraction, originally planned to be 7.0 m.g.d. (31.8 Ml/d) is now up to 10.0 m.g.d. (45.4 Ml/d).
		These notes are generally about the Agreement made between the South West Water Authority and the Wessex Water Authority whereby, in consideration of the construction and operation of the Reservoir by South West Water Authority for the benefit of both Water Authorities, the Wessex Water Authority will pay South West Water Authority some of the costs.
		Wessex may take 10 m.g.d. (45.4 Ml/d) for 183 days a year and 7 m.g.d. (31.8 Ml/d) for the remaining 182/3 days subject to a maximum of 2,555 million gallons (11,614 Ml) per year.

Table 6. (continued)

COST FACTORS CONSIDERED

Under the Scheme the Wessex Water Authority
agreed to make payments to South West Water
Authority:

(a) in respect of the <u>capital cost</u> of "the total
     scheme"
(b) towards the <u>financing cost</u> of "the joint scheme"
(c) towards the <u>running costs</u> of "the joint scheme"

In the above context:

(i)   The "total works" [the reference in the
      Agreement to the phrase 'the total scheme'
      is probably an error] means works set out in
      the First Schedule to the Agreement. They
      comprise [works] goods and services provided
      under some thirteen major contracts for
      geological testing; construction of the dam;
      provision of pumping, electrical and instru-
      mental apparatus; contracts for telemetry; the
      conversion of farm buildings into a recre-
      ational centre; for the acquisition of land;
      site investigations; promotional expenses,
      roadworks, tree planting; etc., etc.

(ii) The "joint scheme" means those parts of the
      total works which are not (under the Fourth
      Schedule to the Agreement) related specifically
      to either Water Authority. (Under the Agreement
      Wessex Water Authority in this way meet the
      whole of the actual electricity costs for
      pumping water from the reservoir to their
      Maundown Treatment Works).

(iii) "Financing costs" consist of interest and
      depreciation.

Table 6. (continued)

Operating Costs		Capital Costs	Administrative Costs
Running Costs	Standing Costs		
Wessex are to pay the South West Water Authority towards the running costs of the joint scheme: 1) In respect of running costs other than for fisheries and recreation: (a) 46% of the annual running costs of the joint scheme: meaning all costs, charges and expenses incurred by South West Water in the operation and maintenance thereof other than for fisheries and recreation; (b) the actual cost of electricity for pumping water from the Reservoir into Wessex's area. 2) In respect of fisheries and recreation running costs: (i) 46% of the excess of running costs over income directly referable to reasonable facilities for fishing; sailing; and quiet enjoyment of lands associated with the reservoir but (ii) if income exceeds running costs, 46% of the excess to be paid to Wessex by way of adjustment.	See column about "Running Costs."	Wessex to pay South West Water, in respect of capital costs: (a) a capital sum being the net total cost of those parts of the total works as relate specifically to the Wessex Water Authorities and (b) 112/191th of the costs of those parts of the total works relating to the supply of water to certain outlying farms and villages # (c) Wessex are also to pay (annually) 46% of the financing costs of the joint scheme based on the capital costs thereof. #Wessex were to have supplied water to certain outlying farms and villages from other sources. When it was decided to supply these from Wimbleball a formula was worked out as to how much this cost should remain to be met entirely by Wessex (112/191ths) and how much should be met jointly (79/191ths).	See column about "Running Costs."

Table 6. (continued)

Other Costs		Review	Arbitration	Termination
Special Items	Rates			
		The Agreement, made under Section 81 of the Water Resources Act 1963 on 5th September 1978 is to be reviewed at the request of either party: (a) if there were at any time var- iation in the terms of the license issued under the Act of 1963 whereby the former West Somerset Water Board were authorized by the Devon River Author- ity to abstract the water they needed from the Reservoir; (b) on a date not less than five years after the date of the Agreement and at intervals thereafter of not less than five years.	The Agreement provided that there should be established for the purpose of resolving any issues concerning the operation of the reservoir a Committee of two officers of each party, with power to co-opt additional officers from either or both. The Chairman of the Commit- tee is to be an officer of the now South West Water Authority. If the committee are unable to agree any issue it shall be referred for the decision of the Chair- men of the South West Water Authority and the Wessex Water Authority acting jointly. Any dispute about the terms of the Agreement is to be referred to an Arbitrator appointed jointly by the parties or, failing agreement, by the Chairman of the National Water Council or his successor. Any such reference to be in accordance with the Arbiration Act 1950.	

Table 7.   Empingham Reservoir Scheme Cost Sharing Analysis

Main Sponsor	Other Participants in Scheme	Title and Outline of Scheme
Anglican Water Authority	Severn-trent Water Authority	**Empingham Reservoir Scheme** (see Plan Apart) Empingham is a large pump storage reservoir on the River Gwash which it impounds. Empington is also fed with water brought by pipelines from two other rivers in adjoining catchments: from the River Nene by an intake and pumping station at Wansford, and from the River Welland by intake and pumping station at Tinwell. The water from these sources arrives below Empinham Dam and the "Reservoir Pumping Station" there pumps the water into the reservoir. Water from the reservoir is first pumped by a [different] "Empingham Pumping Station" to a treatment works and further pumping station at <u>Wing</u>. Some of it is (believed) there taken to supply local Mid-Northampton needs. Some (up to 3 m.g.d.) goes to Severn-Trent River Authority. Associated with the Wing treatment works is a pipeline to and disposal works at Pilton (not shown on the plan). From Wing, water is also pumped by pipeline to <u>Beanfield</u> Service Reservoir and Pumping Station near Corby. Some of it is again (believed) used for local supply. Some (up to 2 m.g.d.) goes to Severn-Trent Water Authority.
Also that the "statutory entitlement" of Severn-Trent Water Authority means an aggregate of 4 m.g.d. of which not more that 3 m.g.d. may be taken at Wing: and not more than 2 m.g.d. taken at Beanfield.		These notes are generally about the cost-sharing Agreement between Anglian Water Authority and Severn-Trent Water Authority; chiefly in respect of the above works. It is however appropriate to mention too that from Beanfield water also goes by pipeline to service reservoirs at Hannington. And that from Empingham Reservoir water is discharged down the River Gwash for abstraction further downstream for South Lincolnshire.

Table 7. (continued)

COST   FACTORS   CONSIDERED

Operating Costs		Capital Costs	Administrative Costs
Running Costs	Standing Costs		
"Running Costs" means costs properly charged to Anglian Revenue Account in relation to abstraction storing, treating and conveying water (including disposal of sludge therefrom); in respect of fuel, electricity or other power, oil and chemicals (including any which may be applied to aqueducts, pipes or tunnels at any pumping station or at any intermediate shaft).	"Standing Charges" means all amounts properly charged to Anglian's revenue account in any year in relation to the abstraction, storing, treating and conveying of water (including sludge disposal resulting therefrom) other than running costs; rates; abstraction charges and administrative expenses.	See other columns	See "Cost Factors How Shared; Administrative Costs."

Table 7. (continued)

COST FACTORS - HOW SHARED

Operating Costs		Capital Costs	Administrative Costs
Running Costs	Standing Costs		

IN CONSIDERATION OF THEIR ENTITLEMENT SEVERN-TRENT TO PAY ANGLIAN:

Running Costs	Standing Costs	Capital Costs	Administrative Costs
See "Wing Running Costs" below. In respect of "Wing Running Costs" which themselves comprise (a) the running costs of the Empingham pumping Station and pipeline to Wing and the Wing Works PLUS (b) that Fraction of the "Reservoir Running Costs" of which the numerator is the quantity pumped to Wing and the denominator the sum of the quantity taken to Wing and that discharged from the Reservoir: Severn-Trent must pay that fraction of which: the numerator is the quantity supplied to Severn-Trent; and the denominator is the quantity supplied from Wing Works. In respect of "Beanfield running costs" (which include costs of pumps and pipeline from Wing and the Beanfield Service Reservoir): That fraction thereof of which the: Numerator is the quantity supplied to Severn-Trent in any year from Beanfield and the Denominator is the quantity pumped to the Beanfield Service Reservoir.	In respect of the "Reservoir Standing Charges" (viz standing charges for inter alia the river intakes; pumping stations; pipelines; reservoir pumping station and Empingham Reservoir): 4/65ths parts thereof (65 m.g.d. being the ascribed yield of the reservoir to public supply). 4/50ths of "Wing Standing Charges" (50 m.g.d. being the rated average treated water output of Wing treatment works. (Wing Standing Charges including for Empingham Pumping Station; pipeline to and treatment plant at Wing; and pipeline to and disposal plant at Pilton). As to "Beanfield Standing Charges": 2/46ths of the charges relating to the pipeline from Wing and the service reservoir at Beanfield. (46 m.g.d. being pipeline capacity): PLUS 2/13ths of the charges as to Wing Pumps. (13 m.g.d. being their rated output).		In respect of Administrative Charges a sum equal to 2½% of the sums payable by Severn-Trent in respect of (a) the Reservoir Standing Charges; (b) The Wing Standing Charges; (c) the Beanfield Standing Charges; (d) The Wing Running Costs; (e) the Beanfield Running Costs; (f) Rates.

Table 7. (continued)

Other Costs		Review	Arbitration	Termination
Special Items	Rates			
	Such proportion of rates levied on Anglian in any year as one half the actual aggregate quantity of water supplied to Severn-Trent in previous year bears to total quantity of water supplied by the Anglian Authority in that calendar year as notified to the Inland Revenue.	Either party may at any time 5 years after the making of this Agreement (which is for a supply of water in bulk) ask by notice for reconsideration; provided that such party has not given such notice in previous 5 years.	Any dispute to be determined by arbitrator appointed by agreement or failing agreement by failing agreement by the President of the Institution of Civil Engineers.	Severn-Trent are are entitled to determine the Agreement by 12 months notice in writing at any time after 1st April 2076.
Pumps, meters, pipes and apparatus at <u>Wing</u> exclusively for Severn-Trent Water Authority to be vested in Anglian; but Severn-Trent to pay Anglian's costs of providing, operating and replacing etc.		But, in any event, notwithstanding the above either party may by 12 months notice ask for all or any of the heads of agreement relating to the fractions or proportions in respect of: (i) the Reservoir Standing Charges' (ii) the Wing Standing Charges; (iii) the Beanfield Standing Charges; (iv) the Rates to be varied on the ground that the circumstances on which they have hitherto been agreed have changed to a material extent. Failure to agree shall be referred to arbitration.		
Pumps, meters, pipes and apparatus at <u>Beanfield</u> exclusively for Severn-Trent to be vested in Anglian: but Severn-Trent to pay Anglian's costs of providing, operating and replacing etc.				

Table 8.   Central Scotland Water Development Board Apportionment
Scheme Cost Sharing Analysis

Main Sponsor	Other Participants in Scheme	Title and Outline of Scheme
Central Scotland Water Development Board	Borders Regional Council Central Regional Council Fife Regional Council Lothian Regional Council Strathclyde Regional Council Tayside Regional Council	**Central Scotland Water Development Board Apportionment Scheme 1976** The Central Scotland Water Development Board was established persuant to Section 3 of the Water (Scotland) act 1967 It supplies water from its "Sources of Supply" - the Loch Lomond and Loch Turret sources operated by the Board under statutory powers - to Constituent Water Authorities being Scottish Regional Councils. The constituent water authorities under the scheme were those assigned under Section 9 of the Act: the Borders; Central; Fife; Lothian; Strathclyde and Tayside Regional Councils. The quantity of water reserved to those authorities under the scheme was by reference to their estimated requirements during the financial year ending 15th May 1988 agreed to be supplied by the Board at mutually agreed supply points. The apportionment scheme of 1976 superseded an original scheme of 1968 subsequently amended in 1970 and amended again in 1975. The reservations of each constituent water authority under the scheme, as set out in a Schedule thereto were: Borders Regional Council 0.00 Central " " 176.20 Ml/d Fife " " 0.00 Lothian " " 40.90 Ml/d Strathclyde " " 181.80 Ml/d Tayside " " 13.60 Ml/d This scheme apportions the aggregate amounts to be requisitioned from the Constituent Water Authorities under the Act. Under the Act any such scheme was required to have regard to the general principle that the amount requisitioned was to be proportionate to the quantities of water supplied or expected to be supplied to each constituent authority by the water development board.

Table 8. (continued)

COST FACTORS CONSIDERED			
Operating Costs		Loan Charges	Administrative Expenses
Running Costs	Standing Costs		

For the purposes of the Scheme the Aggregate Amounts (to be requisitioned by the Development Board from the constituent Water Authorities) was classified into

(1) Operating Expenses (partly comprised of Running Expenses and partly comprised of Standing Expenses)

(2) Loan Charges

(3) Administration Expenses

"Operating Expenses" means such operating costs in any financial year identified by the Board as directly attributable to operation of their sources of supply including without prejudice to that generality salaries, wages, maintenance and local rates.		"Loan Charges" means repayment and interest on monies borrowed in any financial year to meet approved capital expnediture together with appropriate loan expenses.	"Administration Expenses" means all expenses of Board in any financial year other than operating expenses and loan charges.
"Running Expenses" means that part of the operating expenses which bears to the operating expenses the same proportion as the total quantity of water supplied by the board to all Constituent Water Authorities in any financial year bears to the combined reliable yields of the Board's sources of supply for the said year. (See requisition attached to copy Scheme for exemplification of this concept and others related to this scheme). So: the Running Expenses are a proportion of the Operating Expenses. They are the same proportion as Quantity Supplied to All Constituent Water Authorities ―――――― Total Reliable Yield of Sources of Supply.	"Standing Expenses means that part of the operating expenses remaining after deducting the running expenses.		

Table 8. (continued)

COST FACTORS - HOW SHARED

Operating Costs		Loan Charges	Administrative Expenses
Running Costs	Standing Costs		
The amount of Operating Expenses to be apportioned to any Constituent Water Authority for any financial year shall bear the same proportion to the total operating expenses as the quantities of water _estimated_ to be supplied to that Authority in the year bears to the total water supplied to all Constituent Water Authorities. In other words: Operating Expenses apportioned to any Constituent Water Authority: Total Operating Expenses as Quantity Supplied to Constituent Authority: Total Quantity Supplied to all Constituents. PROVIDED That if in any year the total quantity supplied by the Development Board is less that 80% of combined reliable yields of sources of supply the Operating Expenses will be apportioned as follows:		The amount of Loan Charges to be Apportioned to any Constituent Water Authority in any year shall bear the same proportion to the total loan charges as the reservation of such Authority bears to the reservations of all Constituent Water Authorities. [Loan Charges of Constituent Authority = Individual Reservation times Total Loan Charges divided by Total Reservations of all Constituents.]	The amount of Administrative Expenses to be Apportioned to any Constituent Water Authority in any year to bear the same proportion to all Administrative Expenses as the proportion the rate product of the limits of supply of each Constituent Water Authority bears to the aggregate of rate products of limits of supply of all Constituent Water Authorities.
The Running Expenses (only) will be apportioned as set out above, i.e.: Running Expenses (only) to any one Constituent Authority shall bear to total Running Expenses of all Constituents as Quantity of water supplied to that Constituent bears to Total Quantity supplied to all Constituents.	The Standing Expenses to be apportioned on the same basis as Loan Charges are apportioned, viz: The Standing Expenses to each Constituent shall bear same proportion to the total Standing Expenses to all Constituents as the Reservation of each bears to the total reservation of all Constituents.		
Note The combined reliable yield of all sources of supply in 1978/9 was 403 Ml/d. The total taken by all Constituent Water Authorities was 218 Ml/d (54%).			

Table 8. (continued)

Other Costs		Review	Arbitration	Termination
Special Items	Rates			
Where actual quantities supplied to any Constituent Water Authority in any year vary from the quantities estimated to be supplied the charges for operating costs to be calculated on actual instead of estimated basis (see scheme for precise wording). Where a Constituent Water Authority with no reservation (e.g., Fife) requests a reservation then, if approved by the Board, the apportionment of operating expenses and loan charges shall apply and scheme be amended (with approval of Secretary of State) to include reservation. In that event a balancing payment (as to, inter alia, aggregating past loan and standing charges) is also to be paid. See scheme for details. Under the Water (Scotland) Act 1980 water authorities and water development boards "shall in matters of common interest in relation to their functions consult together and collaborate."		Under the Water (Scotland) Act of 1980 the Central Scotland Water Development Board was continued. Area of Board was to comprise limits of supply of Tayside; Fife; Lothian; Central; and Strathclyde Regional Councils. Board may apply for revision of any apportionment scheme (to Secretary of State). Secretary of State may require a Water Development Board to apply for revision of scheme.		

REFERENCES

1. White, R. J., Esq., Personal communication, National Water Council, London, England, 1981.

2. Walker, D., Esq., Personal communication, National Water Council, London, England, 1981.

3. Subcommittee on Evaluation of Standards, "Proposed Practices for economic analysis of river basic projects," Report to Data Agency Committee on Water Resources, U.S. Government Printing Office, 1958.

4. Hill, S. W., Esq., Arthur Collins & Co., London, England, 1981.

5. Water Resources Act, 1963.

6. Water Acts, 1945 and 1948.

7. Public Health Act, 1936.

8. Water Act, 1973.

9. Bray, P. H., Esq., Chief Executive, Anglian Water Authority, Huntingdon, England, 1981.

10. Fraser, A., Esq., Central Scotland Water Development Board Operational HQ, Baltimore, Torrance, Scotland, 1981.

11. Duncanson, J. K., Esq., Policy and Review Group, Southwest Water, Exeter, England, 1981.

12. Noblet, A. F., Esq., Southwest Water, Exeter, England, 1981.

21. EDUCATIONAL NEEDS FOR THE OPERATION OF COMPLEX WATER SYSTEMS

Emanuele Guggino
Universita di Catania, Italy

In recent years the agency for the development of South Italy, Cassa per il Mezzogiorno, and the Regional Government of Sicily have constructed a large number of public works related to resources development and utilization. Huge investments have been made to construct various sets of public works for municipal, agricultural and industrial water supplies, as well as some tens of municipal and industrial wastewater treatment plants. Numerous agencies share the responsiblility for operating these facilities. but their operation does not allow for coordination in decision making, whether from a technical or economic point of view. The decision-making process is difficult and requires the simultaneous availability of competent operators in many fields such as economics, sociology, law, and technology. The decision making is even more difficult since there are no data collecting and control systems.

To this point in time the emphasis has been on construction of these works, not operation. The operators have not been trained, nor have the necessary management institutions been set up. Thus, despite decades of laudable interventions by such agencies as Cassa per il Mezzogiorno, the Ministry of Public Works, and the Sicilian Regional Government, the goals have not been reached to develop multipurpose, municipal, agricultural and industrial water supplies and ensure pollution control. Why? The answer is not simple. The problems are very complex and a multiplicity of causes can be ascribed. The most important, it is asserted here, is the lack of operator training. Programs are lacking for this purpose.

Much has been achieved in planning, designing, financing, and constructing these plants. But, by the same token, little has been achieved in operating and controlling the complex water resource systems of which the plants are only a part. In particular, little has been done to select, prepare, train and bring up to date the operators who are the most essential part of efficient system service.

In Southeast Sicily the complex water resource system, which is inevitably interconnected and interacts with the socioeconomic system, can be considered comprised of the following subsystems:

1. water supply systems for municipal, agricultural and industrial uses;

2. pollution control systems;

3. flood control and soil conservation systems.

On the basis of these experiences we would briefly like to examine

1. the kind of "operators" required to operate complex water resource systems; and

2. the most suitable kind of "training" for the these 'operators';

These water systems encompass one or more sets of works and plants, operators and users, and agencies and institutions. They are mutually connected and interact so as to achieve the goals for agricultural, civil, and industrial purposes. It is important here to understand the terms "water management" and "operation." Water management here means a series of activities which can be grouped together in the following phases: planning, programming and financing; design, construction, operation and maintenance; and control. For the sake of brevity, the earlier stages, i.e., planning, programming and financing, and design and construction, are labelled "planning." Normally the later stages, i.e., operation, maintenance and control, are labelled "operation" and refer to running the system.

These water systems must be managed, and operated using a global approach to achieve, what we may call here, "total water management." The term global infers not only the physical and technological aspects of designing and constructing the works and plants but also the sociotechnological problems of the operators, users, and the institutions. The operators needed to operate these complex water resource systems must be selected therefore and trained using an industrial type approach.

The continuing population increase in southern Itlay, along with the improvement in living standards, means that all natural resources need to be more fully exploited. The delicate relationships of our "spaceship earth" components, to provide adequate levels of air, earth, and water, makes imperative their rational management if we are to provide adequate levels of public health, energy, and flood control, and water supplies. The elusive "quality of life" depends now, and even more in the future, on how these natural resources are managed.

Among these natural resources, water resources are of prime importance. Because water is scarce, its allocation is subject to intense competition. In addition water projects have vast and far reaching socioeconomic effects and, therefore, are often highly controversial. The increasing social importance of water is seen by the increased interest in pollution control. This problem has priority within the recent environmental movement. Within this movement, great importance is attached to not wasting or degrading natural resources. Thus more is being done to reuse water, and, at the same time, reduce pollution.

It is understood already that flood control and soil conservation go hand in hand with a general water policy. But this policy should encompass also the correct management of complex water systems, comprised of plants, operators, and institutions. The term, "operation," is a phase of management which is becoming increasingly relevant. The activity includes administrators, directors, technicians, and teachers. It is therefore necessary to train "operators" who are capable of efficiently managing complex water systems within this total context. The "operators" are both witnesses and creators of the complexity. Our knowledge of the physical and technical worlds, as of the socioeconomic universe, is becoming increasingly marked by this characteristic of complexity. It underlies the decision making of these same "operators." There are two ways to deal with this complexity: (1) by trying to reduce it, (2) by increasing the productivity of the service. Both solutions need operators. It is therefore necessary to train operators capable of reducing system complexity and increasing service productivity.

Complex water systems are characterized by a high level of component integration, and interaction between these and the various subsystems. To correctly manage a complex water system, a "systems approach" is necessary. It is a rational approach based on the three fundamental functions: (1) knowledge, (2) decision, and (3) action. To implement there must be an analysis of significant information collected by an efficient organization.

During these last decades water resource system operation
rules, policies and procedures have evolved toward a greater
sophistication. This has been brought about by technical
innovation stimulated by the increased importance of the so-called
"water industry", an increased sensitivity to the problems of the
environment, and a change in the organizational structures
involving society as a whole. So long as there was a relatively
low percentage of water resources being used, water management and
operation remained fairly simple. But in an industrial type
context it becomes more necessary to face both operational
problems and those arising from training operators. We need
operators who are capable of making decisions swiftly so as to
ensure effective results in reaching short and medium term
objectives. Operators also must be able to get on with connected
activities resulting from their decisions, as, for example,
collecting and evaluating sociotechnological data, or involving
users to employ their water resources efficiently according to
each situation.

In a complex water resource system, operation aims can differ
according to the various sociotechnological situations. At the
same time there may be multiple interests in operation by
different agencies with different aims. In educating operators
one must bear in mind the various socially dynamic problems. The
education of operators is a socially dynamic problem, too. Thus
we must educate both young people and operators already in place.
This means selecting, preparing, training and bringing up to date
directors technicians, managers, teaching staff, administrators
and politicians who are already a part of society. These are the
people who should ensure that the ever more complex water resource
systems operate properly, not only from a technological or from a
sociopolitical point of view, but from a sociotechnological point
of view. The "operators" must be able to employ efficient tools,
such as systems engineering, to operate water systems.

Sicily has been moving in this direction. This has been
happening since 1970 with the Water Resource Management School
(S.G.E.R.I.), since 1972 with the Centre of Economic Studies
Applied to Engineering of Catania (C.S.E.I.), since 1973 with the
Education and Study Centre for Southern Italy (FORMEZ) at Rome,
and with Fund for Southern Italy (CASMEZ). These organizations
have been involved in training "operators" to correctly operate
complex water resource systems. In fact the S.G.E.R.I. has long
been active in the water industry and in environmental engineering. We cannot mention all that has already been done, but we
draw your attention to the philosophy and criteria of the thirteen
methodology courses for water management and the two environmental
engineering courses held at ERICE. Also, in 1973 the COLORADO
STATE UNIVERSITY at Fort Collins and CATANIA UNIVERSITY carried
out research into the applicability of the systems approach to

water resource management. This had the technical and financial support of FORMEZ in Rome. The research resulted in a training course of thirty audiovisual lessons in water systems engineering developed in both English and Italian. Also since 1975 the C.S.E.I. of Catania has been following this same philosophy and criteria to train operatros to operate water resource systems.

To summarize what has been said, we have outlined both the characteristics of the operators needed to operate modern water resource systems, and the best training for these operators. The water resource systems encompass three categories, i.e., (1) water systems for municipal, agricultural and industrial supply and for hydroelectric power, (2) water systems for pollution control, (3) water systems for flood control and soil conservation. The "operators" are administrators, managers and technicians. They must be selected and trained by using an industrial type approach, since complex water resource management is the same as managing an industry, i.e., the water industry in this case. Operators must be trained to reduce the complexity of the systems and increase service production. It is no longer enough to simply plan works and plants. It will be necessary to ensure that these complex systems are efficiently run by suitable "operators." This kind of training leads to a systems management.

PART V: CASE STUDIES

22. MANAGEMENT OF WATER SERVICES IN ENGLAND AND WALES

A. G. Conybeare Williams
South West Water Authority, England

1. INTRODUCTION

This chapter reviews the recent reorganization of the water industry in England and Wales and then in more detail, some specific aspects that have emerged in the management of the South West Authority in promotion of a major water scheme. The reorganization was mandated by the 1973 Water Act (1) which changed radically the administration of our water services. The Act itself was the result of a series of government committees (2), examining the increasing community demands for water, and the need to make the most economical use of the country's water resources according to a planned water economy. The report of the Jeger Committee (3) established that the water demands could not be met without higher quality standards for wastewater effluents. For example, in 1966 it was reported (4) that of 148 maritime local authorities which discharged effluent to the the sea, only 22 had full wastewater treatment works. In addition to public health considerations due to such practices, aesthetic issues were relevant. The implications of recent E.E.C. Directives (5) have to be taken into account in this context.

There were other considerations which persuaded the Government to review the administration of water services. At a time when the local government system was being restructured these bodies were, in 1968-69, responsible for 31 percent of public expenditure which was 15 percent of the gross national product. Some £3,600,000,000 was spent on revenue or current account and £1,412,000,000 on capital account. Of a total working force in England of 21,4000,000 the local authorities were in 1968 employing over 1,360,000 people full-time and nearly 587,000 part-time (6). There was need for a practicable equation to balance control of broad policy at the national level with efficient and economic regional and local control of admini - stration. This was a principal factor underlying the need for reorganization of the water services, in the hands of local government.

2. THE REORGANIZED WATER SERVICE IN ENGLAND AND WALES

The government in 1971 recognized (7) that the fundamental problem was the need to double the output of water conservation works by the year 2000. They concluded that a comprehensive water resources management plan should be drawn up to provide a strategy within which more detailed planning and subsequently executive action could proceed. They determined that for the whole of England and Wales,and their populations of 50 million people, ten regional water authorities should be created, each to be on a river basin concept. Each would undertake to satisfy the entire spectrum of water needs within its respective jurisdiction. The specific objectives (8) are as follows:

1. to provide sufficient water for the community's requirements in terms of public health and a modern economy,

2. to ensure that water is of the appropriate quality for the purpose concerned,

3. to ensure that resultant waste is disposed of without damage to public health or the environment,

4. to make adequate provision for land drainage and for flood protection, including protection from estuary and sea flooding,

5. to achieve the well-being and development of salmon and freshwater fisheries and to safeguard these aspects,

6. to guarantee to the consumer services in all these respects which are reliable in both quality and quantity,

7. to pay due regard to the interests of people and communitites who may be affected by proposals for the development of water resources, such as reservoirs and other works,

8. to recognize the growing importance of amenity aspects and to secure that the recreational use of rivers and other inland waters are developed in a way that is compatible with other needs.

The formulation of economic and financial objectives for the new water authorities has required a considerable amount of "working out." The authorities are required to employ in the most efficient manner both the resources they inherit and the resources they invest in the vital expansion, development and maintenance of water services.

The need for a country to plan the development of its water resources nationally depends both on availability of rainfall - the natural resource - and on the political philosophy of the country. But given an abundance of rainfall it could be argued that a national plan for the development of water resources is not needed (9). This may be the case if there is no conflict between neighboring regions for a specific water resource and so long as regional water policies are broadly in harmony with national economic policy. On the other hand, if the region is arid, the planned development and "control" of water resources probably will be essential. The need for an overall national strategy would be compelling from many points of view.

When the Government determined in 1973 to reorganize the water industry in England and Wales (Scotland has its own statutory and administrative structures and so has Ireland), they considered, in assessing the strategy for water resources, a whole series of assessments made over a period of years (10). The conclusions were:

1. there would be a continuing need to augment water resources on a substantial scale for a long time to come,

2. many of the main areas of growth are in those parts of England and Wales which either have a lower rainfall than elsewhere or where in any case the main sources have already been or are currently being fully exploited.

These circumstances indicated clearly that it would be necessary for the reorganized water industry and for the ten new regional water authorities to pursue schemes which would involve transfers of large quantitites of water from source to point of use.

3. METHODS OF MEETING DEMAND

The reorganized water industry has a great deal of flexibility in meeting demands for water. Some of the less conventional methods are reviewed in the following sections in the context of the reorganized industry.

3.1 Conjunctive Use

The water industry in England and Wales places particular value on "conjunctive use." We use this term to describe schemes which involve the use of two or more sources of supply. Usually one is for normal circumstances and the other is for times of peak demand or of a shortage of water from the normal source. But the term may be used also to connote other kinds of coordinated use of water. For example, and upland reservoir might be used for a water supply and to regulate a river in time of low flow. The objectives may be to:

1. maintain the ecology and amenity in the river,

2. providing an acceptable constant level of flow in the river to achieve adequate dilution of effluents released from sewage works downstream,

3. increasing yield to maintain downstream water supply abstractions,

4. satisfy downstream community water supply demands conjunctively with supplies from independent groundwater resources,

The relationship between water supply and sewage treatment is in England and Wales a close one. The view is, however, that pollution is less a question of general improvement than of attention to specific locations on rivers where a combination of control of effluents, uprating or improvement of particular works, and mastering storm sewage overflows will produce good value for money.

3.2 Artifical Recharge

In many areas of England and Wales water supplies are obtained from water stored naturally underground. The Chalk, Permo-Triassic Sandstones and Jurassic Limestones are the most significant strata in terms of total resources. Groundwater thus represents an important water resource and the industry has acknowledged the need for further controlled development and experiment (11). In certain cases the abstraction of groundwater has, over a period of many decades, exceeded the natural replenishment from infiltration. In order to combat this situation and to enhance the availability of groundwater supplies, studies have been carried out which have proved the feasibility of artificial recharge, by which natural replenishment is supplemented with purified river water introduced into wells or by infiltration from lagoons when there is surplus water in the rivers. Such a prototype scheme for artificial recharge has been installed by the Thames Water Authority in the Lee Valley. Feasibility trials have been carried out by the Severn Trent Water authority in connection with the extensive Bunter Sandstone Aquifers in the Midlands of England.

3.3 Desalination

Desalination is often put forward as a means of meeting new demands for water for supply without the capital - and social - costs of the construction of new reservoirs. This technique is invariably put forward as a viable alternative whenever a scheme for a major impoundment is promoted in England and Wales. In examining such a proposed alternative several aspects arise for discussion. The former Water Resources Board published in 1969 (12) their conclusions, following an examination of the known technology taking into account feasibility and economic studies of the Water Research Association, that the cost of producing fresh water by desalination processes was by several factors greater than developing new water supply by conventional means. Later the Water Resources Board (13) advised with regard to the prospects for the refrigerant process that had been developed by the United Kingdom Energy Authority. Again the conclusion was reached, despite the technical potential of the process that the estimated cost of water from desalination had increased considerably more than that of water from conventional sources.

In 1975 the South West Water Authority, faced with the need to complete its overall resources strategy by the promotion of two further major reservoir schemes and apprehending the likely arguments of objectors, undertook its own specific examination (14) of desalination as an alternative to impounding reservoirs. This investigation was undertaken with the assistance of the United Kingdom Energy Authority. The report concluded that the cost of meeting specified future deficiencies of water by the provision of a desalination plant would be high compared with the costs of satisfying those deficiencies by providing a new conventional reservoir.

Both the capital and operating and maintenance costs of a desalination plant are high relative to costs associated with conventional water supply schemes in Britian. For this reason, when a desalination plant is used to supply water in conjunction with conventional sources, the cheapest solution is found when as much water as possible is taken from the conventional sources and the deficit is made good by the desalination plant.

The introduction of water from a desalination plant into an existing supply network would raise particular problems. Water would leave a distillation plant at a higher temperature than that usually encountered in public supply systems and would be almost free of hardness. The combination of high temperature and lack of hardness would lead to corrosion of the mains. To overcome these problems the product water from the desalination plant would need to be treated in a "dirtying" plant before it was put into the supply. Additional specilized treatment plant would thus be necessary and further capital and operating costs incurred.

Apart from cost of production per unit of water, a desalination plant requires a great deal of maintenance even when it is not producing water. The report (14) showed that difficulty could be encountered in achieving economic arrangements for manning a plant which, although operating at only 25 percent or less of its capacity in an "average" year, would require sufficient skilled operators on maintenance work and standby duty to run the plant at full production during a drought.

These arguments with respect to cost and efficiency in themselves are compelling reasons for concluding that a desalination plant would not be a viable economic alternative to the construction of a new reservoir. One of the strongest lobbies of opposition to the construction of a new reservoir is to be found among those who are concerned to protect and preserve the environment. This lobby of opinion tends to overlook, when it presses desalination as an acceptable alternative, that such plants are themselves open to objection on environmental grounds. Apart from the issues of appearance and the effect of the visual impact of the plant, other factors such as noise, siting and disposal of effluent would certainly, in England and Wales, weigh heavily against this method.

4. RESERVOIR SELECTION

Against this general background describing the reorganized water industry in England and Wales we turn now toward the problem of selecting a site for a new reservoir and of promoting a scheme for its construction. In preparing the case for such a new major scheme, the water authority must account for a full range of factors and uncertainties. But there are two main questions common to any such proposal outlined below.

Is the need verified? This is the first question. To answer this question the reliable yield from an optimization of all existing water sources has to be set against demand estimates for the future. The latter will take account of population distribution and growth, including any peak seasonal patterns.

The second question is, is the site selected and the scheme proposed one which will satisfy the need and which is otherwise acceptable? The problem of site selection is not one of just adopting the answer which may be right or best in quantative economic terms. Nowadays in England and Wales the identification of the optimum schemes of water resource development can no longer be made on simple traditional narrow considerations. A Water Authority is required in formulating proposals to take into account any effect on the environment and amenities, on geological or physiographical features of special interest, and on other defined aspects (1,15). It is difficult, if not impossible, as yet to place quantifiable values upon amenity and similar aspects of the environment.

In the last twenty or so years we in Great Britain have seen the emergence of other criteria - in part supplementary to the two foregoing issues - which have to be examined and satisfied before a major public works project can have any real hope of succeeding. At a period of inflation and of rising costs and when there has been an immense growth in spending in the public sector, more attention and scrutiny than ever is directed to the economic issues and to the cost benefit aspects of schemes. In promoting recent major reservoir schemes the South West Water Authority has had to support each scheme with evidence from its own highly qualified and experienced specialist staff and to provide "supporting" or complementary opinion from emminent and established consultant expert witnesses. The evidence so given has had to deal with such matters as:

1. general engineering evidence as to hydrology, reliable yields, deficiencies, choice of sources,

2. dam structure and design,

3. geology,

4. water chemistry and water quality,

5. population growth,

6. landscaping and environment,

7. recreational use,

8. fisheries,

9. land acquisition and compensation and land use,

10. agricultural economy,

11. highway implications,

12.   financial and economic considerations,

13.   ecology,

14.   archaeology   and   features   of   historical   importance,

15.   evidence of prior consultation with defined parties and
       bodies in the site selection process.

All this evidence has to be produced and given at public
inquiry when, of course, it is subject to test by objectors to the
proposal.   A hope to succeed nowadays in the promotion of a public
works scheme requires virtual technical perfection, specialist
evidence from many disciplines, and a scheme which has economic
viability.   I do not question for one moment that state of
affairs.   On the contrary, it would be irresponsible to deny that
major capital investment should take place otherwise.

Increasingly, in recent years those who oppose major water
resource projects have, at public inquiry, argued that the
proposed scheme would be unnecessary, or could be deferred, if a
more economical use of water were achieved and if this was allied
with household metering.   In 1977 the South West Authority
promoted a scheme for a major reservoir at Roadford which,
yielding some 30.4 million gallons a day, would complete the plan
for the water resource strategy of the South West.   This project
was investigated by government inspectors at a public inquiry
lasting over two months in 1978.   The Authority, by expert
advocates and witnesses, gave all the evidence traditionally
expected.   The objectors attacked the proposals on all the grounds
available to them.   We still await, three years later, a firm
governmental decision.   We have heard from government that the
inspectors who conducted the inquiry were satisfied by the case we
submitted.   The Minister, however, has deferred a decision until,
at his direction, we have looked again at another site (one we
rejected some years ago) and has advised us that we will be
expected, when we report with our latest findings on the alternative site, to take due account of the need to pursue vigorously
efforts to reduce wastage and take all practicable measures to
conserve supplies.

The promotion nowadays of a major reservoir scheme may well
raise as associated issues such matters as energy conservation,
water leakage and control and metering.

5. ENERGY CONSERVATION

Energy costs in the water industry in the United Kingdom
amount to many millions of pounds a year.   The importance of the
subject was stressed in December 1980 at a Symposium arranged by

the Institute of Water Engineers and Scientists. In his opening address at the Symposium, Mr. Tom King MP, the Minister for Local Government and Environmental Services (16) stated the key issue in the following remarks, "It is a sobering thought that it is less than eight years ago that the price of oil was 1.80 dollars a barrel - today it stands at something like 36 dollars" .... "we have a finite resource without any clear indication of where replacements are to come from." He argued the need for each water authority to prepare a "total energy budget."

In the United Kingdom, as in other countries, the substantial increase in the cost of oil and other forms of energy has caused the water authorities to examine two main issues:

1. Is the authority purchasing its energy at the right price and according to the tariff system best suited to its needs and its finances?

2. Is there scope for conservation of energy in the present systems of operating works and plant and, if not, what alternative forms of energy are available for consideration?

The South West Water Authority set up two years ago its own in-house Energy Conservation Review Team. Some measure of the importance which we attach to the work of this team is evidenced by the fact that the price of electricity rose by 17 percent in April of 1980 and a further 10 percent from 1st August 1980. The authority's budget for power has risen in the last financial year to some £2,500,000 out of a total revenue expenditure of just under £50,000,000.

One of the major problems facing the water authority in the present economic climate is to keep the charge it levies for services as near as it can to the rate of inflation. So many of our heads of expenditure in terms of operating costs offer no margin whatsoever for economy. It is important, therefore, to test all possible areas of saving. Energy costs must be so tested. There are nearly 1000 locations in our region where we have works and premises which are being reviewed in this context. We have, for example, some 600 sewage works. Our exercise includes examination of energy consumption by better operational control. A preliminary finding is the need for maximum demand monitoring equipment at major water treatment and supply works. The first report (17) by our energy conservation unit indicates valuable steps towards progress in this field. Further work is being planned in the development of control curves for the operation of resource systems integrating river abstractions with reservoir levels, storage and release; direct measurement of efficiency of pumping; the introduction of variable speed control to existing motors using new technology; and in finding ways to reduce the pumping costs of "unaccounted for" water and sewerage infiltration.

An indication of the possible scale of leakage from mains and the consequences on energy costs is given in a recent paper by two senior officers of the Thames Water Authority (18). They point out that if some leakage estimates for London are correct a reduction of 10 percent of that leakage would save 22 million watt hours in energy and some £500,000 in revenue expenditure each year. The authors accept that the degree of renewal or leakage tracing warranted by capitalizing the saving is, of course, arguable. They believe, however, that the fundamental principle is to ensure any capital expenditure scheme has the energy consideration stated and preferably to seek to have this included in whatverer procedure is used to evaluate the worth of the proposal. Also they suggest that changes in energy economics will have to be taken into account in future developments. Total energy schemes and schemes for waste heat recovery are becoming economically viable where once they were considered hardly worthwhile. In the fuels industries pricing policy is aimed at ensuring efficient utilization. Future work must be directed at having included with designs an overall assessment of their energy implications. Steps have to be taken to ensure the best use has been made of energy sources.

6. WATER LEAKAGE AND CONTROL

Measures to control water leakage are accepted in the water industry, both at national and regional levels, as demanding a very high priority in steps towards the achievement of water economy. The National Water Council (19) asserts that no single leakage control method is economic for all situations. It also recognizes that it is clearly uneconomic to ensure that pipelines and reservoirs will never leak: it is also clear that there is an economic limit to the loss of water that should be tolerated through leakage.

The demand aspects of major water resource schemes have always been a dominant part of the case for the promotion of the projects. Until recently the unaccounted for water component of total demand has attracted little attention. With an increasing awareness of environmental considerations and energy conservation it is now necessary for a promoting authority to prove that due attention is being given to reducing unaccounted for water. It must be acknowledged however, that within the term "unaccounted for" water there is some legitimate usage, such as that for fire fighting or sewer flushing and it is only the leakage component over which the promoters may have control in terms of achieving savings. The National Water Council study (19) indicated that leakage losses ranged from 10 percent to 50 percent of the total water put into supply. Recent thinking, however, inclines to the view that percentage losses may be misleading and that, at least in urban areas, leakage may be better expressed as a flow per property per unit of time based on net night flow lines. In rural areas a loss per kilometer of main per unit of time is likely to be a more practical measure of leakage. Until recently leakage of 20 percent - 25 percent of water put into supply was regarded as generally acceptable but such generalized figures, whether expressed as a percentage or a loss per property, cannot be supported on detailed examination. An acceptable level of leakage can only be determined for individual systems, having regard to the interrelationship between the costs of finding and repairing leaks, the cost of supplying pumping and treating supplies and the overall cost of providing new sources. It can be demonstrated that in an area with high pumping and treatment costs and where the development of new sources is expensive per unit of yield, it is worth pursuing an extensive waste detection program to reduce leakage to about 16 percent or 5 liters/property/hour. This is likely to be the lowest level of leakage which can be obtained economically using an intensive district and waste metering program. Conversely some experienced opinions consider that it may be economic to allow leakage to rise to 50 percent or more, or 18 liters/property/hour (19) where supply costs are low and a new source cheap to develop. This is likely to be the highest level of leakage which will arise if a totally passive approach is adopted and leaks are repaired only when they are reported.

The benefits which can be derived from leakage control fall into two categories: (1) reduction in running costs, and (2) deferment of expenditure on capital works. The benefits arising from reductions in operating costs will take place when a more intensive leakage control program becomes established. On the other hand, the benefits arising from the deferment of captial expenditure may not be immediate: they will begin to apply when the first capital scheme, however small, is able to be deferred or reduced in capacity because of the introduction of a more intensive leakage control program.

7. METERING

When in the water industry in England and Wales metering is discussed, the topic arises in two different but related aspects: (1) household metering as a method of charging, (2) household metering as a method of conservation. Both aspects are discussed.

7.1 Household Metering as a Method of Charging

In England and Wales there has been a long tradition of basing charges for household water services on rateable values. More recently unmeasured customers pay on a dual basis, i.e., a standing charge to reflect the cost of resource works, pumping and mains etc., and a further sum based on the rateable value of the individual property. Commercial users generally pay by rateable value also. However, because the actual use of water made by a large shop or store may be quite disproportionately small in relation to a high rateable value, an abatement factor is applied generally to reduce the rateable value figure to lessen or "cushion" the impact of water charges on this class of consumer. Industrial users - manufacturers and producers - pay by volume for water used.

Prior to water reorganization most households received only one bill for the local services they received. It was common practice for the appropriate council or local authority to act as collecting agents for the local water undertaking and to include the charges for water, based on rateable value, in the general bill for all local authority services, including the cost of conveyance, treatment and disposal of sewage. The actual amounts paid for water and for sewage treatment tended to be "lost" among the total bill for other services. Furthermore, in so far as the sewage cost element was concerned the level of charges was reduced by reason of the general rate support grants received by local councils from central government. In some areas there was an element of subsidy also in respect of water supply.

Since the 1973 Water Act water authorities have been moving towards the introduction of direct billing. The water authorities now send annual accounts for the services they provide to the customers who use those services. Water authorities receive no grants for sewage or water services and therefore, today, the consumer receives a direct bill for the real level of cost. This change of practice and procedure has made the individual customer much more aware of the cost of his water services and at a time of inflation and rising costs, consumers have been loud in their criticism of increasing water service costs. There is a tendency to overlook the inescapable impact on charges caused by new works and new reservoirs, and to forget that higher standards of water quality cost more, and to overlook the ongoing effect of the financing charges inherited from predecessor bodies. Also, there is a failure to remember that by most standards and criteria water is cheap. Most households in England and Wales will spend more each week on newspapers than they pay weekly for all the services provided by the water authority for that household. In the region of my South West Authority, for instance, the weekly bill for all water services for the average household (taking average rateable value to be £168) is of the order of £1.50 per week, e.g., less than the price of two packets of cigarettes.

At a time when people are sensitive to increasing costs, more and more attention has turned in England and Wales to alternative methods of charging. Arguments are advanced on many grounds. For example, that to base water and sewage charges on rateable value is not an equitable method as it takes no account of the level of occupancy and therefore of actual demand upon and use of services. There is a strong body of opinion in favor of the introduction of the alternative method of charging consumers for the water they use as measured by meter. Metering remains a controversial topic upon which there are conflicting views.

7.2 Household Metering as a Means of Conservation

The view of the water industry in England and Wales (20) is that the metering of domestic water use is likely to have two effects: (1) when meters are first installed a permanent reduction in consumption is likely to result from more careful usage, and (2) increases in the price of water to users after meters are installed can lead to further reduction in demand. The possibility of a third effect has been noted, i.e., namely that further price increases might cause the slowing down of the future rate of increase in demand for water.

We are aware of other experiences in Europe and America concerning the reductions in consumption which result due to metering. The percentage reduction in consumption ranges widely between one country and another. In North America analysis since 1900, covering periods before and after metering, shows a reduction between 11 and 83 percent, and typically about 30 percent. More recent work in Europe shows a range of reductions of the order of 10 to 40 percent. In England there has been one established long term domestic metering situation - that at Malvern, and a much more limited exercise at Fylde. Here the evidence is less persuasive. Data from the Malvern situation and experience has been examined by two researchers whose reports (21, 22) are well known. Work on Malvern data between 1954 and 1969 indicates a statistically significant reduction in quantity consumed when price is increased. The later data for the period 1969 - 1974 did not confirm this. It may be that so many economic changes occurred through this period as to preclude the availability of sufficient reliable data to enable separate identification of the effect of the price changes to be made. The water industry considers that a permanent reduction range of 10 to 40 percent is feasible. Recently the industry has moved significantly further towards the introduction of domestic metering. Most of the ten regional water authorities offer, albeit in some cases on a phased basis, the option to domestic consumers to have meters installed and thus to pay by volume.

Two important matters call for comment in this review of metering as a conservation issue. The first indicates that for the "average" consumer the introduction of metering will be of no financial benefit. A calculation, on the assumption of the cost of meters installed internally, shows that in the case of a property of £450 rateable value with a single occupancy and an average annual consumption of 10,000 gallons, the potential annual saving in households water supply charges when compared to full unmeasured standing charges would be as little as £0.23, ignoring the cost of the meter. The significance of this statistic, in considering how many consumers will seek to pay by meter, is highlighted by the data in Table 1. There are approximately 468,000 households in the region of the southwest. These are categorized in Table 1 according to rateable value.

Table 1. Distribution of Domestic Properties
by Rateable Value

Range of Rateable Value	Number of domestic properties	Percentage
up to 400 R.V.	463,000	99.0
400 to 500 R.V.	2,000	0.4
500 to 750 R.V.	2,000	0.4
over 750 R.V.	1,000	0.2
	468,000	100.0

The potential saving as a result of metering is of course very sensitive to the interrelationship between, the rate and the standing charge, the quantity of water actually consumed, the price per gallon. The authority is placing a higher standing charge to more nearly recover the "fixed" proportion of their costs - such as interest, rates, employee costs etc. - which remain substantially constant irrespective of consumption. The £0.23 as quoted above is based on the standing charges being fully phased in - a situation which is unlikely to prevail for another 2 or 3 years. However, if one uses the actual standing charge prevailing in 1980-81 the savings to the household would be £19.00. On the evidence of these figures it would appear that if the authority were to offer to the domestic user payment by meter as an alternative to payment by standing charge and rateable value, the vast majority of consumers would gain no financial benefits, and would therefore be unlikely to avail themselves of this method of paying for water supplied. On this assumption the availability of metering may have to be discounted as ever making possible in reality the percentage saving in demand and consumption referred to in the preceding paragraph.

The second aspect of possible economies in consumption which might derive from the introduction of metering relates to the provision of new resources. The table in the preceding paragraph has indicated that the financial benefits of metering might attract few consumers. If, for the purposes of argument, one were to adopt a saving in a given immediate period of, say, 10 percent of the current water consumption in a defined area of the region where supplies are known to be in deficit, the question to be answered is to what extent such a reduction would affect the authority's plans for the creation of a new major resource. If one were to assume a proposed new resource with a storage of 8120 million gallons, the initial thought might be that to reduce consumption by 10 percent would be of great signigicance. A reduction of 10 percent in demand would have comparatively small effect on the required storage or upon the top water level because of the usual cross section of the reservoir. The top water level over many hundreds of acres might be a number of inches lower, but this would have no practical effect, or saving, in the acquisition of land to the normal limits of deviation. Similarly, the height of the dam would be varied minimally, if at all, and capital costs would be basically the same.

7.3 Metering — A Conclusion

There can be no doubt that whilst to many consumers and politicians the concept of paying for water by measure is attractive, particularly at a time of high and increasing water charges, there are a number of factors which in reality cast serious doubts upon that approach. Quite apart from the significance, of metering to the consumer in relation to the amount to be paid by him, the foregoing paragraphs will have indicated that a water authority's capital program for major resource works is not likely to be affected by a program for the introduction of meters as an aspect of a drive for water conservation.

8. DEVELOPMENT PLANS AND EXPENDITURES ON CAPITAL PROJECTS

It is logical to follow the preceding sections of this chapter with a reference to the legal requirement on each water authority to prepare statutory surveys for the management of water and its use (23) and to create a five year rolling program for capital works schemes. Capital expenditure is strictly controlled by government and in presenting this paper I propose to discuss these aspects and also to indicate some of the major areas of diddiculty facing water authorities as a result of a "restricted" capital allocation and also to outline the methods adopted to ensure that capital projects are selected on rational criteria, including cost benefit and socio-economic considerations.

REFERENCES

1. Water Act, 1973.

2. Central Advisory Water Committee, diverse reports, H.M.S.O., 1943 to 1971.

3. "Taken for granted," report of the Working Party on Sewage Disposal, Ministry of Housing and Local Government, H.M.S.O., 1970.

4. "Survey of Methods of Sewage Discharge," Ministry of Housing and Local Government, 1980.

5. "Quality of Bathing Water," European Economic Community Directive, 76-160, 1980.

6. Employment and Productivity Gazette, H.M.S.O., October 1968.

7. "Reorganization of Water and Sewage Services," Department of the Environment, Circular 92-71, H.M.S.O, 1980.

8. "A background to water reorganization in England and Wales," Department of the Environment, H.M.S.O., 1973.

9. Taylor, L. E., "Water resources planning at national level," Central Water Planning Unit, Department of the Environment, 1978.

10. "The Future Management of Water in England and Wales," H.M.S.O., 1974.

11. "Water Industry Review," National Water Council, London, 1978.

12. Report on Desalination for England and Wales, H.M.S.O., 1969.

13. Desalination, H.M.S.O., 1972.

14. South West Water Authority, report on Desalination as an Alternative to Impounding Reservoirs, 1975.

15. Water Resources Act, 1963.

16. King, T., "Water," opening speech to the symposium on energy use and conservation in the water industry, reported in the Journal of the National Water Council, Number 36, January 1981.

17. "Conservation of Energy," a report to South West Water Authority 18th February 1981.

18. Thomas, G. A. and Wrighe, G. D., "Energy Management: The Thames Experience," paper delivered at Public Works Congress, Birmingham, 1981.

19. "Leakage Control, Policy and Practice," National Water Council, London, 1980.

20. "Paying for Water," a discussion of economic and financial policies for the water services, April 1976, National Water Council, 1976.

21. Rees, J. A., "Factors affecting metered water consumption," a study of Malvern U.D.C. Final Report to the Social Science Council, 1971.

22. Fox, G. T. J., "Universal Metering," British Water Supply, January 1973.

23. Section 24, Water Act 1973.

23. RESERVOIR OPERATION IN THE WUPPER-RIVER SYSTEM

Richardo Harboe
Rhur-University, Bochum, Federal Republic of Germany

1. INTRODUCTION

The Wupper-River Authority owns and operates several sewage treatment plants and reservoirs within the Wupper-River Basin. Although 19 treatment plants have been built, the water quality at a control gage in the city of Wuppertal is very low (3). In order to improve the river water quality, dilution of treated waste waters is necessary. For the purpose of providing such dilution, as low-flow augmentation, the authority operates four reservoirs in parallel and is building a new one in series.

The objective of this chapter is to describe real time optimal operating rules for these reservoirs (1). Besides low-flow augmentation other purposes, such as flood control and recreation should also be considered. To solve this problem a two step approach is used. In the first step, approximate optimal operating rules are obtained; the method is a so-called sequential optimization using dynamic programming models. In the second step, simulation with synthetic data is performed using these optimal operating rules. But they are changed as needed in order to achieve an improvement in real-time operation.

In recent years an increasing interest for finding optimal operating policies of complex water resources systems for design and management purposes has developed. Other authors, as outlined in Figure 1, have studied the problem using operations research techniques such as linear programming, dynamic programming and simulation, with various degrees of success. Hall (4) combines the use of dynamic and linear programming techniques using dynamic programming for single reservoirs and linear programming for the

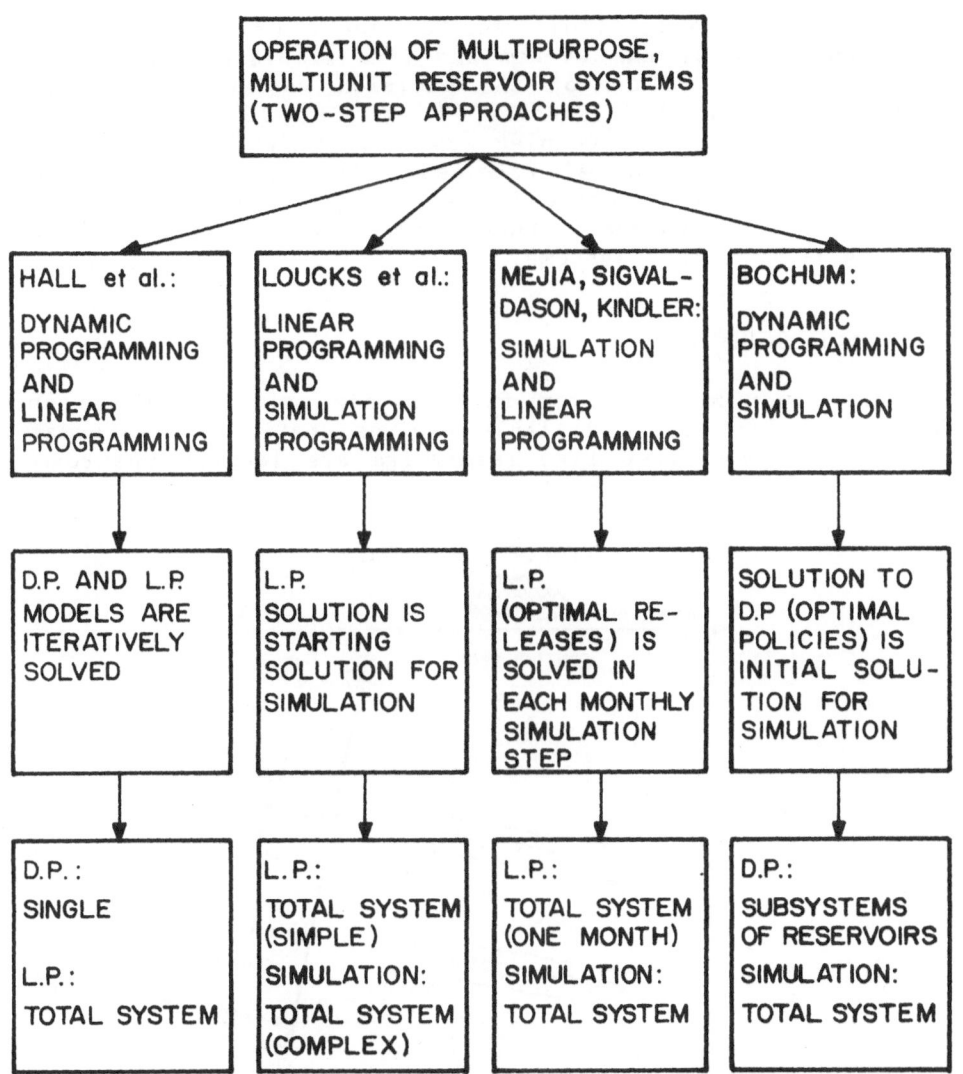

Figure 1. Approaches to Optimization of Complex Reservoir Systems

whole system. Loucks (10) uses a linear program to solve a simplified model of the system and then simulates its operation with the optimal values of the decision variables as starting parameters. Mejía (12), Kindler (9) and Sigvaldason (14) applied a simulation model to a complex system in which a linear program was solved to find optimal releases. The approach described in this chapter has been tested for a relatively simple system (5). It is believed, however, that good results can be obtained extending it to larger systems.

2. SYSTEM DESCRIPTION

The Wupper River system includes six main reservoirs shown schematically in Figure 2; Table 1 gives statistical information about them. The first five reservoirs are located upstream on

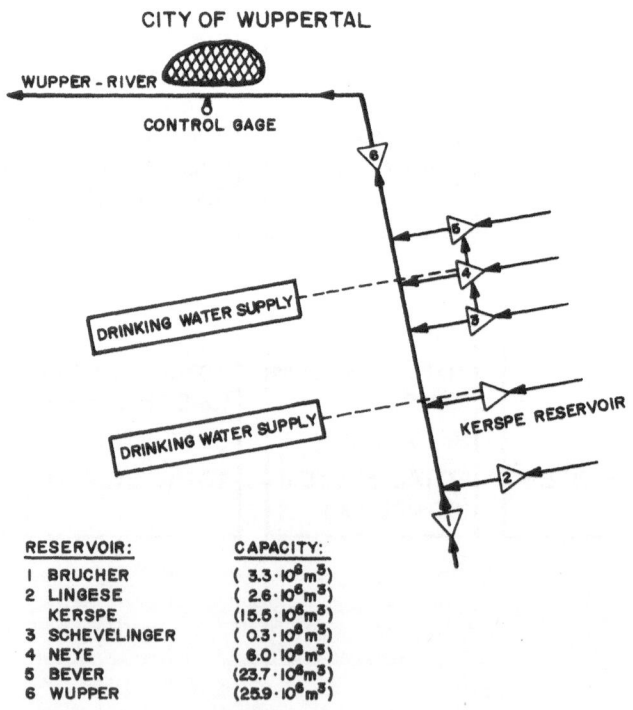

RESERVOIR:	CAPACITY:
1 BRUCHER	($3.3 \cdot 10^6 m^3$)
2 LINGESE	($2.6 \cdot 10^6 m^3$)
KERSPE	($15.5 \cdot 10^6 m^3$)
3 SCHEVELINGER	($0.3 \cdot 10^6 m^3$)
4 NEYE	($6.0 \cdot 10^6 m^3$)
5 BEVER	($23.7 \cdot 10^6 m^3$)
6 WUPPER	($25.9 \cdot 10^6 m^3$)

Figure 2. Wupper River System

different tributaries of the Wupper River, and are thus all connected in parallel. The sixth reservoir, under construction,

Table 1. Characteristics of Wupper River Reservoirs

Reservoir	Capacity [10^6 m^3]	Dead Storage [10^6 m^3]	Flood Control Storage [10^6 m^3]			Mandatory Releases [m^3/s]	Mean Monthly Inflow [m^3/s]
			1.Nov.-31.Jan.	1.Feb.-28.Feb.	1.Mar.-31.Mar.		
1. Brucher	3.340	0	0.400	0	0	0.02	0.16
2. Lingese	2.600	0	1.100	0	0	0.03	0.26
3. Schevelinger	0.290	0	0	0	0	0.007	0.17
4. Neye	6.00	0	1.056	1.056	1.056	0	0.31
5. Bever	23.700	0	5.000	0	0	0.07	0.70
6. Wupper	25.900	0	9.900	decreasing from 9.9 to 5.0	decreasing from 5.0 to 0	1.00	3.12*

*Inflow from intermediate catchment between upper 6 reservoirs and Wupper Reservoir

is located downstream from all reservoirs in the main stream and is therefore in series with the previous ones. The reservoirs serve four purposes, namely low-flow augmentation, flood control, drinking water supply and recreation. Of these purposes the main one is low-flow augmentation for water quality improvement which has to be provided at a control gage located downstream from reservoir 6 at the city of Wuppertal. The natural flows at this control gage are very low during the summer, as shown in Table 2.

Table 2. Characteristics of the Flow at Control Gage

Flow at Control Gage		Monthly Flows (m^3/s). 1946-1977											Mean Annual Flow [m^3/s]
		Nov	Dec	Jan	Feb	Mar	Apr	May	Jun	Jul	Aug	Sep	Oct	
Influenced by Reservoirs	Minimum Monthly Flow	1.39	2.26	1.96	2.54	2.48	3.15	2 88	2.46	2.43	1.76	0.60	0.81	-
	Mean Monthly Flow	7.84	12.16	11.21	10.71	7.82	7.49	4.89	4 78	5.56	5.19	5.24	6.24	7.41
From Intermediate Catchment*	Minimum Monthly Flow	1.38	1.34	1.79	1.54	2.31	1.41	0.81	1.06	0.97	0.78	0.46	0.66	-
	Mean Monthly Flow	6.85	10.88	10.03	9.92	6.97	6.30	3.34	3.11	3.98	3.73	3.59	4.80	6.11

*Between upper 6 reservoirs and control gage (between Wupper Reservoir and control gage it is 49 percent thereof)

Thus when trying to keep a low-flow target of 3.75 m^3/s it is necessary to release supplementary water from the upper five reservoirs an average of five months in each year. In order to reach a target of 5.00 m^3/s, supplementary releases would be needed on the average of 6.5 months each year. Kerspe Reservoir is not considered because it is used completely for drinking water supply, and it almost never spills any water. Reservoir 4, Neye, is used for drinking water supply, but it also stores water diverted from Reservoir 3, Schevelinger. Also it releases it when necessary to Reservoir 5, Bever. Flood control is provided as a fixed constraint in all reservoirs leaving a flood control storage space free between November and January. Recreation takes place mainly in Reservoir 5, Bever, during the summer months. Small mandatory releases have to be made on each reservoir as a constraint with seasonal variation. The inflows to the reservoirs have strong seasonal variation; 30 percent of annual inflows occur during the May-October summer period and 70 percent in winter. Based on 31 years of the reduced historical record, significant lag 1 serial correlation and lag 0 and lag 1 cross-correlation coefficients were measured.

The five upper reservoirs have been operated to achieve a minimum-flow goal of 3.75 m^3/s, reduced to about 2 m^3/s in critical periods. This operating procedure was violated in many months buring historical operation of 32 years. The critical flow series at the control gage can be appreciated with the following statistics:

a.   The flow was lower than 3.75 m^3/s at least once in 19 out of 32 years, or in 72 months in the whole period,

b.   In 32 years there where 15 years in which flow under 3.75 m^3/s had a duration of at least 3 months,

c.   Within the 19 years with flow under 3.75 m^3/s, it was less than 2.25 m^3/s in 10 of those years, i.e., 52.6 percent of the time.

The water quality of the Wupper-River is frequently very low, especially downstream the city of Wuppertal. To improve the quality many sewage treatment plants have been constructed in the watershed and the construction of the Wupper Reservoir having a capacity of 25.9x10^6m^3, has been planned. The main purpose of this reservoir, together with the existing ones, will be to try to increase the low-flow at Wuppertal in order to dilute the already treated wastewaters, and to improve ecological and aesthetical aspects. As can be appreciated from the monthly time series of flows, minimum monthly flow, as influenced by the present reservoirs, was 0.6 m^3/s and occured in September 1959. Low inflows to the reservoirs occur during the May-October summer period, which has 30 percent of annual inflow. This is the period when the water quality conditions are critical.

3. INSTITUTIONAL ASPECTS

The Wupper-River Authority is a public enterprise which owns and operates nineteen sewage treatment plants and seven reservoirs within the Wupper-River Basin (8). The cities, industry and public utilities are members of this authority and provide a portion of the funds. Additional funds come from federal and state government. For example, if the Wupper-River Authority will operate a reservoir leaving part of the storage free for flood control, it receives financial support from the state for building the dam.

Drinking water supply is provided by private enterprises, mostly owned by cities, which have also built reservoirs in the area and facilities to transfer water from the Rhine.

A new reservoir, Dhünn, not included in the system is being built to serve multiple purposes, e.g., flood control, low-flow augmentation, recreation and water supply. The Wupper-River authority will deliver the drinking water to the utilities. The reservoir was not included in the models because it lies on a tributary of the Wupper-River which does not affect the flow of the Wuppertal control gage. The Wupper-River authority was created 50 years ago in order to manage the water resources of the region on a basin wide principle.

4. OPTIMAL OPERATION OF RESERVOIRS

As noted above, the premise of this chapter is to apply a single reservoir optimization model in a chosen sequence to all reservoirs of the system. Thus the computational burden increases only linearly with the number of reservoirs considered, i.e., sequential optimization.

The single reservoir optimization model consists of a computer program written in FORTRAN IV using the deterministic dynamic programming technique. It works backwards in time, using months as discrete time periods. The state variable is the amount of water in the reservoir at the beginning of each period. Flood control is provided by setting maximum allowable storage levels during months with flood risk. Discrete steps, 2.5 percent of the active storage volume each, are used to represent the state variable.

The amount of water in the reservoir at the end of each period is chosed as decision variable. The advantage compared with the choice of reservoir release as decision variable is that there is no need for iterative procedures to compute evaporation losses, since evaporation is an explicit function of water stored at the beginning and end of each period.

The main objective is to maximize minimum flow at the control gage. The objective function for this purpose is:

$$maxZ = min[REL_n + Q_n] \qquad (1)$$

The following recursive equation (1,2) is derived for this objective function:

$$f_n(S_n) = \max_{REL_n} [min\{(REL_n + Q_n), f_{n-1}(S_{n-1})\}] \qquad (2)$$

in which:

S_n = amount of water stored at the beginning of each period n,

REL_n = release from reservoir during month n,

n = periods, numbered backwards n = 1,2,...,N,

Q_n = flow to be augmented at the control gage,

$f_n(S_n)$ = optimum return (maximum of minimum flow at control gage) from periods n through 1.

The state transformation equation can be written as:

$$S_{n-1} = S_n + I_n - REL_n - E_n(S_n, S_{n-1}) \qquad (3)$$

in which:

I_n = inflow to the reservoir during period n,

E_n = net evaporation losses from water surface of the reservoir during period n.

The following constraints are satisfied during the optimization:

1. mandatory release: $REL_n \geq MANREL_n$ $\qquad (4)$

2. flood control : $S_n \leq S_{max\ n}$ $\qquad (5)$

3. spilling : $S_{n-1} \leq CAPACITY$ $\qquad (6)$

As result of the optimization, a target discharge level, $f_N(S_N)$, at the control gage is obtained. This level can be reached with 100 percent probability using historical streamflow record. The optimal releases from the optimization are not used in a forward run as usual, because with the maxmin objective function the target would increase after the critical period due to perfect knowledge of future inflows.

The optimal discharge level is now used as a low-flow target in a forward simulation. The following simple operating rule is applied, which uses the target and constraints.

Release is maximum of:

a. mandatory releases, such as for fish, aesthetical reasons, rights of water users,

b. release necessary for low-flow augmentation up to the optimum discharge level (target) found in backward optimization, $(f_N(S_N))$,

c. release necessary to satisfy the monthly flood control reservation, $(S_n \geqq S_{max\ n})$,

d. release necessary to avoid spilling.

When applying this release rule there is no need for inflow forecasting because it will be possible to meet the low-flow augmentation target with 100 percent probability when using the same inflow record as used in the optimizaiton.

The application of the suggested approach to any reservoir system requires first finding the optimal low-flow target with the deterministic dynamic program and secondly to run a simulation model with the simple operating rule described above. For each reservoir of the system this optimization and simulation steps have to be solved before the next reservoir can be analyzed.

Special attention must be paid to the sequence in which the reservoirs should be treated. The reservoirs first chosen show higher storage levels which may be desired, e.g., for recreational purposes. A flowchart for the single-multipurpose-reservoir (SMPR) program is shown in Figure 3. The results of this sequential application of single reservoir models are the following. First, the optimal low-flow augmentation target of 2.45 m^3/s is obtained for Reservoir 5, Bever. In the simulation run Bever Reservoir increases the natural flow at the control gage, resulting from the so called "large intermediate catchment" which is located between the upper three reservoirs and the

380

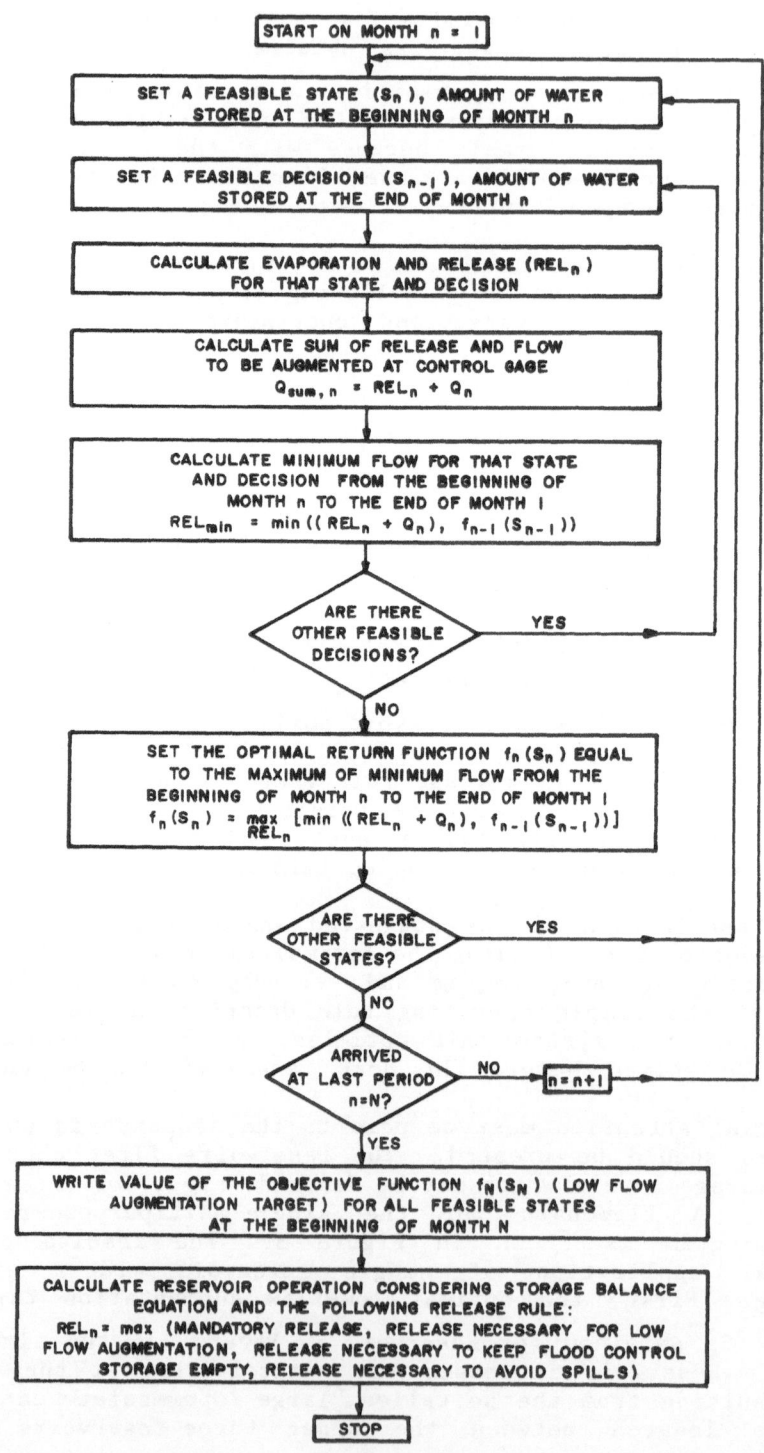

Figure 3. Flow Chart of SMPR Program for a Single Reservoir

control gage. All other reservoirs are thought to be fully closed, which means that they contribute nothing to the flow at the control gage. Reservoir 4, Neye, is not optimized but will be included in the simulation of the whole system.

The second reservoir in the optimization and simulation sequence is Schevelinger Reservoir. An optimal low-flow target of 2.46 m^3/s is computed by the single reservoir optimization model, considering the flow to be augmented at the control gage as the one already augmented by Bever Reservoir.

Similarly, Lingese Reservoir is treated as the third reservoir in the procedure, yielding a low-flow target of 2.62 m^3/s. The fourth reservoir is Brucher with a target of 2.80 m^3/s. The fifth and last reservoir is the Wupper Reservoir, Number 6, because it is located in series and downstream of all previous reservoirs.

The flow to be augmented by Wupper Reservoir is different from that for the other reservoirs, because the uncontrolled portion of the flow now comes from the so called "small intermediate catchment" between the Wupper Reservoir and the control gage. This uncontrollable flow is 49 percent of that from the "large intermediate catchment." The four reservoirs upstream are nevertheless operated as shown before. Inflow to the Wupper Reservoir consists of releases from upstream reservoirs plus flow from intermediate catchment between the upper five reservoirs and Wupper Reservoir. Applying the single reservoir optimization model results in a low-flow target of 3.64 m^3/s for the Wupper Reservoir. This means that if the Wupper Reservoir had existed during the considered 32 years of historical record it would have been possible to guarantee a low-flow of 3.64 m^3/s for all months if the whole system had been operated in the suggested manner.

The results obtained with this sequential optimization certainly represent a local optimum. Nevertheless, as presented in a paper by Boehle et al. (2), it is very close to global optimum. Subsystems of two parallel reservoirs and three parallel reservoirs were optimized using incremental dynamic programming and the results are very close to the results obtained when these subsystems are optimized using single-reservoir models in a sequence. The same optimum was obtained when optimizing an equivalent reservoir which was located at Reservoir 6. Since this reservoir had a capacity equal to the sum of all capacities it represents a most favorable condition, its optimum target being an upper limit for any other optimization.

The advantages of the sequential optimization procedure are the following:

1. It provides a target low-flow augmentation level for each reservoir in given sequence.

2. The targets can be reached in a simulation using historical record with simple operating rules without forecasting of inflows (although the optimization model was deterministic).

3. Using historical record as in the optimization the target can be met with 100 percent security.

4. The computer time required increases linearly with the number of reservoirs considered instead of geometric increase for global systems optimization. Extensions to more complex systems present no problem.

5. It can be applied easily to reservoirs with other purposes, such as power, water supply, irrigation, navigation, for which target levels are to be optimized.

These advantages of the sequential optimization are nevertheless due to two important aspects, which should be present in other systems where this method is to be applied. These characteristics are:

1. The inflows to the reservoirs are highly cross-correllated, thus the reservoirs are simultaneously drawn down in critical periods (small river basins with similar hydrological behavior among its streams).

2. The form of the objective function (maximin) which indicates that no benefit is considered when flows at a control gage are above the optimized targets.

5. SIMULATION OF THE RESERVOIR SYSTEM

A simulation model was developed to use the operating rules developed and historical and synthetic streamflow records. The computer program consists of a main program and several subroutines. The main program SIMULA calls the subroutines in a given sequence for each month of simulation. In particular, the subroutine BILANZ (balance) is called once for each of the six reservoirs in the same order in which they were optimized. SIMULA also calculates the total inflow to Wupper Reservoir, consisting of natural inflow from intermediate catchment, plus all releases from the reservoirs in parallel. The subroutine INPUT reads the data, i.e.: all inflows and flows at control gage, data about the reservoirs (capacity, dead storage, maximum storage due to flood control, lake surface vs. storage), evaporation rates, mandatory releases and low-flow augmentation targets. A subroutine called ORGANI selects from all given data the ones needed for a particular month to be simulated. Subroutine WERTEN calculates and evaluates statistically all failures to meet target low-flows and mandatory releases. Three important parameters are evaluated:

1. Security according to duration. This equals the number of months without failure divided by the total number of months that are simulated.

2. Security according to frequency. This equals the number of years without monthly failures divided by total number of years.

3. Security according to the quantity. This equals the total low-flow target in $10^6 m^3$ minus the sum of all deficits in $10^6 m^3$ divided by the total low-flow target.

The subroutine OUTPUT writes tables with results of simulation (storage levels, releases, increased flows at control gage) and of statistical analyses. A subroutine called PLOTS can draw diagrams with the above mentioned results.

The most important subroutine, BILANZ, performs the monthly water balance for each reservoir considering the operating rule, the flow at control gage already augmented by previously simulated reservoirs and all other constraints imposed upon the reservoir. This subroutine BILANZ performs the following main steps:

1. Calculate the necessary release to increase flow at control gage up to given low-flow target set for each reservoir. This necessary release is considered, meanwhile, as the actual release from the reservoir. (For Neye Reservoir the low-flow target is replaced by drinking water supply demand).

2. Check if mandatory release for each reservoir is satisfied with previously calculated actual release. Otherwise, set actual release equal to mandatory release.

3. Calculate reservoir contents at the end of this month (balance equation without considering evaporation), using deterministic inflow to the reservoir (historic or synthetic).

4. Check if the maximum storage level due to flood control is respected. If not, calculate overflow and increase actual release correspondingly.

5. Check if storage is not below dead storage. Otherwise decrease actual release correspondingly.

6. With new reservoir contents at beginning and end of this month calculate evaporation and then actual storage level at the end of this month.

7. Calculate actual release considering evaporation losses.

8. For Schevelinger Reservoir: Calculate water transfers into Neye Reservoir.

9. For Neye Reservoir: Calculate water transfers into Bever Reservoir.

10. Check for failures to meet mandatory release and target low-flow at control gage.

11. Calculate actual spills due to constraint on maximum controllable release.

12. Calculate increased flows at control gage, after low-flow augmentation through this reservoir.

Due to the water transfer from Schevelinger Reservoir through Neye Reservoir into Bever Reservoir, several iterations had to be performed to obtain the exact releases from the reservoirs. In a first step no transfers (which occur as a spill) from Schevelinger to Neye were assumed. Then release from Neve for drinking water supply was evaluated. Then Bever Reservoir is operated to achieve its low flow target (including transfers from Neye). Finally Schevelinger Reservoir can increase its low-flow target and eventually spilling into Neye is obtained. If spilling into Neye occurs, the mass balance for Neye and Bever have to be repeated. Iterations proceed in this way until no change in release or spills (transfers) occur.

This simulation model is designed to use any sequence of hydrologic inputs, however, if the same historical record as was used in the optimization is considered, all target low-flows and mandatory releases will be satisfied. As a result, 100 percent security is obtained in all parameters. When a higher low-flow augmentation level as obtained in optimization or synthetic hydrologic sequences are used this result will change.

The first analysis with the simulation model was to use the reduced historical record of 31 years and several low-flow targets. When a target higher than 3.64 m^3/s (obtained in optimization) for Wupper Reservoir was used, all targets for the other reservoirs were increased proportionally, i.e., a target of 5.00 m^3/s for Wupper Reservoir implies a target of (5.00x2.80)/3.64 for Brucher Reservoir and so on as presented in Table 3.

Table 3. Low-Flow Augmentation Targets from
Optimization and Increased Targets

Reservoir	Low-flow augmentation targets of individual reservoirs in m^3/s for different target of Wupper Reservoir			
	3.64	4.00	5.00	6.00
Bever[1]	2.45	2.69	3.36	4.04
Neye	Drinking water supply			
Schevelinger[1]	2.46	2.70	3.38	4.05
Lingese[1]	2.62	2.88	3.60	4.32
Brucher[1]	2.80	3.07	3.85	4.61
Wupper[2]	3.64	4.00	5.00	6.00

[1] Natural flows at control gage from intermediate catchment up to upper reservoirs have to be augmented

[2] Natural flows at control gage from intermediate catchment up to Wupper reservoir have to be augmented

The results are shown in Table 4. As can be seen, the number of months with deficit increases from 0 to 69 in a total time series of 372 months when low-flow target is increased from 3.64 to 6.00 m^3/s. The security according to the quantity is reduced to 92.72 percent, that is, the deficit is only 7.28 percent when low-flow tartet is up to 6 m^3/s. Figure 4 shows part of the hydrographs at the control gage, which includes both natural flow (from intermediate catchment between upper reservoirs and control gage) and the flow augmented by releases from Wupper Reservoir.

Table 4. Results of Simulation with Historical
Streamflows Record

Low-flow target for Wupper Reservoir [m³/s]	Hydrologic sequence (No. of years)	Total failures		Securities according to:			Security for mandatory release (duration) [%]	Minimum flow at control gage [m³/s]
		Months	Years	duration [%]	frequency [%]	quantity [%]		
3.64	historical (31)	0	0	100.00	100.00	100.00	100.00	3.64
4.00	historical (31)	4	2	98.96	93.55	99.69	99.74	• 1.51
5.00	historical (31)	30	10	92.19	67.74	97.52	98.70	0.856
6.00	historical (31)	69	15	82.03	51.61	92.72	95.83	0.493

The releases from all other reservoirs were considered as inflow to Wupper Reservoir together with natural inflow from intermediate catchment. Therefore, a second regulation of these flows was possible. The low-flow target for Wupper Reservoir was 5 $\frac{m^3}{s}$ (12.942 x $10^6 m^3$/month in Figure 4). As can be seen in Figure 4, during the 7 years out of 31 years considered failures of different magnitude occurred in 18 months, the greatest failure in 1976 during a dry period when all reservoirs are empty. Releases occur also during high flows because of constraints on reservoir operation and spills.

The second analysis with the simulation model was performed using crosscorrelated synthetic streamflow records which were generated at six river gaging stations (13). A multivariate streamflow generation model originally formulated by Matalas (11) and extended for monthly flows by Young and Pisano (16) was used. Ten synthetic series (I through X) of 51 years were generated and their characteristics are presented in Table 5.

Two sets of simulation runs were performed namely one with the 3.64 m^3/s target for Wupper reservoir, which has 100 percent security with historical records. The results are included in Table 6. It can be seen that on the average the security according to frequency (number of years of failure divided by total number of years) is only 98 percent with 10 synthetic records. Only 50 years were used for the evaluation since it was appropriate to count failures in years starting in April and ending in March and not according to hydrologic year, i.e., November to October.

The same simulation runs were repeated with a 5 m^3/s target for Wupper Reservoir (all other target increased proportionally). The results are included in Table 7. The securities are somewhat

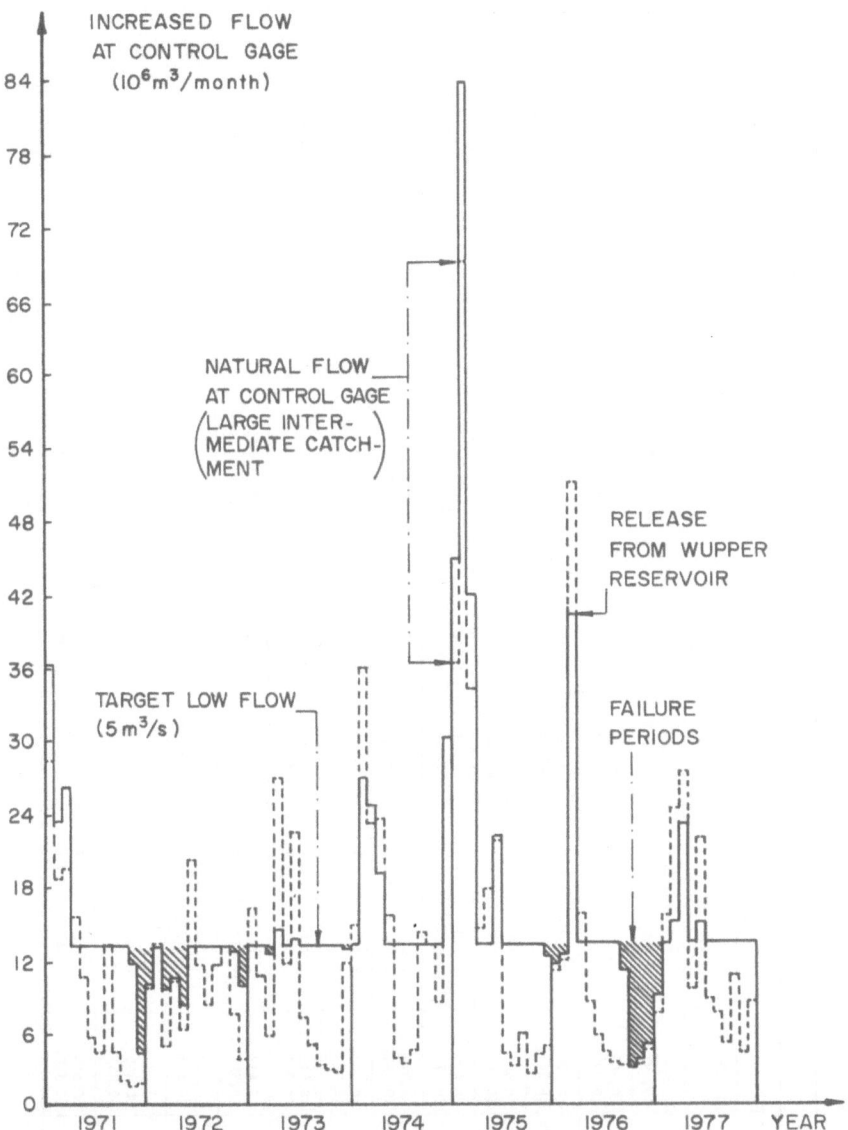

Figure 4. Increased Flows at Control Gage Showing
Years with Failures

Table 5. Annual Mean Values and Standard Deviations of 10 Synthetic and 1 Historic Records at Six Gaging Stations

Annual mean value and standard deviation in million cubic meters per month

| Record | Inflows to: | | | | | Flow at control gage |
	Brucher (1)	Lingese (2)	Bever (3)	Neye (4)	Schevelinger (5)	(6)
Synthetic I (51 years)	0,439 (0,369)	0,725 (0,680)	1,911 (1,680)	0,829 (0,725)	0,444 (0,629)	16,480 (12,801)
Synthetic II (51 years)	0,400 (0,323)	0,672 (0,565)	1,785 (1,515)	0,781 (0,669)	0,417 (0,610)	15,256 (11,656)
Synthetic III (51 years)	0,386 (0,345)	0,653 (0,612)	1,718 (1,675)	0,749 (0,713)	0,405 (0,693)	15,541 (13,254)
Synthetic IV (51 years)	0,403 (0,342)	0,668 (0,585)	1,697 (1,559)	0,766 (0,704)	0,427 (0,644)	15,133 (12,288)
Synthetic V (51 years)	0,421 (0,372)	0,708 (0,652)	1,857 (1,771)	0,828 (0,813)	0,466 (0,699)	16,256 (13,937)
Synthetic VI (51 years)	0,414 (0,345)	0,695 (0,612)	1,816 (1,650)	0,799 (0,714)	0,436 (0,635)	15,894 (12,778)
Synthetic VII (51 years)	0,303 (0,324)	0,663 (0,567)	1,739 (1,512)	0,780 (0,705)	0,421 (0,637)	15,818 (12,375)
Synthetic VIII (51 years)	0,419 (0,362)	0,720 (0,667)	1,910 (1,822)	0,865 (0,832)	0,464 (0,653)	16,606 (14,277)
Synthetic IX (51 years)	0,419 (0,353)	0,721 (0,637)	1,881 (1,718)	0,813 (0,760)	0,469 (0,716)	16,364 (13,583)
Synthetic X (51 years)	0,457 (0,394)	0,773 (0,679)	2,010 (1,802)	0,909 (0,819)	0,486 (0,675)	17,672 (14,406)
Mean value of all 10 synthetic records	0,415 (0,353)	0,700 (0,620)	1,832 (1,670)	0,812 (0,754)	0,444 (0,659)	16,100 (13,140)
Historic $\bar{X}$ (s) (31 years)	0,413 (0,359)	0,694 (0,349)	1,818 (1,620)	0,810 (0,717)	0,442 (0,592)	16,047 (13,310)

Table 6. Results of Simulation with Targets from Optimization Models

Target for Wupper reservoir [m³/s]	Synthetic record	Monthly failures after Wupper (out of 612)	Yearly failures after Wupper (out of 50)	Securities according to:			Flow at control gage (50 years)		Lowest monthly flow at control gage		
				Duration [%]	Frequency [%]	Quantity [%]	Mean value [10⁶m³/mo]	Standard deviation [10⁶m³]	without reservoirs [10⁶m³/mo]	with reservoirs [10⁶m³/mo]	[m³/s]
3.64	I	2	1	99.67	98.00	99.97	16.480	12.801	0.082	8.253	3.19
3.64	II	0	0	100.00	100.00	100.00	15.256	11.656	0.608	9.424	3.64
3.64	III	0	0	100.00	100.00	100.00	15.541	13.254	0.084	9.424	3.64
3.64	IV	1	1	99.84	98.00	99.99	15.133	12.288	0.143	8.688	3.36
3.64	V	6	3	99.02	94.00	99.78	16.256	13.937	0.022	5.605	2.17
3.64	VI	2	1	99.67	98.00	99.97	15.894	12.778	0.101	8.229	3.18
3.64	VII	0	0	100.00	100.00	100.00	15.818	12.375	0.281	9.424	3.64
3.64	VIII	4	2	99.35	96.00	99.82	16.605	14.277	0.027	2.720	1.05
3.64	IX	4	2	99.35	96.00	99.95	16.364	13.583	0.162	7.967	3.08
3.64	X	0	0	100.00	100.00	100.00	17.672	14.406	0.067	9.424	3.64
3.64	Mean value of 10 records	1.9	1.0	99.69	98.00	99.95	16.102	13.136	0.158	7.916	3.06
3.64	historical	0	0	100.00	100.00	100.00	16.071	13.327	1.206	9.424	3.64

Table 7. Results of Simulation with Target of Wupper Reservoir of 5 m³/s

Target for Wupper reservoir [m³/s]	Synthetic record	Monthly failures after Wupper (out of 612)	Yearly failures after Wupper (out of 50)	Securities according to:			Flow at control gage (50 years)		Lowest Monthly flow at control gage		
				Duration [%]	Frequency [%]	Quantity [%]	Mean value [10⁶ m³/mo]	Standard deviation [10⁶ m³]	without reservoirs [10⁶ m³/mo]	with reservoirs [10⁶ m³/mo]	with reservoirs [m³/s]
5.00	I	23	8	96.24	84.00	98.99	16.480	12.801	0.082	0.231	0.089
5.00	II	28	12	95.42	76.00	98.91	15.256	11.656	0.608	2.433	0.940
5.00v	III	45	19	92.65	62.00	97.96	15.541	13.254	0.084	2.175	0.840
5.00	IV	49	13	91.99	74.00	98.02	15.133	12.288	0.143	1.127	0.435
5.00	V	45	11	92.65	78.00	97.37	16.256	13.937	0.022	0.016	0.006
5.00	VI	32	12	94.77	76.00	98.22	15.894	12.778	0.101	1.837	0.710
5.00	VII	40	11	93.45	78.00	97.76	15.818	12.375	0.281	2.523	0.975
5.00	VIII	35	12	94.23	76.00	98.13	16.605	14.277	0.027	0.013	0.005
5.00	IX	32	9	94.77	82.00	98.43	16.364	13.503	0.162	0.524	0.202
5.00	X	24	8	96.08	84.00	98.84	17.672	14.406	0.067	3.962	1.529
5.00	Mean value of 10 records	35.3	11.5	94.23	77.00	98.26	16.102	13.136	0.158	1.338	0.517
5.00	historical	30 (out of 384)	10 (out of 31)	92.19	67.74	97.52	16.071	13.327	1.206	2.215	0.856

higher with synthetic record than with historical record, but some individual runs showed worst results. It can be concluded that the security according to the frequency lies between 67 and 77 percent. In all cases, the security according to the quantity was very high.

6. CONCLUSIONS AND RECOMMENDATIONS

The first results of an attempt to simulate a complex reservoir system with operating policies that were obtained by optimization have been described. In order to test the quality of these operating rules several simulation runs with synthetically generated inflows were undertaken. The six flow records were generated simultaneously so as to maintain serial correlation for each station and crosscorrelations (lag 0 and lag 1) among all stations. The operating rules were based on the results of a sequential application of dynamic programming models for single reservoirs. The objectve function was to maximize low-flow at a control gage while satisfying several constraints representing other purposes. In this way, a low-flow augmentation target was obtained for each reservoir in a given sequence. This same sequence and the same targets or increased targets, all in the same proportion, were used in simulating the operation of the system with historical record and 10 synthetic records. The results of the simulations with synthetic records have on the average higher security of achieving the targets than the corresponding result with historical record. Individual runs show nevertheless worst conditions.

The application of sequential optimization to this system leads to near global optima for the operation policies of each reservoir. These policies can be applied in real-time operation (without forecasting of inflows) as was shown in the simulation model. They could even be applied on a daily or hourly basis as needed.

The two-step approach, which is similar to the ones developed by other authors, provides a good basis for developing optimal operating rules for complex reservoir systems of more than three reservoirs, when a global model (even using incremental dynamic programming algorithms) would fail due to computer storage and time.

Future research along this line includes improvement of the synthetic data generation techniques, optimization step also using synthetic data, improvements in the operating rule which can be tested only in simulation.

REFERENCES

1. Boehle, W., Harboe, R. and Schultz, G. A., "Low-Flow augmentation for water quality improvement in a river system in Germany," Vol. 2, Proceedings, XVIII Congress of the IAHR, Cagliari, Italy, pp. 63-72, 1979.

2. Boehle, W., Harboe, R. and Schultz, G. A., "Sequential optimization for the operation of multipurpose reservoir systems," Proeedings, International Symposium on Real-Time Operation of Hydrosystems, Waterloo, Ontario, Canada, June 1981.

3. Brechtel, H., "Stand und Entwicklung der wasserwirt-schaftlichen Planung des Wupperverbandes," Das Recht der Wasserwirtschaft, 22, 1979.

4. Hall, W. A. and Dracup, J. A., Water Resources Systems Engineering, McGraw-Hill, New York, 1970.

5. Harboe, R., "A stochastic optimization and simulation model for the operation of the Lech River system," Institut fuer Hydraulik und Gewaesserkunde, TU München, 21, 1976.

6. Harboe, R., "Introduction to dynamic programming in water resources planning and operation," CNR-Istituto di Ricerca per la Protezione Idrogeologica Nell' Italia Centrale, Perugia, Italy, 1980.

7. Harboe, R., Schultz, G. A. and Duckstein, L., "Low-flow and flood control: Distributed versus lumped reservoir model," Proceedings, IFAC Symposium on Water and Related Land Resource System, Cleveland, USA, pp. 313-321, 1980.

8. Harboe, R., Schneider, K. and Schultz, G. A., "Simulation of a reservoir system with optimal operating policies," Vol. 4., Proceedings, XIX Congress of the IAHR, New Delhi, India, pp. 193-204, 1981.

9. Kindler, J., "The Monte Carlo approach to optimization of the operation rules for a system of storage reservoirs," Hydrological Sciences Bulletin, 1977.

10. Loucks, D. P., "Surface water quantity management models," in A. K. Biswas (Ed.): Systems Approach to Water Management, Chapter 5, McGraw-Hill, New York, 1974.

11. Matalas, N. C., "Mathematical assessment of synthetic hydrology," Water Resources Research, 4, 1967.

12. Mejía, J. M., Egli, P. and Leclerc, A. "Evaluating multireservoir operating rules," Water Resources Research, 6, 1974.

13. Schneider, K. and Harboe, R., "Anwendung des Young-Pisano Modells zur Erzeugung gleichzeitiger künstlicher Abflußreihen für verschiedene Stellen in einem Einzugsgebiet," Wasserwirtschaft 7/8, pp. 219-225, 1979.

14. Sigvaldason, O. T., "A simulation model for operating a multipurpose multireservoir system," Water Resources Research, 2, 1976.

15. Wupperverband, 50 Jahre Wupperverband, Wuppertal, F. R. of Germany, 1980.

16. Young, G. K. and Pisano, W. C., "Operational hydrology using residuals," Journal of the Hydraulics Division, HY4, 1968.

24. OPERATION OF WATER SYSTEMS IN EASTERN SICILY

G. Rossi, E. Guggino, S. Indelicato
Universita di Catania, Sicily

1. INTRODUCTION

This chapter is a case study which refers to an extensive region in eastern Sicily, shown in Figure 1, called here the Simeto River basin. It includes extensive urban areas such as Catania and Syracuse, developed agricultural areas and large industrial districts. The availability of water is not adequate to meet the demands of the users in the region. This is true especially in the summer when river flows are low and the use rate in the municipal and agricultural sectors is highest. Three major operation problems are prominent in this case. The first has to do with effective operation in the storage of the natural flows. The second is related to the coordinated utilization of various sources of water regarding both time distribution and quality characteristics of sources and the requirements of various user sectors. The third problem is to integrate discharges from waste-water treatment plants into the total water system.

The water supply in the region comes from several sources, used conjunctively. They include: surface waters of the Simeto river and its major tributaries, the Dittaino River and the Gornalunga River, and the other minor streams of the eastern drainage of Sicily; ground waters of the Mount Etna volcanic aquifer and of the Syracuse aquifers; and treated wastewaters, which are expected to be available at the major wastewater treatment plants such as Catania, Syracuse and Magnisi.

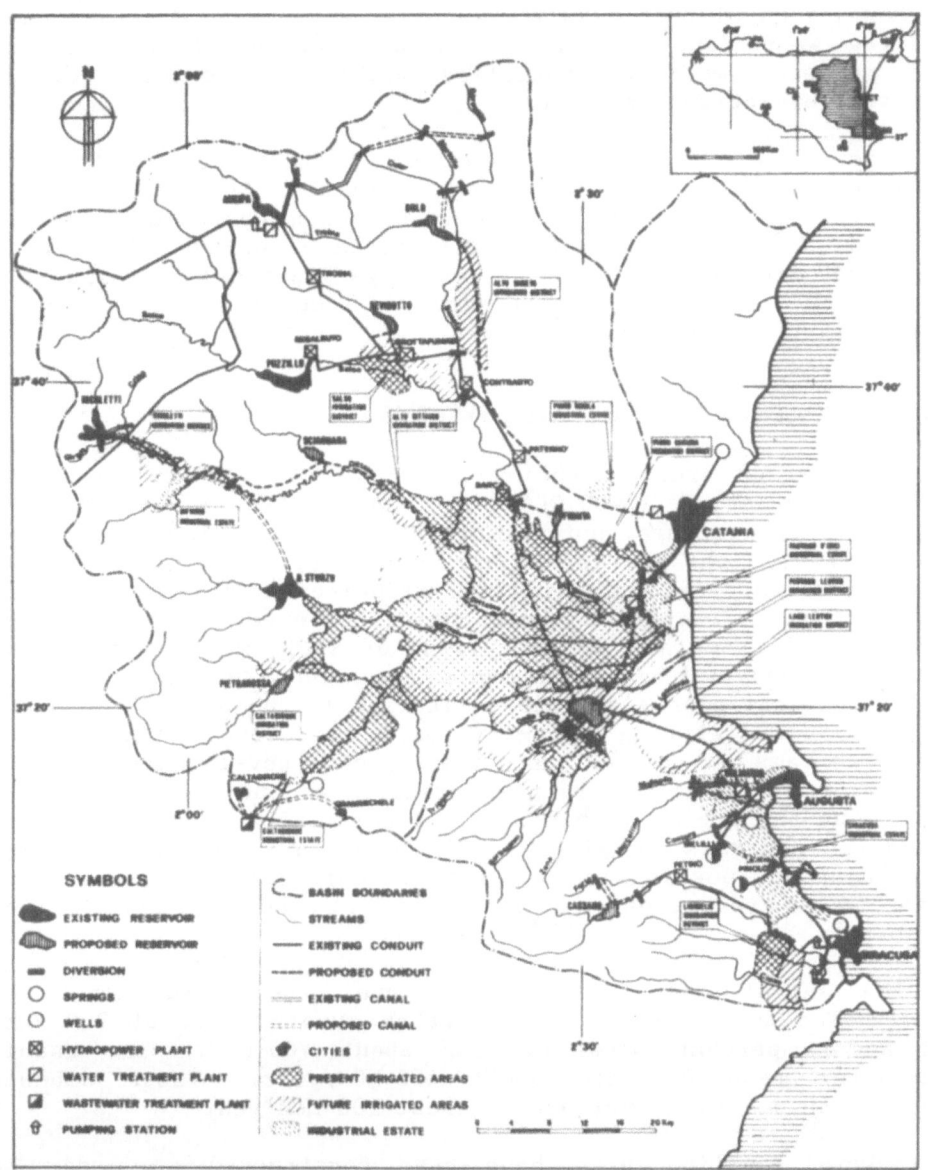

Figure 1.   Water system in eastern Sicily.

Management of such a system requires a multi-purpose approach involving water supply, flood control, and pollution control. These purposes must be satisfied within the context of an integrated operation scheme.

This chapter has three sections. The first deals with identification of existing water systems, the second with development perspectives, and the third with operation problems and strategies for coordinated utilization of water resources in normal and drought conditions. The organizational and institutional needs related to improvement of decision processes are outlined in the conclusions.

2. IDENTIFICATION OF THE WATER SYSTEMS

The surface area of the region is about 7,000 km^2, and its population is about 1,500,000 inhabitants. The major economic activity is agriculture, but during the previous decades there has been a shift of the working population towards the industrial and service sectors. Also agricultural practices have changed due to construction of irrigation works, which has occurred in the Piana di Catania district and in the plains of the Simeto tributaries. The main industrial establishments are along the coast. They include chemical plants and oil refineries in the Augusta-Syracuse area, and manufacturing and electronics in Catania. Due to the action programs of the national and regional governments, further development is occurring in the agricultural and industrial sectors. Thus the region is becoming a leading area for the economic and social progress of Sicily. The major problems constraining the development possibilities relate to more effective management of the water resources and protection of the environment from pollution.

The annual mean precipitation over the region is about 700 mm. The mean annual surface water runoff is about 900 MCM, with an 80 percent occurrence of about 700 MCM. The usable groundwater resources are about 450 MCM/year. These amounts compare with a total water use of about 500 MCM/year.

Several public works have been constructed in order to provide the water supply for the municipal, agricultural and industrial sectors, with financing from the "Regione Siciliana" and "Cassa per il Mezzogiorno." The Rome agency, Cassa per il Mezzogiorno, is involved through two of its Special Programs. The Special Program No. 2, approved in 1971, is intended to provide the infrastructures for the development of the industrial area of Syracuse and Gela. The Special Program No. 30, approved in 1976, is intended to develop a multipurpose water scheme for all of Sicily.

Several initiatives have been developed for pollution control. First, some wastewater treatment plants have been built in recent years. Also, a regional committee has been involved in developing a pollution control plan. However, the management problems of water resources cannot be considered solved completely, and in particular those related to the operation of the developing complex water systems.

Actually, the present state of water utilization is a step in an evolution process which dates back to the most ancient times. The Galermo Aqueduct in Syracuse, for example, goes back to the Greek period; it is still working. Also near Syracuse are the ruins of Kolimbethra Dam which stored the flows of the Alabo River, today the San Cusimano River, for irrigation of the Megara Plain. Then at Santa Maria di Licodia are the ruins of the Roman aqueduct, 16 miles long, for municipal supply of Catania. The memory of techniques that the Arabs used for groundwater pumping and storage in small reservoirs for irrigation, are still evident in some dialectal expressions. A dam which has transformed the Lentini marshes into a lake, and was intended probably to function as a fish nursery goes back to the Suevi age. The bridge of Biscari Aqueduct on the Simeto River, built in the eighteenth century, is more recent.

Ideas for the development of a reservoir system for the storage of surface flows of the Simeto river and its tributaries were proposed more than a hundred years ago. Their implementation was initiated only after the Second World War. In 1866, an engineer of the Department of Agriculture, Industry, and Commerce investigated the possibility of developing six reservoirs in the Simeto and Salso River Basins to irrigate the Catania plains. These were in addition to two reservoirs built on the Anapo River to irrigate the Syracuse area. Following this, a government committee in 1887 restricted the analysis to four of the proposed reservoirs on the Simeto River to be used for irrigation. According to the proposals made by Ing. Vismara in 1908, in behalf of the Eastern Sicily Electricity Company, the irrigation use should have been combined with hydroelectric use. Later, in 1947, these proposals were reviewed and used as a basis for the "General Plan of Exploitation of Salso and Simeto River Basins", developed by the Sicilian Electricity Agency immediately after its establishment. Water supply in the urban areas of the region is provided by a number of aqueducts which use groundwater.

The municipal, agricultural and industrial water supply of eastern Sicily, shown in Figure 1, is provided presently by several plants using both surface waters and groundwaters. These facilities are both public and private, with the former predominating. Limiting attention to the public facilities, but leaving out the municipal acqueducts which generally provide

single municipalities or groups of municipalities, it is possible
to identify three main subsystems in the study area: Salso-
Simeto, Dittaino-Gornalunga and Lentini-Ciane. They are identified in Figure 2.

The Salso Simeto system comprises Ancipa and Pozzillo
reservoirs: Santo Domenica, Contrasto and Barca intakes on the
Simeto River, and six hydroelectric plants. The main characteristics of the plants are presented in Tables 1 and 2. The Ancipa
Reservoir has a net capacity of 28 MCM and it regulates both the
flow of the direct basin and the flows of other tributaries which
are connected through a diversion canal. A small portion of the
Ancipa waters is used to supply a 70 Km aqueduct for the supply of
several municipalities in central Sicily and the remaining portion
is for hydroelectric and irrigation uses. The hydroelectric use
takes place in the two power plants of Troina and Grottafumata.

The Pozzillo Reservoir has a capacity of 103 MCM which is
filled to some extent by sediments and its waters are used mainly
for irrigation. The diverted waters in the irrigation period flow
through the Regalbuto power plant. A small amount is delivered to
a minor irrigation district, while the substantial portion is
routed for further hydropower production and irrigation. The
Simeto flows diverted by the San Domenica intake together with the
discharge from Grottafumata and Regalbuto power plants are
conveyed to the Contrasto power plant and then are discharged into
the river. Then the flows are diverted downstream at Contrasto
intake and are conveyed to Paternò and Barca power plants. During
the irrigation season they are conveyed to the irrigation system
of Piana di Catania irrigation district.

A flow of 1 m^3/s from the Simeto River is reserved for supply
to the industrial district of Syracuse in winter only. This flow
is conveyed through one of the irrigation canals of the Piana di
Catania Irrigation District and through other pumping and conveyance facilities which were built within the context of the Special
Program No. 2 for the infrastructures of the industrial area of
Syracuse.

The Dittaino-Gornalunga system, which includes the Nicoletti
Reservoir on the Dittaino River has a capacity of 19,5 MCM and
irrigates the Dittaino Valley. Don Sturzo Reservoir, capacity 110
MCM, on Gornalunga River is used for irrigation also. This
reservoir also regulates the flows of Dittaino River downstream of
the Nicoletti Reservoir.

In the past, the water supply for the irrigated areas and for
the industrial areas of Syracuse originated from the groundwater
of the Priolo-Augusta area. In order to reduce pumping from the
aquifer, and to prevent the seawater intrusion, a surface water

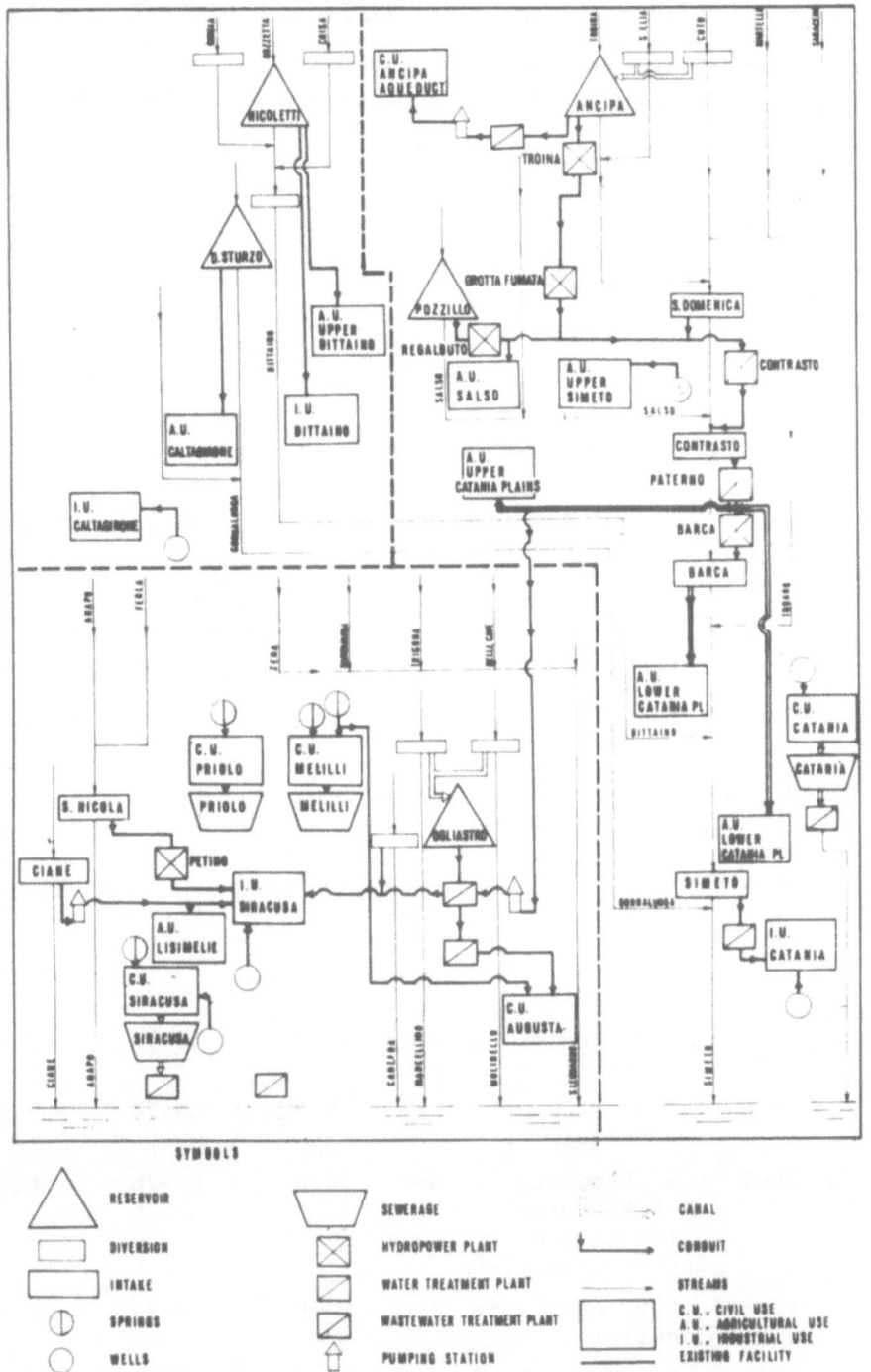

Figure 2.   Schematic of present water system.

Table 1.   Reservoir characteristics in Simeto River basin.

Reservior or Intake	Operated	Water Course	Basin Area (Km^2)	Net Capacity (hm^3)	Uses
Ancipa Reservoir	1953	Troina	51+44	28	Municipal, (E.A.S.)
					Irrigation (Catania District) Hydro-electric, (ENEL)
Pozzillo Reservoir	1962	Salso	577	140	Irrigation (Several Districts)
					Hydroelectric (ENEL)
S. Domenica	1966	Simento	557	-	Irrigation (Catania District)
					Hydroelectric (ENEL)
Contrasto Intake	1966	Simento	308	-	Irrigation
					Hydroelectric (ENEL)
Barca		Simeto		-	Irrigation (Catania district)

Table 2.   Hydropower palnts in Simeto River basin.

Power Plant	Operated	Diversion by	Power (KVA)	Max Discharge (m^3/s)	Annual Mean Energy Period	Annual Mean Energy KWH
Troina	1954	Ancipa Riservoir	33	9.0	54-77	45,000
Grottafumata	1958	P.P.Tronia outlet	18	10.1	58-76	23,000
Regalbuto	1964	Pozzillo reservoir	6	16.5	64-76	7,000
Contrasto	1967	S. Domenica + P.P.Grotta-fumata and Regalbuto outlets	49	36.0	67-76	50,000
Paterno	1963	P.P. Contrasto outlet	19	24.0	63-77	37,000
Barca	1963	P.P.Paternò outlet	12	24.0	63-76	18,000

system has been developed over the past decade. The system includes the Ogliastro Reservoir, capacity 4.5 MCM, which regulates the flows of Mulinello and Marcellino Rivers, the Cantera intake, and the Ciane aqueduct, and, since 1980, the above mentioned diversion of winter waters from the Simeto River.

The study area includes a large number of wastewater treatment plants which were built according to varying design criteria, and upon local initiatives, not related to any general plan for environmental protection. Only a small number of these facilities are operating.

Within this situation, and due to the limited amount of water in some areas, there is reuse of untreated municipal waste waters. Such reuse usually is performed without official sanction and without control of the effluent quality.

All of the existing wastewater treatment facilities have been designed to serve single municipalities. Today, however, there is a tendency to establish wastewater treatment districts, which would provide for joint treatment at a common plant serving groups of municipalities.

The present status of wastewater treatment in Catania and Syracuse province is as follows. The seventeen facilities in the study area serving 30 municipal communities could serve a population of about 320,000, which is about 20 percent of the present population. But only four of the seventeen facilities are in operation. The level of treatment does not meet the quality standards required by the recent pollution control acts, e.g. national laws no. 319 of 1976, and no. 850 of 1979. The other thirteen facilities need to be enlarged or improved in operation. Eight additional facilities are to be constructed in the forseeable future.

It should be noted that the water systems within the region have been affected by several hydraulic works in streams, and forestry measures for soil conservation. In particular the banks of the Simeto River and a few of its tributaries were reinforced. Then a basin-wide flood control and soil conservation plan was approved recently. The plan includes the establishment of a flood warning system and it reserves a portion of the capacity of the reservoirs for flood storage and routing.

3. PROJECTED WATER SYSTEM CONFIGURATION

Analysis of development prospects for eastern Sicily suggests that the water demands will increase rapidly in the coming years. Two planning documents, the General Plan of Municipal Aqueducts and Special Program No. 30 projected the municipal demand for the year 2000 to range from 150 to 180 MCM/year. In anticipation of this growth, several new municipal aqueducts were proposed; the construction of several are underway now.

Projection of agricultural water demand at year 2000 is based upon about 100,000 hectares of irrigated land with a corresponding water demand of 400 to 450 MCM. Sprinkler and drip irrigation is the technology to be used.

The projected industrial water demand is 150 to 200 MCM. To meet these projected demands, several initiatives are currently under investigation to develop the surface water system of the Simeto River. These are indicated in Figure 3 and include:

1. Construction of an intake on the Simeto at Barca di Paternò and a pipeline to convey the winter stream flows from the Simeto to the Lentini Reservoir. The flows from San Leonardo River and from its minor streams will be stored in this reservoir.

2. Construction of Lentini Reservoir, capacity 130 MCM, having the purpose to satisfy water demands of the industrial districts of Syracuse and Catania and the irrigation demands of the land reclamation districts of Lentini.

3. Implementation of the transbasin diversion to Ancipa reservoir. This extension would increase the municipal supply and reduce the risk, in dry years, of irrigation deficits.

4. The construction of three new reservoirs on Simeto river: Bolo, Revisotto, Finaita.

The Bolo Reservoir, of capacity 70 MCM, has been proposed for two uses 1) municipal, to stabilize Catania's water supply; 2) irrigation to supply the districts bordering on the Upper Simeto Valley, now irrigated by Etna groundwater, and to increase the available volume to the Plain of Catania. Regulated flow in Revisotto Reservoir, capacity 22 MCM, and Finaita Reservoir, capacity 7 MCM, should be dedicated to irrigation use, too.

The development of the Dittaino-Gornalunga system involves:

1. the transbasin diversions to Nicoletti Reservoir;

2. the Sciaguana Reservoir, capacity 10 MCM, on the Dittaino, to supply irrigation districts and industrial areas;

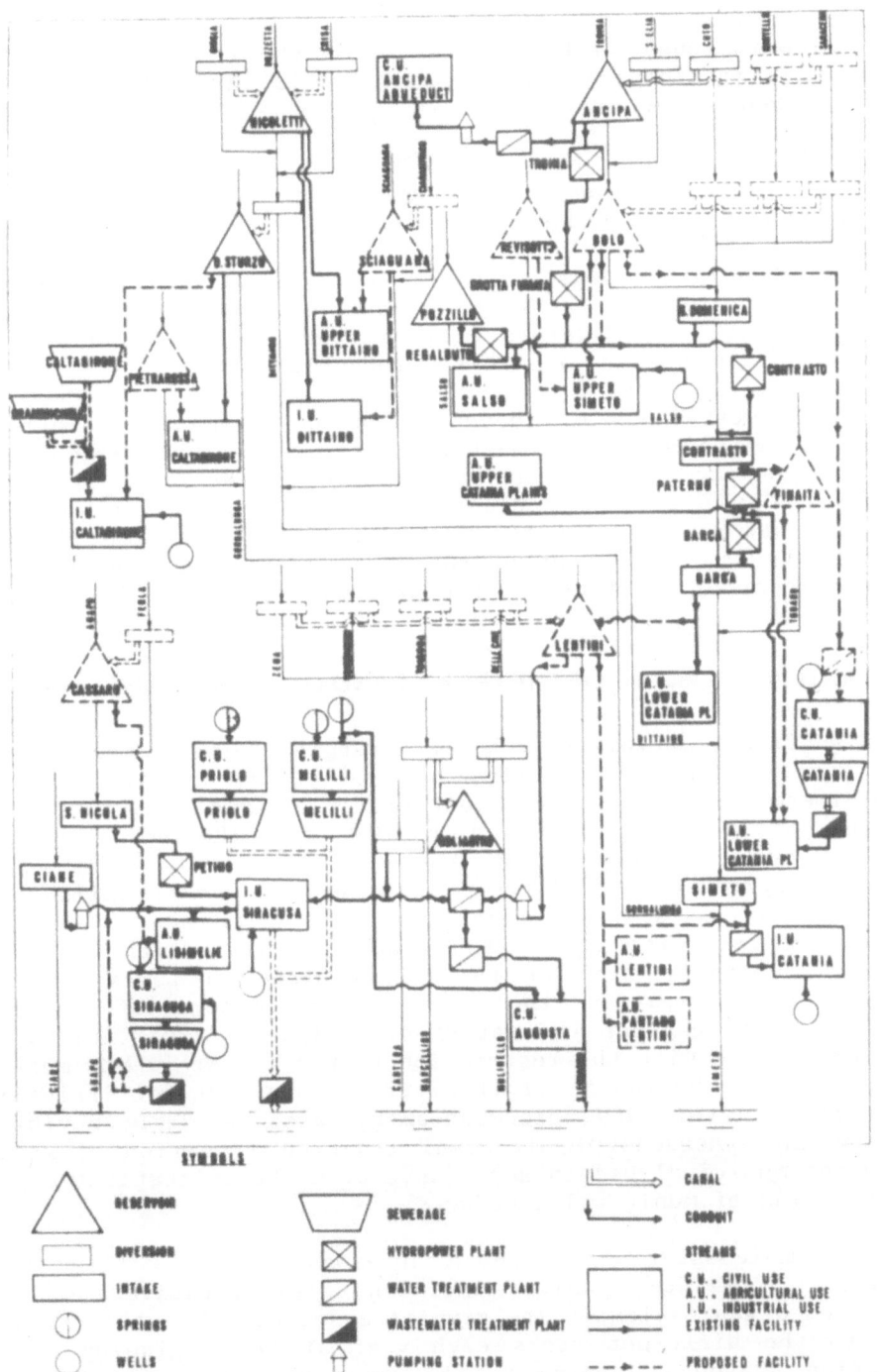

Figure 3. Schematic of the future water system in eastern Sicily.

3. Pietrarossa reservoir, capacity 20 MCM, on the Gornalunga tributary, to allow the enlargement of the irrigation district which today is served by Don Sturzo Reservoir; and

4. the use of treated municipal wastewater from the towns of Caltagirone and Grammichele for supply of the Caltagirone industrial district.

The Lentini-Syracuse system, in addition to Lentini Reservoir, may include the Cassaro Reservoir, capacity 34 MCM, on the Anapo River. Its waters will be used in the industrial and agricultural sectors and for power production.

The effectiveness of these various development alternatives for the system, which would meet the water demands at various time horizons was investigated in the context of the Special Program No. 30 of Cassa per il Mezzogiorno. The results are given in terms of water availability 100 percent of the time for the municipal users, 90 percent of the time for the industrial users, and 80 percent of the time for the agricultural users.

4. OPERATION PROBLEMS OF THE PRESENT AND DEVELOPING SYSTEMS

Significant operation problems occur in the management of the water system of eastern Sicily. presently, the operation of the Salso-Simeto system is carried out mainly by ENEL, which provides for the control of dams and intakes for the power plants. Operation criteria depends on use priorities which were established by agreements. At first an agreement was established in 1961 for such uses among Cassa per il Mezzogiorno, ERAS, Ente Siciliano Elletricità (today ENEL) and the irrigation districts (Piana di Catania, Lago di Lentini, Pantano di Lentini) which reserves to irrigation the use of the waters of Pozzillo Reservoir. It reserves the waters of Ancipa Reservoir for hydropower production with the restriction that at least 14 MCM must be retained at the beginning of the irrigation season for irrigation use. Then a later 1969 agreement among Cassa per il Mezzogiorno, ENEL and Ente Acquedotti Siciliani reserved a portion of Ancipa waters for municipal use in accordance with the indications of the national plan of municipal aqueducts.

The implementation of these agreements is controlled by a management committee which includes representatives of the interested organizations. The committee decides the allocation of water to the different users each year at the beginning of the irrigation season according to the available storage and to the forecasted inflows. The decisions about the source of the releases at the various steps of the operation is given to ENEL.

The responsibility of operation of the Dittaino-Gornalunga system is split among various organizations. ESA controls the Nicoletti reservoir, the Reclamation Agency of Altesina-Alto Dittaino manages the irrigation networks fed by the same reservoir, and the Reclamation Agency of Caltagirone deals with the operation of Don Sturzo Reservoir and of the connected irrigation system. Even in this case the coordination does not exist.

In the Syracuse area, a part of the industrial supply system is operated by an industrial company (Montedison), but most of the facilities are operated now under the direction of the Syracuse Industrial District. The necessity of performing the water treatment provides an incentive to concentrate the responsibility of the operation.

The control of groundwater withdrawal should be performed by the Office of the State Engineer which is invested with authority to allow permits for searching and using groundwaters in force accordance with the 1933 water act. However, the State Engineer is not informed of the actual number of wells and, due to lack of personnel and tools, is not able to check whether the withdrawals conform the the permits or appropriation rights. As a result of this situation, the Priolo-Augusta aquifer, which between 1951 and 1980 has been lowered 70 meters, is in a serious danger of seawater intrusion.

The empirical operation scheme, used to the present time for less complex systems, will no longer be effective in the future as the number of links among the system components and among the systems increase. The link between the Salso-Simeto system and Syracuse area causes problems already.

A strong competition between agricultural and industrial uses of the Simeto flows arose during the construction of the facilities which were proposed to convey the Simeto water to supply the Priolo-Syracuse industrial district. The Piana di Catania Irrigation District opposed the flow diversion for industrial use, pleading the ditch maintenance needs and the acquired water right. The conflict became even stronger in April 1980, following a problem in Syracuse, in which the State Engineer restricted groundwater withdrawals by the Priolo-Syracuse industries. This was to stop the drawdown of the aquifer and the increasing danger of seawater intrusion. The industrial users urged an immediate use of the Simeto River waters. They threatened to close several factories which also inspired reaction from the trade unions. Moreover, the agricultural users were afraid they might lose part of their appropriated resources which already are inadequate in dry years and definitely are inadequate to meet the increased demands of the future. A temporary solution was adopted which consisted of delaying the withdrawal restrictions and allowing, during the 1980-81 winter, the 1 m^3/s diversion of Simeto waters to the industries. The beginning of new irrigation season is reopening the problem. In addition to the above situation, other severe problems concerning the operation of multi-sources and multiple-purpose plants are arising in the developing system.

Effective innovations have to be introduced to improve the operation of the future complex system in eastern Sicily. First of all it is necessary to allow a distribution of releases from various sources to various users beyond the constraints which are assumed at planning level. This will allow pursuit of both cost minimization and improved performance. As an example the Priolo-Augusta aquifer should be considered in the future as a supplementary source of water, to be used only in drought years.

This operational direction is presecibed now in Italy by the "General criteria for the correct and rational use of water" which were issued in compliance of the N.L. n. 319 of 10.5.1976. These criteria include mention of the "better overall effectiveness which could follow to the use of a group of water sources which integrate each other in order to meet the time-varying demands of various use sectors." In particular in case of droughts or occasional dismissing of service of the facilities, it is necessary to reallocate the available water and to distribute the shortages in order to limit the overall economic losses. Furthermore, in critical situations access should be made possible to alternative or more costly water resources. In particular the over-exploited Priolo-Augusta aquifer should be considered in the future as a supplementary source to be used in dry years only. Furthermore, the treated urban wastewaters (in particular the effluents of Catania and Syracuse sewage treatment plants) should be included among the possible source of supply. However, their reuse would stand as a normal practice only if the treatment cost for reuse does not exceed the total cost of discharge. Otherwise the wastewater reuse should be considered only as a emergency provision. The implementation of such an operational scheme requires an adequate sharing of costs among the users within the overall system beyond the present framework of management organizations and water rights.

The integrated management of complex water systems requires the establishment of either an unique authority or of an appropriate coordination structure which should have access to real time information on water availability at the various sources and water demands in the various use sectors. The organization should be able to make decisions about water releases, from all sources, to each user, and to provide for cost-sharing and equitable compensation of the economic losses following the water shortages due to the implementation of the water reallocation. At present, such integrated management is constrained by the legal and institutional setting, but it is the major efficiency factor in the functioning of the future system.

5. REMARKS

Notwithstanding the many positive actions toward the construction of facilities, an adequate level of water management has not been achieved yet, due mainly to an actual lack of effective operation. However a few favorable tendencies begin to be perceived. The coordination of projects and operating schemes of the single purpose water management organizations is recognized as a present problem. But ancient rivalries among local water management organizations are yet a social problem to be solved in order to allow construction of aqueducts for common service, and to allow the establishment of wastewater treatment districts for the operation of common treatment facilities. The evolution process is slow, but the designers and the funding agencies influence the situation. Cassa per il Mezzogiorno, for example, is now giving funding preference to those projects which have economies of scale. Then the present water management organizations are becoming more aware of the importance of the operation. Operating strategies are being included as a design factor in the construction plans for new facilities and updating those existing. For example, the need of reducing the leaks and wastes has led the Piana di Catania irrigation district to redesign its distribution system. Second, the coordination of projects and operating schemes of the single purpose water management organizations is recognized as a present problem. But ancient rivalries among local water management organizations are yet a social problem to be solved in order to allow construction of aqueducts for common service, and to allow the establishment of wastewater treatment districts for the operation of common treatment facilities. The evolution process is slow, but the designers and the funding agencies influence the situation. Cassa per il Mezzogiorno, for example, is now giving funding preference to those projects which have economies of scale.

Also at the operation level an organizational structure allowing coordination of actions is required in order to overcome the splitting of management responsibilities among different organizations working in the various sectors of water supply and use. The effectiveness of the future management structure is related also to an appropriate training of the personnel involved in decision making and operation. Besides the basic educational services which could be offered by schools and universities, a need exists to establish training courses to provide opportunity for updating of the professional skills of the personnel. The computer facilities, which are becoming available, even to the local water management organizations and which are used at present mainly for trivial administrative services, should be used as aids for decision making and real time control of operation. Finally, the growing social concern about the decisions related to water services demonstrates that a need exists to match effective operation with democratic participation. This is necessary in the definition of the objectives and in control of their implementation.

ACKNOWLEDGEMENT

The contributions of Ing. B. Reitano and Ing. C. Modica in the preparation of this paper are gratefully acknowledged.

REFERENCES

1. VISMARA E.: "Gli impianti idroelettrici nella Sicilia oriental in relazione allo sviluppo industriale dell'isola". Atti del VII Congresso Geografico Italiano Palermo (1911).

2. GUGGINO PICONE E., GRAMIGNANI M., INDELICATO S.: "Contributo allo studio con moderni metodi della situazione a fini multipli di bacini idrografici (Studio sulle disponibilità idriche per l'irrigazione della Piana di Catania)." Tecnica e Ricostruzione, n. 6, 1968.

3. BOSSOLA A.: "L'approvvigionamento idrico intersettoriale della fascia costiera orientale siciliana." Atti del Seminario della 1^a Sezione dell'Associazione Italiana di Ingegneria Agraria. Catania, maggio 1969.

4. MANZI E., RUGGIERO V.: "I laghi artificiali della Sicilia." Istituto di Geografia e Geografia Economica dell'Università, Napoli, 1973.

5. ROSSI G.: "Criteri ed esperienze nello studio di un sistema di risorse idriche superficiali." Giornate di studio su 'Problemi di modellistica a controllo nella gestione delle risorse idriche.' Università della Calabria, Cosenza, marzo, 1975.

6. INDELICATO S., ROSSI G.: "Methodological aspects of water resources planning by simulation techniques." Proceedings of the Second World Congress of International Resources Association, New Delhi, December 1975.

7. ROSSI G.: Application of 'System Analysis'. The Upper Simeto basin System. Lecture 6 Audiovisual Course on 'Water Management a System Approach.' For Collins, 1976.

8. LOTTI C. & ASSOC.: "Studio idrologico delle risorse idriche superficiali della Sicilia occidentale e orientale." Roma, 1977.

9. CIRIEC: "Aspetti normativo-istituzionali e sistemi tariffari nel quadro del piano delle acque della Sicilia." Casmez, Roma, 1977.

10. PENTA P.: "Anlisi della possibilità di utilizzazione ad uso promiscuo di acque già destinate per scopi idroelettrici in Sicilia." Casmez, PS 30, Roma, 1978.

11. FABI F., SIMONELLI G.: "Mathematical models of water resource systems. Some recent realizations as the 'Cassa per il

Mezzogiorno.' International Symposium on logistics and benefits of using mathematical models of hydrological and water resources systems, Pisa, October 1978.

12. ROSSI G., INDELICATO S.: "Analisi e definizione dello schema delle opere per l'approvvigionamento idrico intersettoriale della Sicilia centro-meridionale: Casmez, P.S. 30, Roma, 1979.

13. INDELICATO S., ROSSI G.: Analisi di alternative dei sistemi di risorse idriche della Sicilia centro-orientale." XVII convegno di Idraulica e Costruzioni Idrauliche, Palermo, ottobre 1980.

14. AA. VV.: "Difesa dell'ambiente ed assetto del territorio della Sicilia sudorientale." Convegno sotto gli auspici dell'ANDIS, Siracusa, novembre 1980.

25. WATER MANAGEMENT INSTITUTIONS IN EASTERN SICILY

Emilio Giardina and Salvatore E. Battiato
Istituto delle Finanze, Universita di Catania

1. INSTITUTIONAL FRAMEWORK

The water resources in Italy, both surface or underground, are considered public property. As such they are administered by either the state or the region. Any person wanting to use these waters must apply to the public administration for a diversion license.

This ruling is potentially able to assure efficient and equitable distribution of water. But this is not achieved in fact because there is no general vision applied in its administration. Rather the applications are interpreted only within the narrow limits of convenience indicated by the applicants. Also, in evaluation economic factors are not given adequate attention. In recent years, however, legislation has been introduced which is the administrative instrument for the efficient use of water. In 1968 the regulations for aqueduct planning were put in force. These regulations provided a workable system of financing construction and established criteria for drinking water supply service, a priority that can be of importance for the pre-existing licenses.

In 1976 law n. 329 for the protection of water from pollution was passed. This law also concerns the quantity of water resources. Also introduced were criteria for the more efficient use of water. Water conservation and water reuse were major concerns. In addition, it was decided to continue development of regional plans for water quality management. In these plans the need for aqueducts, sewerage, and sewage treatment must be determined, and the factors determining their construction must be defined. And finally, criteria must be specified for defining optimal territorial ranges for the management of the mentioned services.

In Sicily, where a special autonomous government is in force, the water development plan must be drawn up in conjunction with the general plan for the protection of the environment and in 1977, as administrative functions were transferred from the state to the regions, it was decided that priority must be given to projects which are to be multiple use.

2. THE MANAGEMENT ORGANIZATIONS

In eastern Sicily the organizations and the institutions which take care of the collection, transport and distribution of water, that is of the management of water resources, may be classified according to the scheme in Table 1. Besides these

Table 1. Organizations Responsible for Water Management

A. Irrigation water

   1. Private irrigation firms

   2. Consorzi di irrigazione di diritto privato

   3. Consorzi di miglioramento fondiario

   4. Consorzi di bonifica

   5. Amministrazione dei canali demaniali

   6. Ente di Sviluppo Agricolo (E.S.A.)

B. Municipal water supply

   7. Private firms

   8. Servizi acquedottistici comunali

   9. Aziende municipalizzate

  10. Consorzi acquedottistici

  11. Ente Acquedotti Siciliani (E.A.S.)

C. Industrial water supply

  12. Consorzi per le aree di sviluppo industriale

  13. Ente nazionale per l'energia elettrica (ENEL)

organizations, water management is done also by the land owners without the mediation of any other organization.

Some of the organizations in Table 1, e.g. 1 and 7, are commercial firms whose aim is profit. They may be either individual firms or companies, and the latter may be either unlimited companies or limited companies. These enterprises have the powers and the limits that our law system permits in the practice of commercial activities. Also, they are subject to a regime of administered prices.

The remaining organizations are non-profit. They work either in the interests of their owners or, if they are public organizations, in the public interest. This latter is the far more predominant part, and it is the one that strongly shapes the water industry.

Some of these organizations, e.g. 2 and 3, are of a private type, in that they are owned and operated by the owners of the estates where irrigation is done. The others are public organizations which are designed to provide water supply service, e.g. numbers 5, 9, 10, 11, or to develop such a service in the whole of the public aims that they pursue, e.g. numbers 6, 8, 12. The consorzi di bonifica (reclamation districts) occupy a special position. Although they gather owners of the interested estates, they are public organizations since they are entrusted with the task of putting into effect the reclamation plans of their hinterland. Their public feature has been growing more and more, as a consequence of the interventions of the public administration in the formation of their decisions, and of the public financing which they have been granted.

ENEL fits into the picture as an organization which produces, under a monopoly system, hydroelectric power and therefore manages water storage and regulation, and distribution to other users.

We must also mention those organizations which take care of public service such as sewage and treatment of used waters. In general they are the same public organizations that distribute water for industrial or potable use, e.g. Comuni, union of Comuni, and Consorzi for the development of industrial areas. In Sicily, as in other regions with a special statute, another body is also foreseen, the Consorzio, which is created by the region.

3. CHARACTERISTICS OF THE ORGANIZATION SYSTEM

As seen in the foregoing section, the water resources in eastern Sicily are managed by a variety of organizations, each endowed with a decision-making autonomy of its own. The bodies generally restrict their activity towards only one type of use. The commercial firms, at times, take care of domestic water supply. The Consorzi for the industrial development areas, in the limits of the land assigned to them, develop service for industrial use and for potable uses.

These organizations don't always take care of all the phases of the water provision processes, e.g. collection, storage, diversion, transport, and distribution to consumers. The collection and transport over long distances in some cases are done by one organization which gives the water to other organizations for further transport and distribution. This, for example, occurs in the case of the Salso-Simeto system. ENEL takes care of the operation and maintenance of dams, crossings, and some tracts of canals, and then gives the water to the Consorzi di bonifaca. Also, Ente Acquedotti Siciliani sells water to some municipalities which distribute it to consumers.

The relationships between supplier and transferee may be of two types. The amount of water to be transfered, as in the case of the Salso-Simeto, may be determined at the moment of the concession of license, or in later agreements. In such a case the expenses for the service done are paid to the transferer, and these expenses are shared among all the subjects interested, according to the volume of water which has been used. The amount of water to be transfered, on the contrary, can be the object of a real selling act. In such a case, it must be negotiated and the right price paid. This is defined by an administrative authority, on the basis of the costs to be met by the larger organization.

Regarding the cost of the water for the purchasing organization, there are no substantial differences between the two systems. In both cases prices are determined according to the costs of the service for collection and transport. But the latter system is more favourable for the buyer because of the delays which the administrative authority takes in bringing the water prices up to date.

The actual management of the water system of eastern Sicily is divided into a multiplicity of organizations which have a public character. This does not assure the best conditions for exercising control of such a system.

Above all, the character of the prevailing organizations must be taken into account. Given their public or cooperative nature, these bodies are likely to maximize the welfare of their consumers in terms of quantity of utilizable water. This necessitates a tendency to buy up the greatest volume of resources, which also makes every organization reluctant to yield any water to other parties. Therefore situations have arisen where valuable resources are destined for nonpriority uses, while some users are facing shortage.

In the case of public bodies, there is the possibility that they must apply for external financing. The administrators then may try to gain the support of the electors and the users by providing lower-than cost water tariffs. This causes expansion of the demand and waste of the resource.

Some effects of the current system of financing water systems are yet to be considered. For a large part the financing is done through external sources such as Cassa del Mezzogiorno, Regione, Ministeri, and no burden weighs on the management body. No sum of the capital share is asked to be refunded annually, nor is a debt paid. Therefore there is no market price for the capital on whose basis the management body must evaluate the worth of the project, as occurs when financing is done with a loan. The responsibility for such economic evaluation is left entirely to the financing bodies. This causes the need for coordination of other bodies whose initiatives affect the same water system. Up to now the responsibility for such coordination has been assigned to the financing bodies. But the evaluations done by these bodies are not commensurate with the operational complexity of the water systems involved. The financing body, even if it can take into account all the relevant factors required in the evaluation of the project to be financed, does not control the strategic variables of the water system. Some of these variables are in the decision space of other water bodies, and must be considered as constraints. The result is that investments may be approved that are clearly inferior in comparison to other alternatives.

With the adoption of regional quality management planning, a more suitable basis for evaluation of the projects has been created. This is because of the general powers given to the region regarding the water. Thus with a regional planning basis greater attention can be given to the possibilities for reuse of reclaimed water. This, of course, involves the competence of many bodies, and requires the authority for global programming of the system.

The external financing system causes fractioning of water functions among many management bodies. This, in turn, causes great difficulty in setting appropriate water prices. The consumers served by public bodies, which receive external financing, may enjoy, without any merit, lower prices than those charged to the consumers of private commercial firms. This situation in Sicily is realized in both municipal and agriculture uses. Such low prices are made possible in particular to those agricultural users served by <u>Consorzi di bonifica</u>. Farmers served by private firms and private institutions are subject to higher

unitary prices only because their supplier is not a
Consorzio di bonifica.

4. MEDIUM AND LONG TERM PROPOSALS

In order to improve the operation of the water system of
eastern Sicily several innovations are required in the area of
management.

Some improvements may be achieved in the short term, as they
don't require reorganization of the structure of the managing
bodies. Essentially, the proposal is to assure better coordination between them. This requires the creation of an institutional
forum for coordinated programming.

This forum could take the form of a consortium among the
responsible planning bodies. But because such a consortium has a
voluntary character, there are manifold difficulties which must be
addressed if there is to be a cohesive interaction between the
bodies. Thus far, our experiences with voluntary consortia of
local bodies is not satisfactory. If the consortium is to have
binding decision powers regarding the participants, the important
and delicate problem arises regarding their respective voting
weights. They are so heterogeneous in character that the task is
difficult. An alternative solution is to utilize the programming
powers which belong to the region. The coordination task may be
given to the region as a means to develop a regional plan for
water quality management. To do this an operative structure for
regional administration must be constituted. This body would then
be entrusted with the task of preparing the proposals for the
water system of eastern Sicily which would fit within the framework of a regional plan.

For the long term, a major reorganization of the managing
bodies must be undertaken to create a single planning body for the
whole territory. One possibility is to assign this body the
programming and managing functions pertaining to the production
and wholesale distribution of the water. The existing organizations would then distribute the water to the consumers.

This solution provides unity in decision making in the most
important matters relating to the operation of a complex system,
but would avoid the necessity of facing the problems of the
subtraction of the water services to some of the present bodies
and of the abolition of the others that have aqueduct functions
only. This is necessary, however, if we want to assign total
responsibility to the unitary body for all phases of water supply
planning and operation. In a word, we must create a basin agency
with general competence.

Both approaches must be evaluated, taking into account that the water system of eastern Sicily is interconnected with other systems on the island, and that many bodies are involved in operation of the system. These include organizations such as Ente Acquedotti Siciliani, and Ente di Sviluppo Agricolo, which have jurisdictions over the whole island. Given the political barriers, it seems difficult that a radical organizational reform of the managing bodies could be approved, a reform limited to only one side of the island. To decide which solution is best the alternatives must be examined within the context of a comprehensive general policy for the Sicilian region.

For the long term the problem of organizational unification of the water management authorities must be addressed. It simply is not rational to divide the various water management tasks among so many single function organiztions without coordination. The authority to issue licenses for the diversion of surface waters, defining the standards of used waters, control of water quality, planning of capital investments, and regulation of wastewater discharges, should be coordinated by a single body, whether a consortium or a single all-powerful organization.

26. WASTEWATER REUSE STUDIES APPLIED TO THE MEZZOGIORNO

Fulvio Croce
Hydro-Triad, Ltd., Denver, Colorado

In recent years, water has become an increasingly scarce resource in many regions of the world. Comprehensive water planning for all water resources has been recognized as the most appropriate approach to deal with such problems. Water reuse is becoming more prominent within the comprehensive planning approach.

Southern Italy, also called the "Mezzogiorno," is a case in point in which water reuse could have an important role. Agriculture is still the main commercial activity, and the availability of water is often the major limiting factor. At the same time the per capita production of wastewaters by major municipalities is about 150 cubic meters per year. Thus Palermo, for example, produces about 100 million cubic meters annually, which is discharged into the sea. This amount of water is sufficient to irrigate about 20,000 ha. Thus such a scheme could increase the total supply of water, with a concurrent increase in crop production; another significant benefit is in the area of pollution control.

Implementation of reuse requires consideration of many variables, both technical and nontechnical. The mix of these variables determine whether there is a viable market for reclaimed wastewater.

Water reuse planning requires a systematic protocol. The major steps are enumerated as follows:

1.   determine the available alternatives for reuse,

2.   delineate a set of critical decision-making parameters,

3.   outline a demonstration program,

4.   define problem-solving strategies, e.g., management and monitoring programs,

5.   develop recommendations for full-scale implementation.

These steps are expanded in this chapter to outline a general procedure for water reuse planning with special reference to the critical aspects of the decision-making process (1).

1. PRESENT CONTEXT

Because fresh water is a limited resource, the cost of each new large increment of supply is increasing exponentially due to the remoteness of new sources, escalating energy and delivery costs, and environmental considerations. In many industrialized countries, including Italy, regulations now require a minimum of secondary treatment of all wastewaters before discharge (2). This creates an unique opportunity for making the reuse of higher quality waters economically attractive for meeting present and future demands.

In Germany, India, Israel, South Africa, and the United States, especially California, up to twenty percent of wastewater discharges are already reclaimed and reused, especially for crop irrigation. Several hundred projects are involved (3). Some of these projects, e.g., Melbourne, Australia, and Lubbock, Texas, have collected more than forty years of data. Thus a great deal of research and experience has been accumulated on treatment, environmental impacts, and health effects for reuse. From these experiences, appropriate analyses and comprehensive field investigations could provide the technical basis for planning new schemes.

A second phase in implementation of reuse planning, in addition to the technical aspects, is to set standards for reuse. Only California has a comprehensive set of reuse regulations (4) sufficient for implementation. Table 1 summarizes the California Standards. Reliability of treatment is the most critical issue, even in the technologically advanced California (5).

Table 1. California Standards for Reuse of Reclaimed Wastewater

	Minimum Requirements	
Planned Wastewater Reuse	Treatment	Disinfection (Coliform Count, MPN/100 ml)

Irrigation

-Fodder, Fiber, and Seed Crops	P	No requirement
-Produce eaten raw:		
surface irrigated	S	2.2
spray irrigated	T	2.2
-Processed produce:		
surface irrigated	P	No requirement
spray irrigated	S	23.0
-Vineyards & Orchards:		
surface irrigated (no fruit picked from the ground)	P	No requirement
-Landscapes, parks, etc.	S	23.0

Recreation impoundments

-Lakes (aesthetic enjoyment only)	S	23.0
-Restricted Recreational lakes	S	2.2
-Nonrestricted recreational lakes	T	2.2

Note: P = Primary treatment; effluent not containing more than 1.0 ml/liter/hr setteable solids.

S = Secondary treatment and disinfection.

T = Coagulation, filtration and disinfection after secondary treatment; effluent not containing more than 10 Turbidity Units.

2. RECONNAISSANCE PHASE

A reconnaissance study is the first phase of a reuse evaluation. The primary objective of the reconnaissance study is the preliminary assessment of the overall feasibility of a reuse plan, at a given location, and in an existing context. To evaluate the impacts to the society and to the environment in a comprehensive, systematic manner, a team of experts must be formed. They should represent the disciplines of environmental engineering, political science, public administration, economics, agronomy, chemistry, sociology, and public relations. Table 2 outlines the tasks and objectives of this team, working in the reconnaissance stage. It shows the many considerations that must be examined from the very beginning. Detailed investigations will follow only if favorable conclusions can be drawn at this stage.

Table 2. Team Tasks at the Reconnaissance Stage

Tasks	Objectives
Tasks assignment	Assessment of the program major objectives Distribution of responsibilities Start-up of the team work
Collection of available data	General knowledge of the system Base line assessment, i.e., system "as it is"
Preliminary water analysis	Assessment of water suitability for reuse Preliminary selection of type of reuse Selection of related group for users
Identification of entities involved	Assessment of the administrative, political, and social environment

Assessment of the political and public level of interest	Evaluation of the general political
	    feasibility and suggestions for the
	    program strategy
	Identification of selected parameters for
	    alternatives evaluation

Assessment of the economical and financial status	Evaluation of the general economic
	    feasibility and suggestions for the
	    program strategy
	Indentification of selected parameters for
	    alternatives evaluation

3. GENERATION OF ALTERNATIVES

Successive approximations and group interaction are required to produce compromising alternatives, with minimal adverse impacts. The scope of work entails the following:

1. select the most encouraging alternative,

2. prepare a detailed experimentation and demonstration program,

3.   identify the hypothesis to be checked,

4.   describe the scenarios to be evaluated,

5.   design a pilot test program.

The major planning elements of the feasibility invertigation are described in Table 3.

The investigations outlined in Table 3 are the basis for the planning and design of reuse facilities. In addition, effective team management is needed to insure cooperation and coordination among the team experts.

4. DEMONSTRATION PROGRAM

The complex interaction among the variables, and the financial effort involved in a reuse plan make essential a period of pilot experimentation in the field. The outputs of the demonstration phase are:

1.   concepts are tested,

2.   data are obtained to support decision making,

3.   quantification is provided for all parties,

4.   direction is provided for management and monitoring of the system,

5.   critical parameters to be checked during routine operation are identified.

The demonstration program should be performed under supervision of the team. It should last a minimum of two years, during which treatment and disinfection processes are tested, along with crop patterns, and agricultural and irrigation practices. The parameters previously identified, and affecting the system, are the focus for measurement.

An example of an experimentation and demonstration program is the Monterey Wastewater Reclamation Study for Agriculture (6) soon to be implemented in Monterey, California. In order to design a reclamation project for irrigation on high-value agricultural land, investigations, surveys and analyses will be performed on a wide variety of parameters over a five year period. The results of the study may be applicable to similar projects.

Table 3. Detailed Feasibility Studies

Elements	Characteristics	Parameters
1. Political Feasibility	Reuse may be an impopular issue; Additional water supply may be attractive to farmers; Proper incentives and information are recommended.	Political profitability; Involvement and interest of political parties; Coalitions strength; Future trends.
2. Financial Feasibility	Parallels political feasibility; It is influenced by water price policy; A "pay-back" system is recommended for efficiency and reliability.	Financial profitability; Capital availability; Present and projected price of water.
3. Economic Feasibility	The greater B>C, the greater the chances of success; A sound sensitivity analysis is recommended.	All economic indexes; Costs and values of inputs and outputs; Future trends.
4. Legal Feasibility	A detailed contract will protect suppliers, users, and public; Water quality and quantity, controls and penalties must be fully defined.	All legal clauses; Irrigation status and crops grown; Management responsibilities.
5. Institutional Feasibility	Agencies capabilities must be assessed and/or improved; A detailed administation scheme must be developed; Simplicity and flexibility of operation is recommended.	Existing system and competences; Distribution of costs and responsibilities.

Table 3. (Continued)

Elements	Characteristics	Parameters
6. Technical Feasibility: (i) Water quality	The level of reclaimed water quality directly influences cost, crop pattern, safety; Consequently, it has direct political and economic implications.	Detailed water quality standard, depending on the intended reuse; Degree of dilution with "conventional" water.
(ii) Sanitary	Public safety must be ensured; Proper disinfection and high-level reliability is imperative; An experimentation period is recomended.	Site characteristics; Irrigation and agricultural practices; Degree of dilution with "conventional" water; Storage and monitoring provided.
(iii) Local	The impacts on surface and groundwaters must be studied.	Local geology and geomorphology; Groundwater data; Nitrate pollution potential.
(iv) Pedology	The impacts on the soil must be evaluated; The nitrogen and phosphorus balance must be controlled.	Soil Characteristics; Water mineral content; Agriculture and irrigation practices; Specific heavy metals and toxic compounds.
7. Crop Pattern	This issue has very complex implications; Market trends must be studied; Farmers attitude and acceptance must be assessed.	Baseline data; Indications from other investigations; Public attitute; Enforcement capabilities.
8. Public Acceptance	Public information plays a determinant role; A demonstration program is recommended	Public attitude; Crop pattern; Information and incentives; Monitoring program

5. MANAGEMENT AND MONITORING

The design of a management system with the authority, the competence and the elasticity needed to deal successfully with all the dynamic uncertainties involved with the treatment and reuse of wastewater, should also be responsibility of the team. The same can be said for the monitoring program.

Many of the parameters to be monitored are routine for all wastewater treatment plants. Others may require a more highly specialized personnel, as well as a relatively larger financial effort. Three impact areas requiring special consideration are: public health, toxic substances, and nitrogen compounds. The word "impact" should be considered in broad terms with respect to the food chain, soil-crop interactions, and long term effects. Table 4 illustrates some of the aspects of a monitoring program, related to the foregoing concerns.

Table 4.   Monitoring Scheme (7)

Analysis Recurrence:	Daily	Weekly	Annually or Semi-Annually	
Origin	-Effluent -Return flows -Monitoring wells	-Effluent -Return flows -Monitoring wells	-Water supply wells	-Soil
Para- meters	-Chlorides -Pathogens -Specific conductivity -Other indicators of a change in quality	-pH -Total Hardness -Alkalinity -Ammonia Nitrogen -Nitrite Nitrogen -Nitrate Nitrogen -Total Phosphorus -Methylene Blue Active substances (detergents) -COD and SS -Any critical heavy metal or other toxic compound	-Boron -Cadmium -Chromium -Copper -Lead -Nickel -Zinc -Mercury -Arsenic -Any other potentially harmful element or compound	

(7)   USEPA, "Application of sludge and wastewater on agricultural land," Report No. EPA/MCD-35, Cincinatti 1978.

426

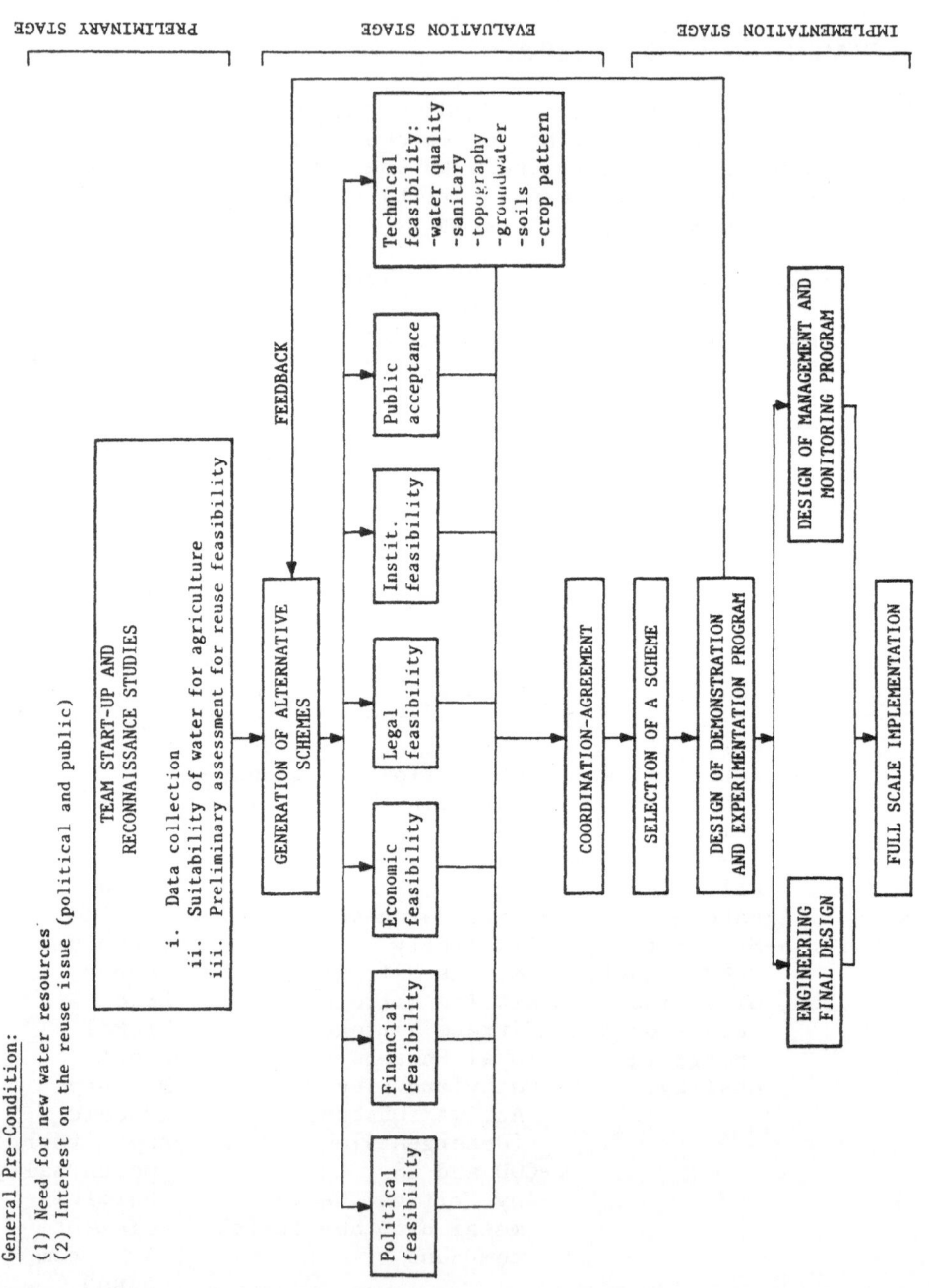

General Pre-Condition:

(1) Need for new water resources
(2) Interest on the reuse issue (political and public)

Figure 1. Appraisal Procedure for Reuse Planning

The vegetation produced may provide the most meaningful item for monitoring. Some toxic substances will accumulate in the plant. Analysis will ascertain possible hazards to the food chain and to the plant itself (7).

If crops are grown for direct human consumption, investigations in the pilot plant stage are needed to determine standards for sanitary safety. If sprinkler irrigation is used, pathogens and viruses must be checked in the water, soil, products, and plants. Marketing of vegetables must be allowed only if all standards are met. A clear set of standards should be prepared for operationing guidance.

6. CONCLUSIONS

The approach presented for planning feasibility studies of wastewater reclamation and reuse for crop irrigation would utilize the experience and the research that has been accumulated. Of special interest is the potential for application in southern Italy.

Conclusions from this study are:

1. technical capabilities for implementation of reuse plans are well developed,

2. past experience shows that nontechnical elements, e.g., reliability, public and political support, and management, are often critical,

3. comprehensive and detailed investigations are required for a successful implementation of a given reuse scheme,

4. an extensive experimentation program, and a public education program are recommended before the final design,

5. management schemes must be developed to administer the project for a continuous, and reliable service.

The rationale of the recommended reuse planning procedure is summarized quite simply in Figure 1. What is not so simple is the implementation of the process described.

REFERENCES

1. U.S. Office of Water Research and Technology, "Water reuse: A bibliography," U.S. Dept. of Interior, Washington D.C., 1976.

2. Repubblica Italiana, "Norme per la tutela delle acque dall' inquinamento," Roma, 1976.

3. Shuval, H. I., "Water renovation and reuse," Academic Press Inc., N.Y., 1977.

4. California Administrative Code, Title 22, "Wastewater reclamation criteria," State of California, Sacramento, 1972.

5. U.S. Environmental Protection Agency, "Survey of facilities using land application of wastewater," Report No. EPA/430/ 19-73-006, Cincinatti, 1973.

6. American Water Work Association, "Municipal wastewater reuse news," Sept. 1980.

7. U.S. Environmental Protection Agency, "Application of sludge and wastewaters on agricultural land," Report No. EPA/MCD-35, Cincinatti, 1978.

27. THE SANTA CLARA VALLEY WATER DISTRICT

Lloyd C. Fowler
Goleta Water District, Santa Barbara, California

1. INTRODUCTION

Santa Clara County is located at the south end of San Francisco Bay. Its location within the State of California is seen in Figure 1. It is a water-deficient area in terms of its present water needs. Water management in this 3,100 square kilometer county is the responsibility of the Santa Clara Valley Water District, an agency created by the California Legislature and controlled by the residents of the district.

When the area was first settled in the late 1700's, water was obtained in limited amounts from springs and creeks. The area was used for production of livestock and dry-farm grains. In the mid-1800's, orchards and other crops were planted which required irrigation during the dry summers.

Some of this irrigation water was derived by diversion of streamflows from the surrounding hills, but this supply was subject to the variations of the climate, and was not dependable. The grounwater basins were tapped first by artesian wells and later by deep well turbine pumps. The expanding use of water caused decline of groundwater tables. In the late 1800's attempts were made by individuals and groups of farmers to recharge the groundwater basins by diversions of winter flood flows.

As the irrigated farming continued to expand with corresponding increases in population, numerous water surveys and studies were made to develop plans and programs for an adequate water supply. These studies and surveys all indicated that the winter flood flows should be stored and used to recharge the

430

Figure 1. Map of State of California Showing Portion of
Water System Related to Santa Clara County

groundwater basins. But it was not until 1929 when a severe drought occurred, that attempts to form a water management agency were successful. By 1936 six reservoirs had been constructed in the mountains to store winter flood flows. The associated facilities were constucted also to recharge the groundwater basins with the stored waters.

Increasing population caused commensurate increase in water use and although additional storage reservoirs were constructed in the 1950's, it was clear that the local water resources would be inadequate to meet the water needs of the area. As a temporary measure the full use of local surface supplies and excessive draft of the groundwater basins, i. e. "mining", made available sufficient water supply for population of over 500,000, and for irrigation of more than 40,000 hectares. The result was that by 1960 the groundwater basins were being overdrafted at the rate of 50,000,000 cubic meters per year.

2. PROBLEMS OF GROUNDWATER OVERDRAFT

The groundwater levels in the basin then receded below sea level, creating the threat of saltwater intrusion. Over the years, saltwater did intrude in the surface zones but has not yet appeared in the deeper aquifers which are the main production units. In other areas of the groundwater basins, lowered ground-water levels resulted in movement of connate saline water into wells. This placed a limit on the amount of groundwater that could be extracted in local areas without major degradation of water quality.

The lower groundwater levels also caused land surface subsidence over an area of 63,000 hectares. Between the years 1912 and 1967 a maximum depression of about 3.6 meters occurred. Figure 2 shows the amount of land surface depression which occurred between 1934 and 1967. Since cessation of overdraft in 1969, and replenishment of groundwater levels, there has been no significant land surface subsidence. This is illustrated in Figure 3 which shows both the depth to water and land elevation at selected index locations.

During the period of active land surface subsidence, water well casings buckled, sanitary and storm drainage sewers lost capacity as a result of changes in slope, roads and railroads had to be raised to stay above floodwaters, and new levees had to be erected and old ones raised to protect developed areas against flooding from San Francisco Bay. Depression of the land surface below sea level required installation of pumping stations to dispose of urban stormwaters and treated sewage.

Figure 2.   Subsidence Contours in North Santa Clara Valley

These damages resulting from land surface subsidence are estimated to have amounted to over $130 million. In addition, there has been an immeasurable loss of aesthetic value to those lands near the Bay which were above sea level and because of land surface subsidence are now at or below sea level.

Receding groundwater levels also result in hidden costs. The lower groundwater levels require additional energy to lift the water, adding millions of dollars to the cost of pumping. This was reflected in higher costs for agricultural products and in higher water bills for the consumer.

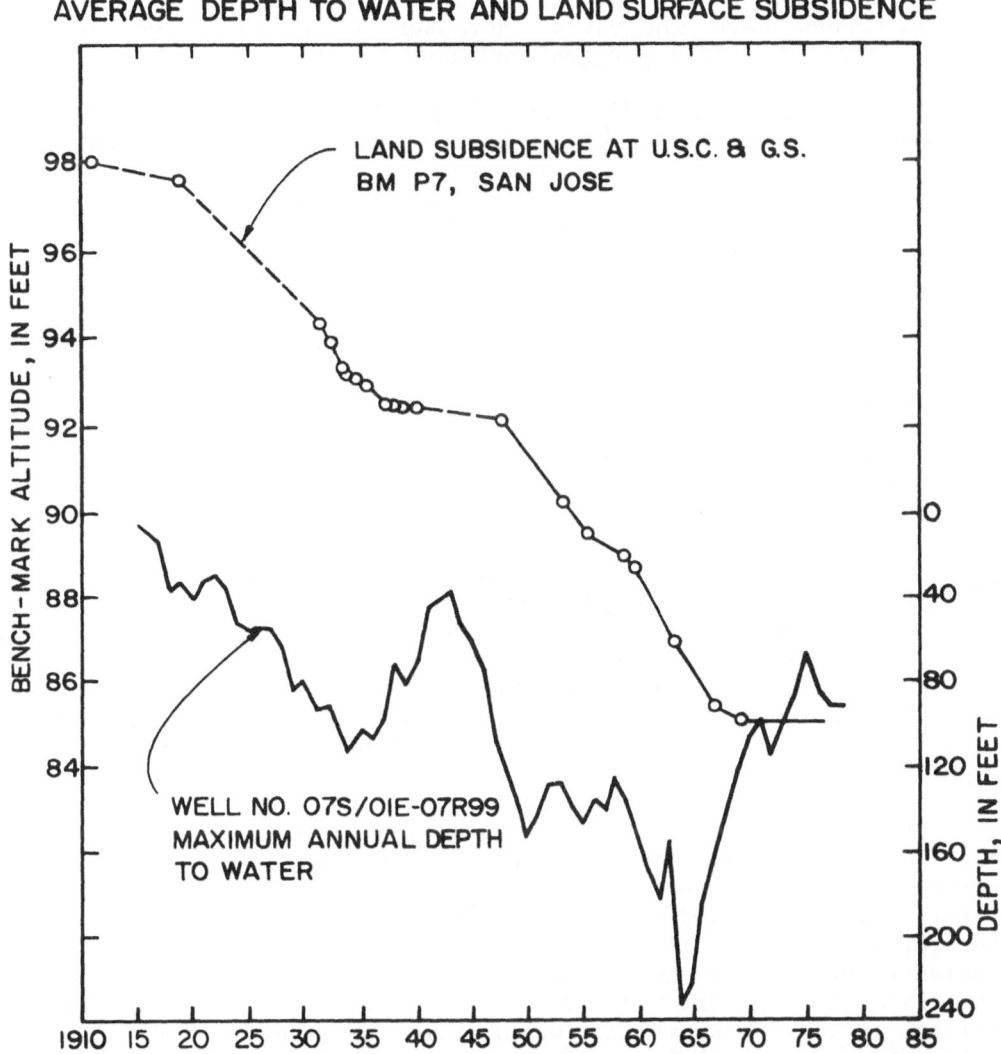

AVERAGE DEPTH TO WATER AND LAND SURFACE SUBSIDENCE

Figure 3.   History of Depth to Water and Ground Elevation at Index
            Locations within Santa Clara Valley Water District

3. WATER SUPPLY PROJECTS

In an attmept to provide additional water for the area, the
Santa Clara Valley Water District undertook a weather modification
program in 1955.   This effort consists of seeding appropriate
cloud formations with silver iodide crystals from ground-based
generators in an attempt to increase rainfall.   Comparisons

between the target area within the district and adjacent nonseeded control areas have indicated increases of up to 15 percent in annual rainfall. This program continues to this date.

The attempts to fully utilize local resources succeeded in improving the available water supply but did not increase the total water resources of the area. As the need for water increased beyond the limits of the local supply, a search began for water supplies from outside the county. Many alternative programs of water importation were considered. From this consideration three supplemental water supplies have been developed as part of the water management program for the area. These include deliveries from the Hetch Hetchy Aqueduct, the South Bay Aqueduct, and the San Felipe Aqueduct; all are shown in Figure 1. The Hetch Hetchy Aqueduct of the City of San Francisco, which delivers water from the Sierra Nevada Mountains to San Francisco, also supplies water for municipal and industrial use to northern Santa Clara County. These supplies have been received since 1962. In 1961 a contract was signed with the State of California for water from the South Bay Aqueduct of the State Water Project. The State Water Project transfers surplus northern California waters to water short areas in central and southern California. Deliveries from this source began in 1965; the water is used for both groundwater recharge and, after treatment, for domestic service. The third source of imported water has been obtained by the district from the San Felipe Division of the Central Valley Project of the U.S. Bureau of Reclamation. Deliveries are programmed for 1986.

As part of its overall water management program the district also has constructed a water reclamation facility to use highly purified water from domestic sewage for injection into the ground-water basin to create a barrier to saltwater intrusion. This barrier will be in the shallow aquifers and will be operated in conjunction with a series of extraction wells, located landward of the injection wells. Initially, the extracted water will be saline; after this saline water wedge has been removed, the extracted water will be primarily injected water and will be used for irrigation of recreational areas. Programs for the development of other water reuse projects which will furnish water for irrigation of landscaping and agricultural crops are currently underway. These programs are expected to result in adequate water supplies for the future, i.e., beyond the time when the current import programs are supplying water to their full capacity.

Another important direction, the district has embarked on a comprehensive water savings program in an attempt to reduce water consumed in households, commerce, industry, and agriculture. Projections of potential water savings indicate that 10 percent of the normal anticipated water needs can be saved by reasonable care

on the part of all water users. The primary aspect of this water
savings program is public education. It is expected to have its
major effect in the future.

4. PRICING POLICY

The water management program of the district is supported by
revenues from taxation and water charges. The pricing policy
embodies a collective concept in which all supplies are seen to
contribute to the general water supply for the entire area. All
costs for the various supplies are accumulated and distributed on
a uniform unit price for water in accordance with the type of use.
Some portions of the supply that require extra costs, such as
treatment to make it potable, or special pumping plants for
delivery, are added charges to those consumers benefiting
directly. This collective price concept is extended throughout
the service area of the district. The charge for agricultural
water is about one-fourth that for municipal and industrial water.
Groundwater extracted from the groundwater basins of the area is
charged for as though it were surface delivered water. All water
services are metered, as are all larger production wells. Since
the water needs of the community are fully provided for, there has
been no question of prior rights to groundwater or of charges
placed on waters that may be generated by local rainfall.

5. WATER SUPPLY CAPABILITY

The combination of water sources, e.g. storage and diversion
of local surface waters, groundwater, and imported water, provide
for a 1979 population of over 1,200,000, and for irrigation of
20,000 hectares. The total supply capability was about 490
million cubic meters in 1979; about one-third of this is imported
water. Additional imported water will be available in 1986.

6. WATER RESOURCES MANAGEMENT IN THE SANTA CLARA VALLEY

6.1 Water Resources Management

The long term goal of the district is to accomplish total
water resources management. The word "management" implies an
executive function that includes planning, organizing, coordinating, directing, controlling, and supervising any project or
activity with responsibility for results. The heart of the
definition of management is responsibility for results. There is
also an "objective function" that is part of this definition.
That is, water resources are to be managed so as to insure
continued and maximum beneficial use.

Water resources management can be divided into three areas. First, there must be knowledge of the physical facts, particularly with regard to surface and groundwater hydrology. There never seems to be enough of the right kind of data, but planning and project development must proceed in the best manner possible with what information is available. As data collecton systems are established and data are accumulated, changes can be made in project plans. But as the emphasis moves to the operational stage the data needs change. This need to change should be reflected during the establishment of the data collection system.

Second, a water supply plan that will provide for the needs of the area must be developed. At one time it was relatively easy to say that whatever the projected needs were, they would be supplied through some formulation of projects. This is much more difficult today than it was in the past, and will probably become more difficult in the future. So, the water requirements must be evaluated carefully. They must be adjusted by conservation (the saving of water), and by the reuse of water. There must be a plan indicating either: 1) how supplies are to be obtained to satisfy demands, or 2) the limits to be placed upon demand. In other words, supply and demand must balance.

Third, there must be some agnecy, hopefully a local agency, with the powers to integrate water supplies and use. The current problems stem from the fact that the water supply and use are not integrated. This results in overdraft, mismanagement, and all the associated problems. History shows that when problems exceed the capability of an individual to correct them some kind of an organization is developed to assist in their resolution. The solution to problems of water management require organized agencies, especially local ones because of the physical and social complexities involved.

6.2 The Management Agency

What are the characteristics of an appropriate agency for water resources management? First, the agency should be one that encompasses the entire service area, and preferably one that includes the supply area as well. The latter is especially important where groundwater basins are part of the supply. It is very difficult to manage groundwaters without including the total area of the basin because what happens in one area affects what happens in the adjacent area. It is difficult to have control of the supply located in some other agency area unless a management agreement is obtained. The best approach is to have one agency covering the entire watershed, including the entire area of the groundwater basin.

Second, the local agency should have the authority to regulate water use, especially groundwater use. To manage the water resources successfully, it is necessary to control the amount of water extracted, where the water is extracted, and where the water is used.

Third, the local agency should have authority to finance supplemental water supplies. If the resources of semiarid or arid areas are to be utilized, it normally means local water supplies will be fully developed and supplemented by imporation projects. This means there has to be a means of financing supplemental water supplies. However, if the management plan is not to develop supplemental water supply projects then the water use must be regulated to fit within the supplies available.

Fourth, the local agency should have the ability to regulate and to integrate whatever local supplies exist with any supplemental imported supplies, and to integrate the surface and underground systems. Unfortunately the integration of surface and underground systems is not easily accomplished.

Of the four areas of needed authority for a water management agency, the greatest problem is in groundwater regulation. Everybody likes to have the ability to use water, when, how, and where, they see fit. The solution to that, of course, is to provide whatever water is needed. Thus, there should be a plan to satisfy the legitimate needs for water. Where the needs cannot be supplied, an adjudication may follow where everybody sues everybody else, the lawyers gain, and everybody else seems to lose. There has yet to be an adjudication that has developed any additional water. It succeeds in placing limits on what might be done, and it has the potential to reduce or eliminate the overdraft. The important concept is to manage the water resources of the area by providing either supplemental water or means to regulate the amount of water use.

To manage water resources, the surface and underground water supplies must be integrated. Groundwater basin management cannot exist without managing the surface water supplies. And surface water management is less optimum if groundwater basins are not included. It should be water management as a total picture. Total water management will be the final solution to the water management problems both nationally and worldwide. This requires the development of management agencies that encompass the entire basin area, and that have the authority to regulate water use, to finance supplemental water supplies, and to integrate local and imported surface water supplies with the groundwater supply.

6.3 Management and Planning with the Aid of Mathematical Models

In the Santa Clara Valley Water District most of the water comes from wells. It reaches the consumer by municipal water agencies, by private water companies, and by individually owned wells. With the advent of imported water, major deliveries of treated surface water to the retail water agencies were possible to supplement their groundwater extractions. The question was how much of the imported water should be treated and delivered directly to the retail water supply system and how much should be used to recharge the groundwater reservoirs which served most of the existing water distributing systems from wells? Unfortunately the Santa Clara Valley groundwater basin can neither accept nor deliver the full amount of the future water needs. The limitations are imposed by the physical characteristics of the groundwater basin; therefore surface water distribution facilities are required to supplement the groundwater facilities in use.

In order to aid in developing an optimum plan for water resources management in Santa Clara County mathematical models of the groundwater basin and of the major surface distribution systems, such as reservoirs and pipelines, were developed. Both models have matching operational areas, i.e., units or nodal areas, for comparative analyses. With these models, and information on future water requirements for the unit areas, integrated or coordinated operation of the groundwater facilities and the surface water delivery facilities was studied. These studies attempt to determine a system that allows maximum utilization of the water supply at minimum cost.

Generally the water supply characteristics of an aquifer, e.g., recharge location, recharge rate, storage capacity, transmission, and in some cases the extraction location and extraction rate, are fixed by its geology. It is possible, however, to correct limitations in the capabilities of a surface distribution system by the investment of sufficient capital. Accordingly the real limitation of the surface distribution system is the level of investment required to meet the desired delivery standards.

In applying this plan to the service area of the Santa Clara Valley Water District the procedure was to develop surface distribution facilities that provided adequate hydraulic gradeline elevations throughout the study area, while providing the desired amount of water to the various unit areas within the area being studied. By superimposing future water demands and delivery schedules on the capabilities of the surface distribution facilities it was apparent where these facilities needed to be supplemented. After determination of the investments needed and the operation and maintenance costs of the supplemental surface facilities required, a comparison was made of the cost of providing all or part of the water delivery by groundwater facilities. In this manner it was possible to develop a water supply distribution system utilizing both the groundwater basin and surface water supply facilities. This scheme would minimize the cost of meeting the projected water supply needs of the area.

It may have been possible to integrate the surface and groundwater models and develop an optimization program that would give the minimum cost solution. This would have required computer capabilities beyond that available to the district and considerable additional programming costs. Therefore the district elected to operate the two mathematical models separately developing many alternative plans relating the groundwater recharge, transmission, and extraction to the capabilities and costs of expanding surface distribution facilities. The end result was a plan for coordinated operation of the surface and groundwater facilities that approximates an optimum combination representing the minimum investment necessary to meet the expanding needs for water. The developed plan is utilized for scheduling of future facility needs and budgetary planning. The major facilities for water conveyance, treatment, and distribution developed in this manner for the Santa Clara Valley Water District are shown in Figure 4.

The development of such optimum plans relate to the water needs of the community. These water needs are programmed in accordance with the desires of the communities following the general plan of each community for its existing and future land uses. If the proposed land uses change it is possible that the optimum plan will change. This requires a continuing communication between the water supplier and the communities with jurisdictions over land use within the water supply area.

The application of an operational program developed by mathematical procedures will be difficult if the appropriate institutional and legal aspects are not considered. But the first step is to develop the operational plan utilizing engineering and economic aspects. Once the physical alternatives have been delineated, these can serve as a basis for evaluation of the legal problems. In the case of the Santa Clara Valley Water District the institutional problem was resolved by the existence of a master water agency with authority to implement the proposed water supply program. It does need the assistance of the various communities served in the delineation of the land use plans upon which the water needs of the area are based.

7. SUMMARY

Historically, water management in Santa Clara County, California, responded incrementally from the need to resolve problems which reached crisis proportions. It is only in recent years that water management has employed objective planning procedures. This mode of management has evolved in the last forty

440

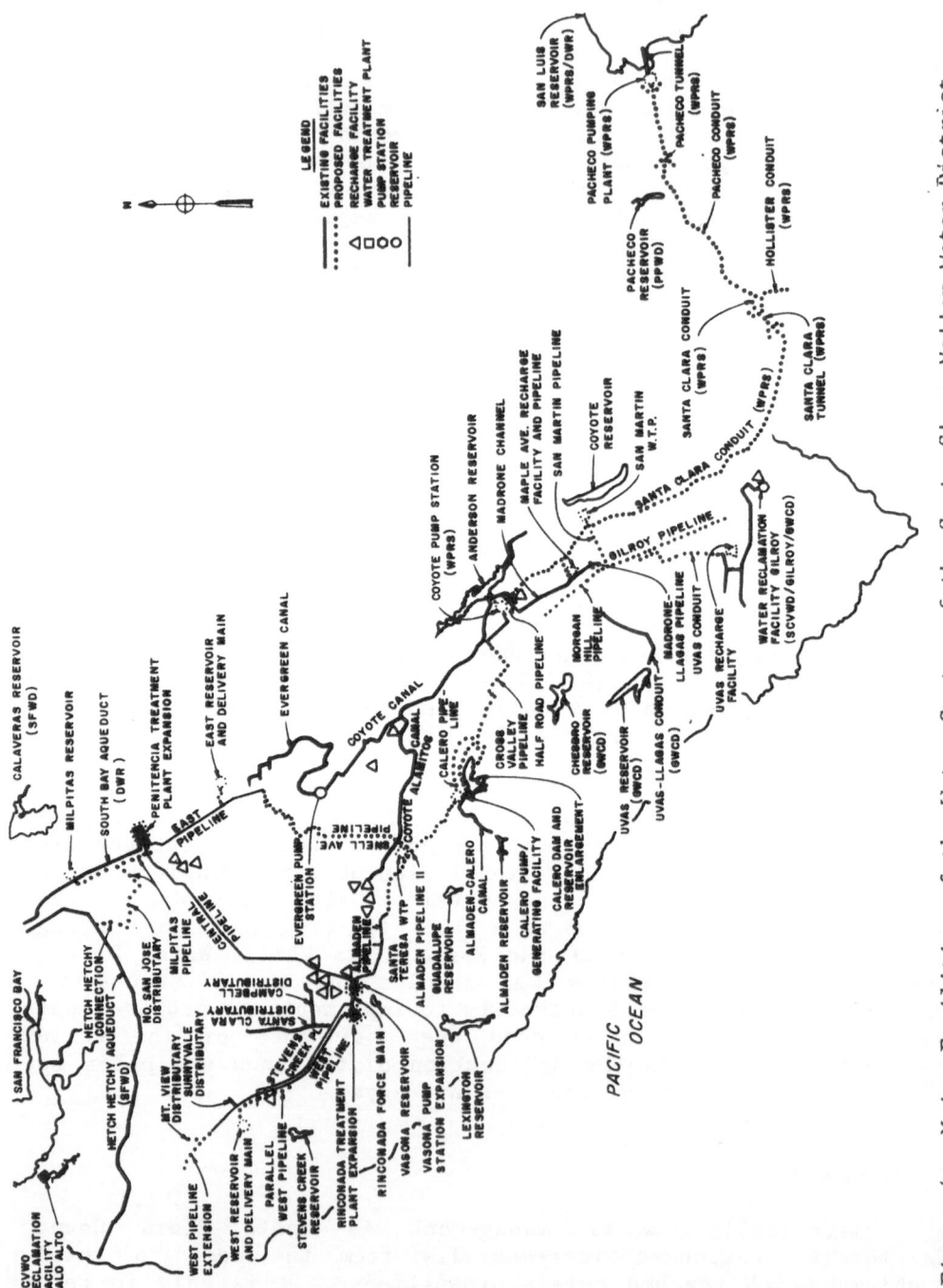

Figure 4. Major Facilities of the Water System of the Santa Clara Valley Water District

years, becoming a sophisticated program designed to meet the current and future water needs of the communities in Santa Clara County.

Water management by the Santa Clara Valley Water District utilizes the many modern mathematical tools and managerial approaches available. The management goal is to obtain maximum utilization of the available water resources at minimum cost. This goal is generally being achieved in spite of a deficiency in authority to control the uses of the water supply.

The district encompases the major watersheds and groundwater basins within the northern part of its area. It does not, however, encompass the entire southern groundwater basin, nor the downstream portions of the watershed in the southern part of the county. The problems presented by this lack of territorial jurisdiction have not yet resulted in water management difficulties. It is possible that such difficulties might materialize in the future as adjacent and downstream jurisdictions attempt to optimize their use of these water resources. Future success in water management in these watersheds and basins will require cooperation with the adjacent jurisdictions.

As the primary water management agency for Santa Clara County, the Santa Clara Valley Water District does have the authority to finance and implement an integrated water system for its area. It can contract with others for supplemental water supplies and it can distribute these supplies in an optimum fashion within the county. This includes the ability to integrate the local surface water and groundwater supplies with supplemental imported water supplies to achieve an economic water supply management sytem.

The district does not, however, have the authority to regulate water use within the county. Especially, it does not have the authority to control groundwater use. It does have authority to charge for extractions of all groundwater put to use within the district but it cannot regulate the amounts or locations of groundwater extraction. The regulation or control of the location and amount of groundwater extraction is a desirable authority, appropriate to full management of the water resources of the area.

REFERENCES

1. Poland, J. F. Subsidence contours in North Santa Clara Valley.

28. INCREASING THE YIELD OF CACHUMA RESERVOIR

Lloyd C. Fowler
Goleta Water District, Santa Barbara, California

The Goleta Water District is a political subdivision of the State of California and was organized under Division 12 of the California Water Code. It was formed by a vote of the people in the District on 17 November 1944. It's mission is to furnish the community it serves with an adequate supply of water as to both quantity and quality in an efficient and economical manner.

This District encompasses an area of about 13,360 hectares. Figure 1 is a map showing the boundaries of the district and Lake Cachuma. Domestic water service is provided to a population of about 75,000. Agricultural water service is provided to about 1,215 hectares. There are nearly 14,000 active water accounts.

One of the original purposes for the formation of the district was to establish a legal entity representing the Goleta Valley area which together with the City of Santa Barbara and the Montecito, Summerland and Carpinteria County Water Districts could enter into contracts with the Santa Barbara County Water Agency for an imported water supply. The County Water Agency had contracted with the U. S. Bureau of Reclamation for a supply of water and repayment of the costs of construction of the Cachuma Project on the Santa Ynez River. This project was constructed to conserve waters of the Santa Ynez River for use in the Santa Ynez Valley and the South Coast area of Santa Barbara County. The initial water service from the Cachuma Project to the South Coast began in 1956.

The present sources of water for the district are surface water from the Cachuma Project and groundwater from wells in the Goleta Valley basin. The safe yield of the Cachuma Project is

443

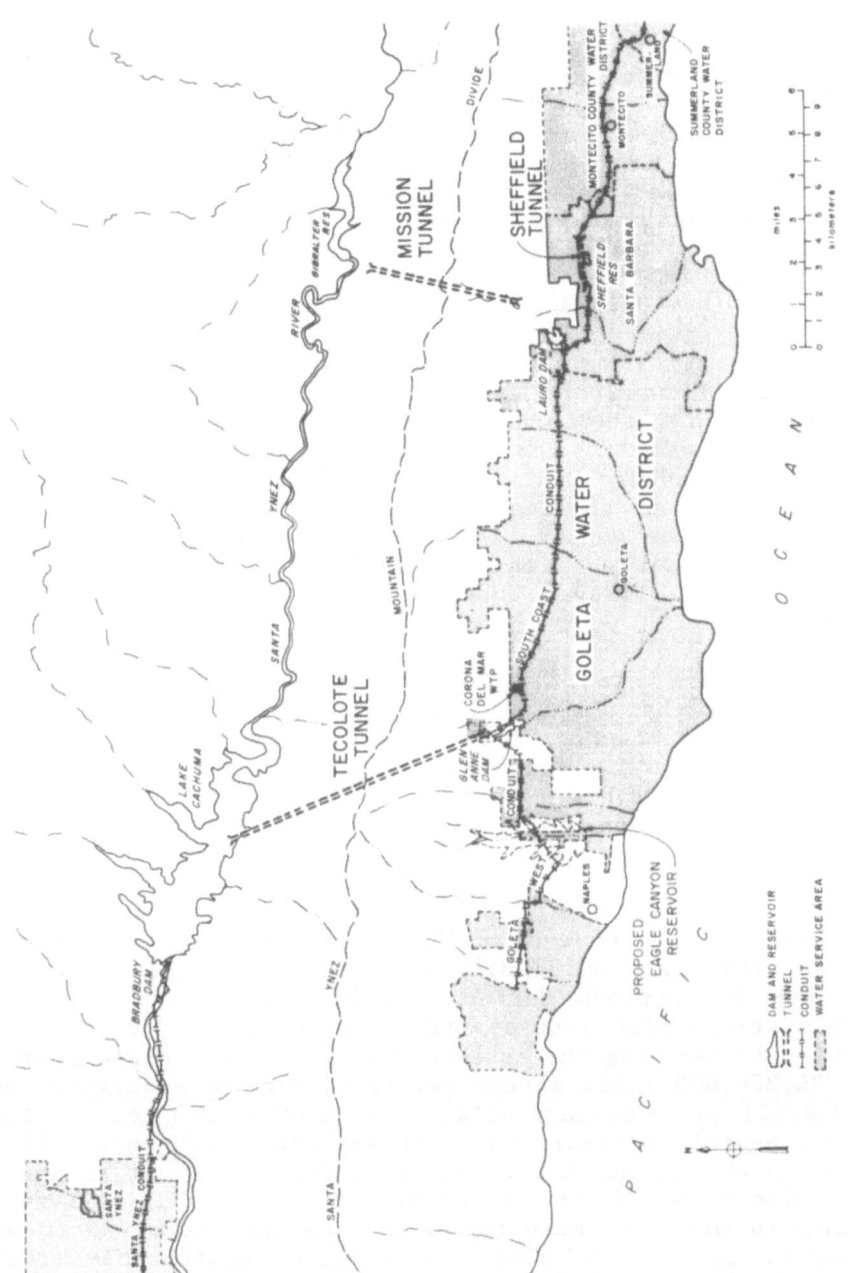

Figure. 1. Goleta Water District

estimated to be 34,300,000 cubic meters per year. The District's share of this yield is 11,750,000 cubic meters per year during the current 1980-85 period; in 1990 the allocation will have increased to 12,440,000 cubic meters per year. Unless the safe yield is increased, this is the maximum share of water the district can obtain from the Cachuma Project.

The safe yield of the Central Subbasin of the Goleta Ground-water Basin, the area where District wells currently extract groundwater, has been estimated to be 4,200,000 cubic meters per year. The subject of groundwater extractions in the Central Subbasin has been under litigation since 1974. It is expected that a decision in this case will delineate user shares in the groundwater. Other groundwater basins in the district may furnish limited amounts of water if wells were developed in them.

Cachuma spill water is the water that flows over the spillway at Bradbury Dam when the Cachuma Reservoir on the Santa Ynez River is full and the inflow to the reservoir exceeds its diversions and losses. Since Bradbury Dam was completed in 1956, Cachuma Reservoir has filled and spilled ten times. This spill has always occurred between the months of January and July, but the duration of the spill is always less than the seven month period. The duration of spill averages about 90 days, or a little less than three months.

The amount of water that has passed over the spillway during a spill season has varied from 5,900,000 cubic meters to 583,000,000 cubic meters. For the ten times the reservoir has spilled, the amount of Cachuma spill water has averaged 133,500,000 cubic meters. Since the reservoir does not spill every year, the average annual amount of spill water during the 23 year history of operation is 58,000,000 cubic meters per year.

In the 23 year period of reservoir operation, it appears that Cachuma Reservoir fills and spills on an average of about every 2.3 years. This is a misleading statistic since there have been no extended dry periods during this 23 year period of record. Based upon stream flow records over a period of 66 years shown in Figure 2, and operating the reservoir at its safe yield withdrawal rate of 34,300,000 cubic meters per year, Cachuma Reservoir would fill and spill on an average of about once every 3.3 years. There have been several intervals in which the reservoir would fill and spill each year and then there have been dry periods of as long as 16 years during which the reservoir would not fill and spill. Therefore, in order to make use of the Cachuma spill water as a firm supply, it will be necessary to have considerable storage space to store spill waters during the short time they are available and hold them for use during extended dry periods.

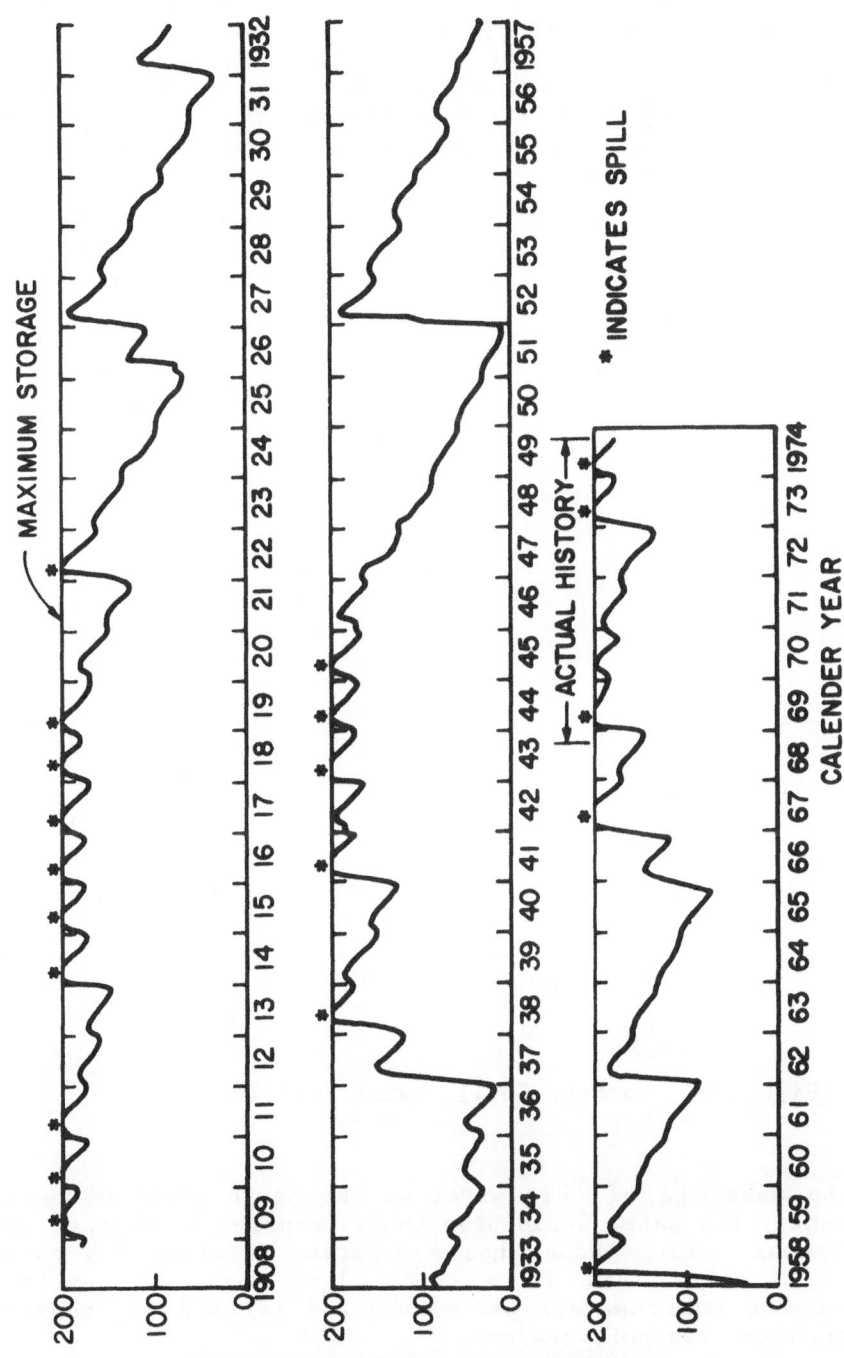

Figure 2. Cachuma Reservoir Storage History Showing Spill Periods

446

In order to simplify a discussion of the amount of Cachuma
spill water that can be used by the Goleta Water District, consider what might have happened during the 1979-80 water year.
During this water year, about 143,200,000 cubic meters of water
spilled from Cachuma Reservoir during the period February to May
1980. The duration and volume of water are shown in Figure 3.

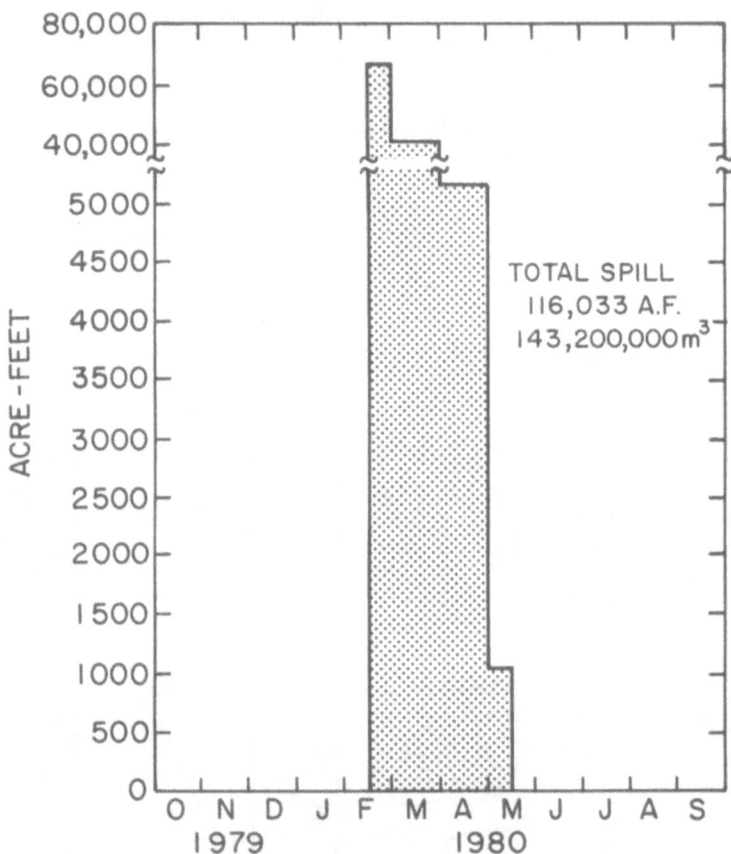

Figure 3. Cachuma Spill, Water Year 1979-80

In order to make use of this water on the South Coast of Santa
Barbara County, the water would have to be transferred through the
Tecolote Tunnel. This tunnel has a capacity of about 2.5 cubic
meters per second. This is a little less than the designed
capacity of 2.8 cubic meters per second and is used for conservative hydraulic transport reasons.

The amount of water that can be transported through the
Tecolote Tunnel for each month is shown in Figure 4.

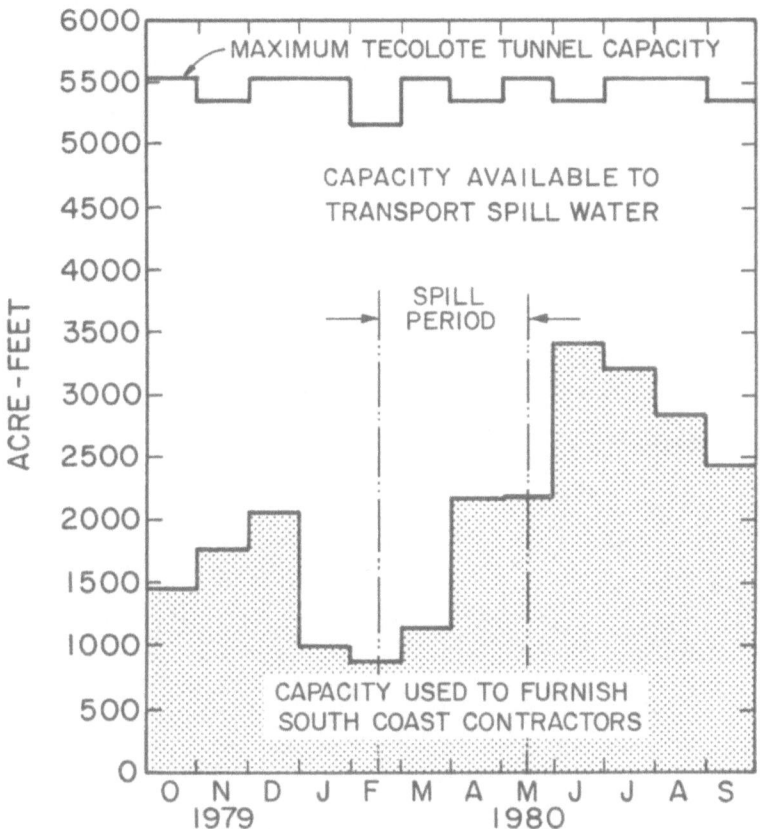

Figure 4.   Capacity of Tecolote Tunnel

Also plotted is the amount of water utilized from Cachuma Reservior by the south coast water agencies. The difference between
the capacity of the tunnel to deliver water and the amount of
water received by the south coast water agencies is the capacity
of the Tecolote Tunnel to transfer Cachuma spill water to the
south coast area.   In 1979-80, this excess amounted to about
13,600,000 cubic meters, which is considerabley less than the
total amount of water that flowed over the spillway at Bradbury
Dam from Cachuma Reservoir. The volume and time distribution of
transferred Cachuma spill water is shown in Figure 5.

In order to make use of this transferred Cachuma spill water,
the Goleta Water District would have to consume it or place it in
storage. If it is to be considered as an addition to the firm
water supplies of the district, it would have to be placed into
storage.   The storage could be either underground or in some
surface storage reservoir or both.

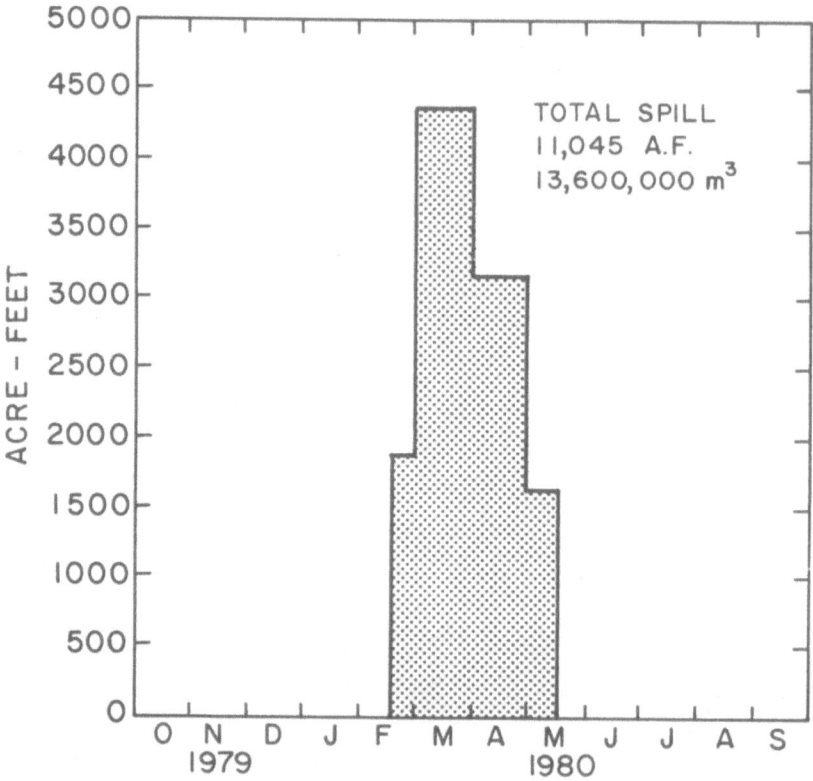

Figure 5.   Spill Water Transportable by Tecolote Tunnel

In order to place the transferred Cachuma spill water in the underground basin, it would be necessary to treat the water prior to injection.   Untreated spill water could be placed in surface spreading basins from which the water could infiltrate into the forebay areas of the groundwater basin.   Unfortunately, such recharge areas are limited and not readily available to the district.   Therefore, for purposes of this evaluation, the ability of the district to treat water is considered the limiting factor on the amount of Cachuma spill water that can be placed into the groundwater basins.

The capacity of the Corona del Mar Water Treatment Plant to treat water for service within the Goleta Water District is shown in Figure 6.   These capacities do not consider the potential plant outages required for maintenance or caused by breakdown as these are considered to be relatively small. The amount of treated water

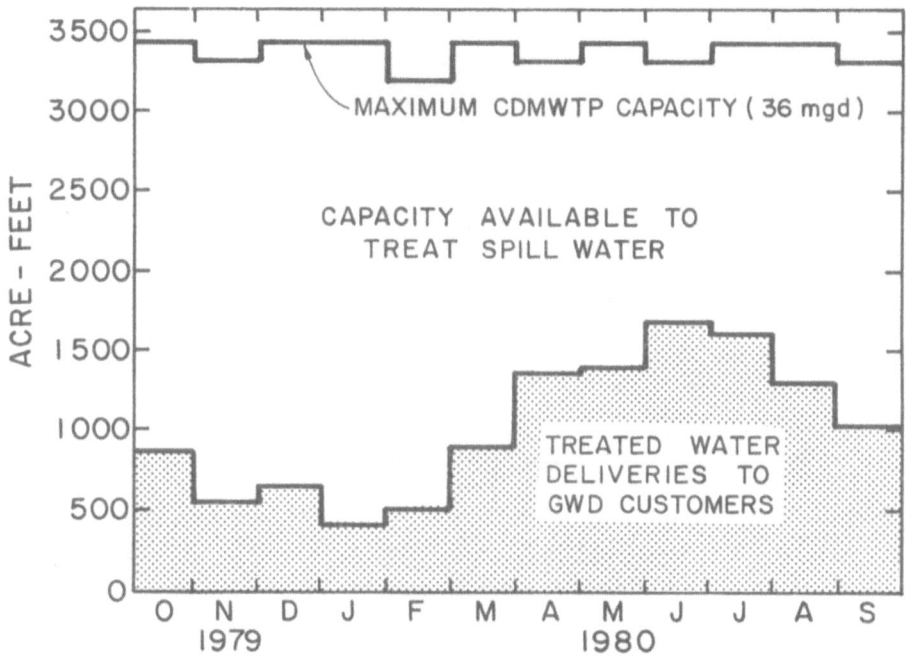

Figure 6. Capacity of Corona del Mar Water Treatment Plant

required to serve the district's customers is also shown in Figure
6. The difference between the capacity of the plant to treat
water and the amount required to service the district customers is
the capacity of the plant to treat Cachuma spill water for
injection into the groundwater basin. The amount of transferred
Cachuma spill water that could be treated for injection purposes
is shown in Figure 7. During the 1979-80 spill period, this
amount of water was 8,240,000 cubic meters. In each month, the
amount treated is equal to the amount of spill water transferred
or the capacity of the plant to treat water, whichever is least.

Surface storage of transferred Cachuma spill water may be
complimentary to or an alternative to storage in a groundwater
basin. Assuming that a dam and reservoir of sufficient capacity
could be constructed, transferred Cachuma spill water could be
placed into storage in such a surface reservoir through the Goleta
West Conduit. This conduit has a capacity to move about 0.57
cubic meters per second into such a reservoir. Part of this
capacity would have to be utilized to service customers along the
Goleta West Conduit. The capacity of the conduit to move water is

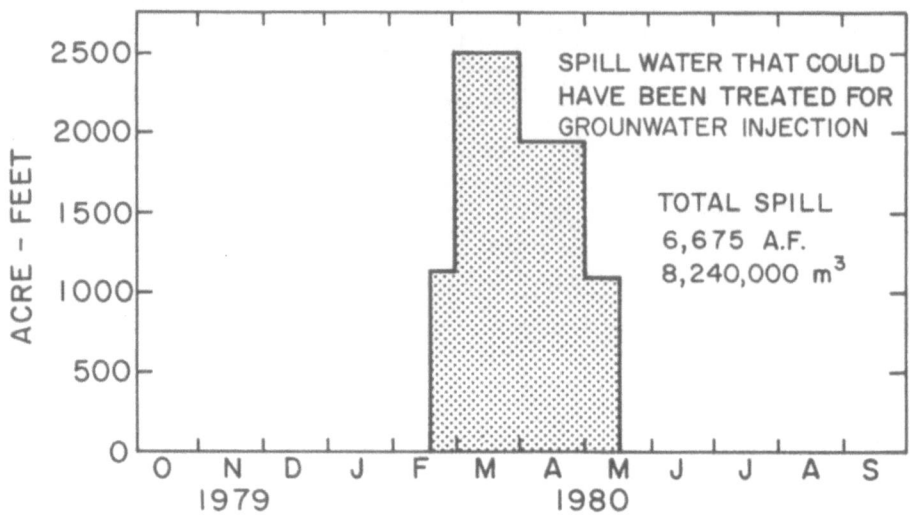

Figure 7.   Capacity for Treatment of Spill Water

shown in Figure 8.   Also shown in Figure 8 are the deliveries to customers along the Goleta West Conduit.   The capacity of the conduit to move transferred Cachuma spill waters to a surface

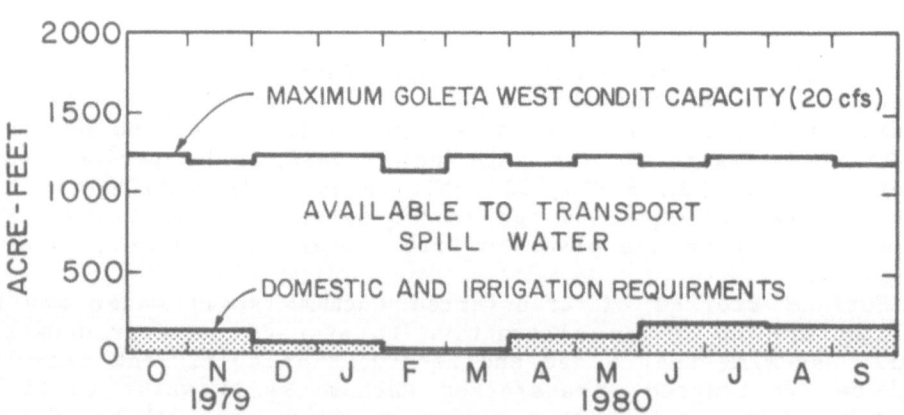

Figure 8.   Capacity of Goleta West Conduit to Transport Spill Water, Water Year 1979-80

storage reservoir is the difference between the capacity of the conduit and the use of water from the conduit. The amount of Cachuma spill water that could have been transferred is shown in Figure 9. It is equal to the capacity of the Goleta West Conduit

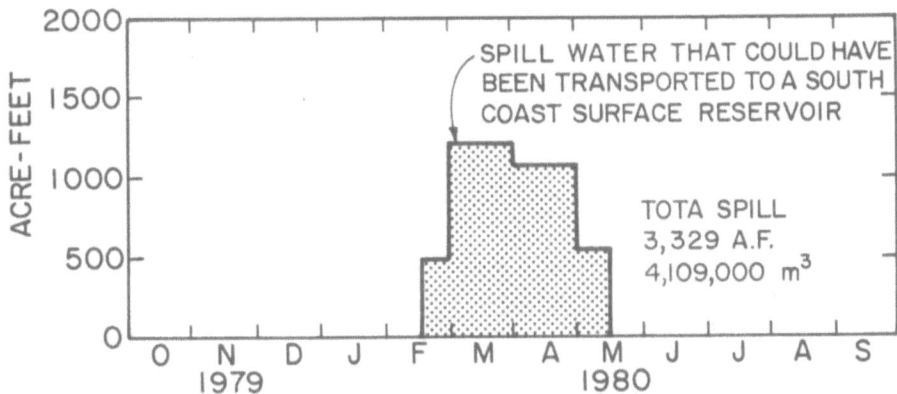

Figure 9.   Spill Water Transportable by Goleta West Conduit, Water Year 1979-80

to move spill water or the amount of spill water transferred, whichever is least. It amounts to 4,109,000 cubic meters for the period February to May 1980.

The amount of transferred Cachuma spill water that could have been placed into storage in 1979-80 by the Goleta Water District by injection into the groundwater basin and by transfer to surface storage is the sum of the amounts shown in Figures 7 and 9. This total is shown in Figure 10. It is only 10 percent less than the amount of potentially transferrable Cachuma spill water.

The 12,350,000 cubic meters of Cachuma spill water that could have been placed into storage by the district in 1979-80 is not a firm yield. It is a one year event that cannot be counted upon as occurring every year. In order to evaluate the long-term annual amount of Cachuma spill water that could be made available to and utilized by the district, the analyses for 1979-80 described above can be applied to the records of the annual amounts of spill water that have occurred since Bradbury Dam was completed. During the 23 years that Cachuma Reservoir has been in operation, water has flowed over the spillway ten times. By summing the estimates of the Cachuma spill water that could be transferred and utilized by the district during these ten spill years, and dividing by the 23 years of reservoir operation, it appears that the district could have added to its water supply an average of about 6,050,000 cubic meters per year.

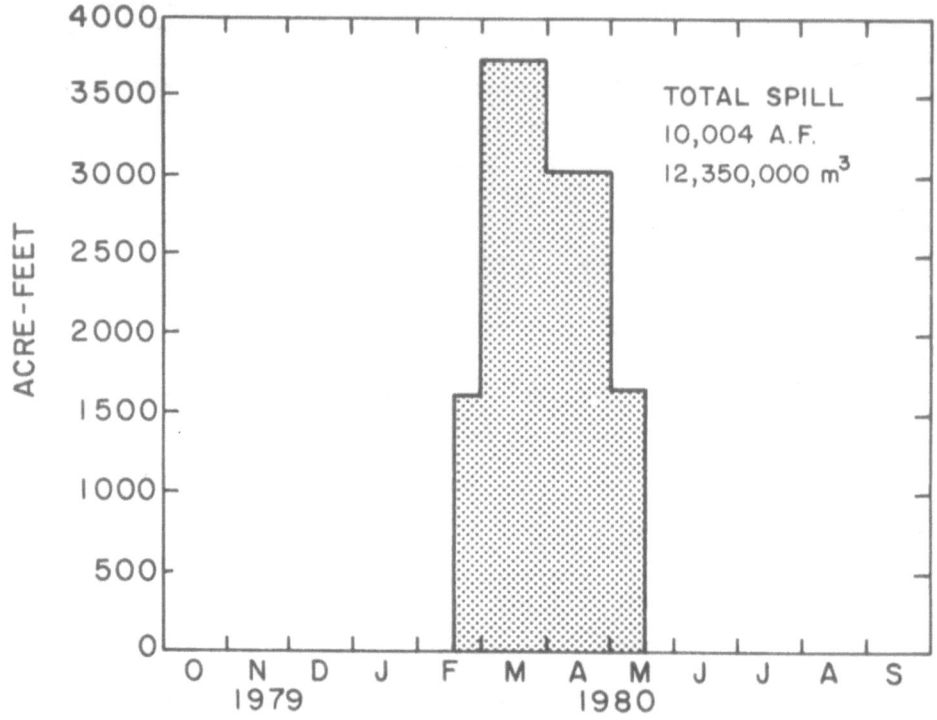

Figure 10.   Treatable and Storable Spill Water,
Water Year 1979-80

As a note of caution, however, the hydrologic period from
1956 to 1980 is not representative of the long-term average for
the Santa Ynez River Watershed. Comparing this shorter more
recent record with a 66 year average hydrologic period of record
(as shown in Figure 1), it is estimated that the firm annual yield
of Cachuma spill waters would amount to around 2,470,000 cubic
meters per year.

The review described above does not consider the possibility
of future increases in community needs for water. As these needs
increase, the capacity of the Tecolote Tunnel, the Corona del Mar
Water Treatment Plant, and the Goleta West Conduit to transport
Cachuma spill waters would be reduced. However, the Tecolote
Tunnel can be pressurized. Since Cachuma spill waters would be
available only when the reservoir was full and spilling, there
would be considerable head available to move greater quantities
of water through a pressurized tunnel. It is possible to increase
the size of the water treatment plant so that it could maintain
its capacity to treat transferred Cachuma spill waters. The

capacity of the Goleta West Conduit to transport Cachuma spill waters could be increased by the addition of pumps. For these reasons, it was not considered necessary to evaluate the potentail effect of increased future water needs on transfers of Cachuma spill waters.

The possibility of a long-term firm yield of about 2,470,000 cubic meters of water per year warrants further detailed evaluation of the use of Cachuma spill waters. Therefore, it is recommended that the district investigate in detail, including legal and institutional concerns, the underground and surface storage of transferred Cachuma spill waters to increase the firm yield from the Cachuma Project.

29. WATER SUPPLY IN ORANGE COUNTY

Neil M. Cline
Orange County Water District, California

1. WATER SUPPLIES IN SOUTHERN CALIFORNIA

Orange County, California is located on the coastal plain of Southern California south of Los Angeles. The current population is 1.9 million. A combination of good climate and employment opportunities in the area continues to stimulate an annual growth rate of nearly 6 percent. The urbanization causes commensurate continuing increases in water supply requirements. In 1980 the water use within Orange County was 380,000 acre-feet.

The average rainfall is twelve inches per year, which is adequate to supply about 25 percent of the current water requirements. To supplement the local supply, Orange County contracts with the Metropolitan Water District of Southern California (MWD) for imported water from the Colorado River and from Northern California. The distances of transport is 240 miles and 400 miles, respectively.

Water importation has been going on in Southern California for the past sixty years as a means to augment the scanty local water supply. The first major import scheme was the Owens Valley Aqueduct, constructed by the City of Los Angeles in 1913. In the 1920's MWD was organized by the California Legislature as a part of the Boulder Canyon Project Act to transport Colorado River water from the Colorado River to the coastal plain of Southern California. To accomplish this 240 mile transport MWD constructed and presently operates the Colorado River aqueduct designed to deliver 1.2 million acre-feet per year. Then in 1960 MWD contracted with the State of California for delivery of 2.0 million acre-feet of water from the California Water Project, which transports water from Northern California.

Both of these massive import programs have been accomplished in spite of enormous political opposition from the areas of origin. The Arizona vs. California litigation over allocation of the Colorado River was in court for 36 years; it was finally settled in 1964 in Arizona's favor. This was 15 years after the completion of the aqueduct. Under the Supreme Court decree California will lose by 1985 more than half the volume it is currently importing from the Colorado River. The waters of Northern California have not been subject to such intense litigation, but the concept of transferring several million acre-feet from north to south has been the subject of heated debate within the State. This has been going on since the 1920's when the idea was first considered by California's water planners. It was brought to fruition in 1960 when the voters of the state approved a bond issue to construct the California Water Project. The project developed a comprehensive system of water storage and statewide water transfer. The key element in permitting improved efficiency of the California Water Project is a canal system called the Peripheral Canal, which would link the Sacramento River to the California aqueduct. It has never been completed due to local opposition and statewide concerns about environmental protection.

Construction of the Peripheral Canal has been and continues to be the biggest single resource issue in the State of California. The voters of California will be asked to reconsider the Peripheral Canal construction in future ballot. At this time there is no assurance that there will be approval of this much-needed facility. What is clear, and both supporters and opponents of the water transfer system recognize, is that alternative methods of matching supply with demand must be found. It is not surprising therefore that in Southern California it is anticipated that, in part, the area's future water requirements will be satisfied through increased water conservation and wastewater reclamation programs. In Orange County, as a result of the long-standing physical and institutional problems which have caused limitations of water availability, it has been determined that the region's water supplies will be composed ultimately, of fully developed local resources, to be used conjunctively with a combination of imported water from Northern California and the Colorado River, and reclaimed wastewater and desalted brackish water. It is estimated that within the boundaries of the Orange County Water District the demand for water will reach 500,000 acre-feet annually by the mid-1990's.

Water reclamation is becoming increasingly prominent in this overall multifaceted approach to water planning. The Orange County Water District has developed one such water reclamation project in which reclaimed water is injected into aquifers connected to the sea in order to hydraulically prevent seawater

456

intrusion in the area's invaluable groundwater reserve. A location map showing Orange County and the boundaries of the Orange County Water District is shown in Figure 1.

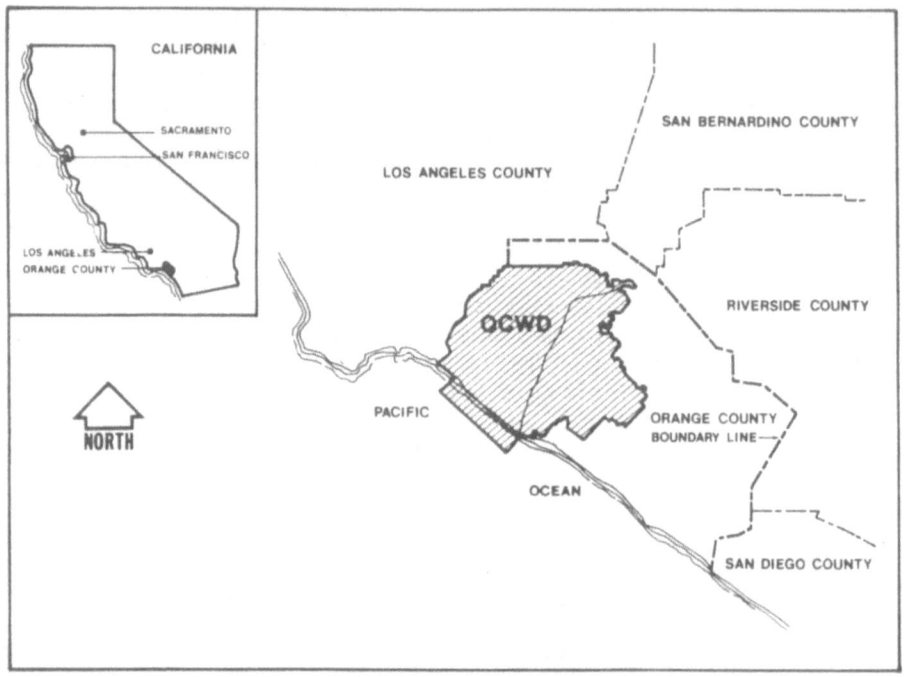

Figure 1.   Orange County Water District Location Map

2. THE COASTAL AQUIFER AND THE EFFECTS OF SEAWATER INTRUSION

Tradtionally, water systems evolve with need. Orange County is typical. The early settlers, before 1900, relied on the surface flow of the Santa Ana River and artesian springs. As demand increased due to the population influx from the East, wells were constructed. By 1921 over 5,000 wells were operating in the county, resulting in a heavy groundwater overdraft, which in turn caused seawater intrusion. The Orange County Water District (OCWD) was formed in 1933 to manage groundwater reserves, and to pursue litigation against upstream diversions of the Santa Ana River. To offset the large overdraft, which was inducing seawater intrusion along the Orange County coast, the District implemented a recharge program, in which large volumes of surplus imported water were placed into groundwater storage. Since 1949 the District has placed over 2.4 million acre-feet of imported water into subsurface aquifers. In addition it has captured nearly

90 percent of the natural runoff of the Santa Ana River. This program benefits directly the 23 cities in Orange County which presently pump about 70 percent of their annual supply. The balance of their requirements are obtained from the import systems. Of the groundwater produced, about one-half has its origin from the natural percolation of the Santa Ana Rver and local rainfall; the remaining half is imported water, artificially recharged to groundwater storage. The groundwater basin functions as a distribution system, and for operational and emergency storage. It is considered invaluable by the cities served. As is frequently the case with coastal aquifers, especially those composed of sedimentary material, sea water intrusion is a frequent condition. It occurs when groundwater levels are reduced to elevations below sea level causing a negative hydraulic gradient toward land. The groundwater basin in Orange County is the depositional plain of the Santa Ana River. In general, the aquifers of the area are composed of sand, distributed fine to coarse in grain size, and separated by silt and clay aquicludes. The Talbert aquifer is the principal zone of production in Orange County, and it is in hydraulic contact with the sea. This zone is of recent age and overlies the Pleistocene deposits within the gap created by the Santa Ana River. The Talbert aquifer is the only one in direct contact with the ocean. The lower three zones of local production are subject to intrusion only because they are in contact with the Talbert.

Figure 2 is a general cross section drawing of the ground-water basin showing the major aquifers. In mid-Orange County the depth to the base of the fresh water bearing sediments is over 4,000 feet. The fresh water base rises to an elevation of 200 feet along the coast and in the Santa Ana Canyon area. Recharge of groundwater is accomplished through percolation in the forebay, and in recent times from injection wells along the coast. Figure 3 is an enlarged cross section drawing showing the coastal geology where seawater intrusion has occurred.

Early settlers of Orange County and the coastal plain found an ideal location for agriculture provided by the moderate climate and artesian wells. Soon supplemental pumping by agriculture exceeded the safe yield of the groundwater basin. Consequently in the early 1920's Artesian pressure levels were lowered below sea level and by 1931 the Talbert aquifer was intruded by about 1.4 miles inland by seawater. The seawater intrusion continued, moving steadily inland, forcing the City of Laguna Beach to abandon their wells in 1947. Newport Beach had to abandon their wells in 1953, and by 1963 seawater had moved as far as 3 1/2 miles inland from the ocean. Although the OCWD managed to prevent further and more widespread damage through massive percolation of imported water into the forebay, the need for a coastal barrier system was obvious. Preliminary studies indicated that to develop

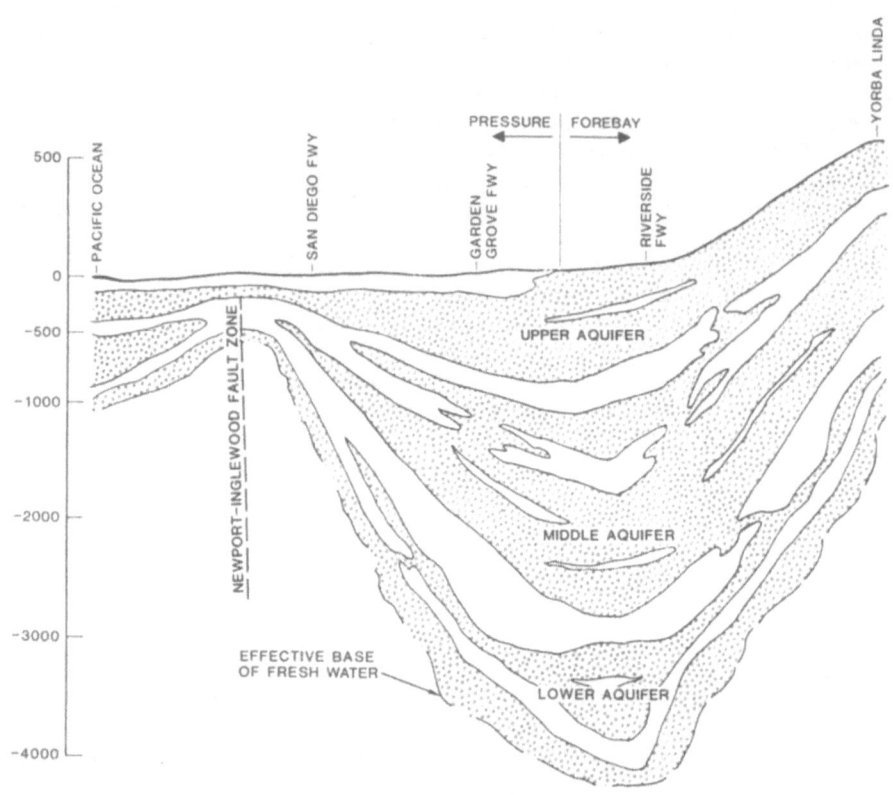

Figure 2.   Cross Section of Orange County Groundwater Basin

an efficient hydraulic barrier to maintain intrusion control would
require 30,000 acre-feet of water per year, a substantial quantity
in water-scarce Southern California.   After weighing all the
factors, including consideration of supply availability and
reliability, environmental problems, water quality, and program
costs, it was determined the best source for barrier injection
would be reclaimed water.   Since 1962 OCWD has been in the process
of developing its coastal barrier project to utilize salvaged
wastewater.

The project involved pilot studies on a variety of processes
of wastewater reclamation.   These studies were conducted by the
district during the years 1962 through 1970.   Pilot injection
programs indicated, for reasons not clearly understood, that
injected reclaimed water remains en masse when flowing out from
the point of injection.   Because a number of coastal community

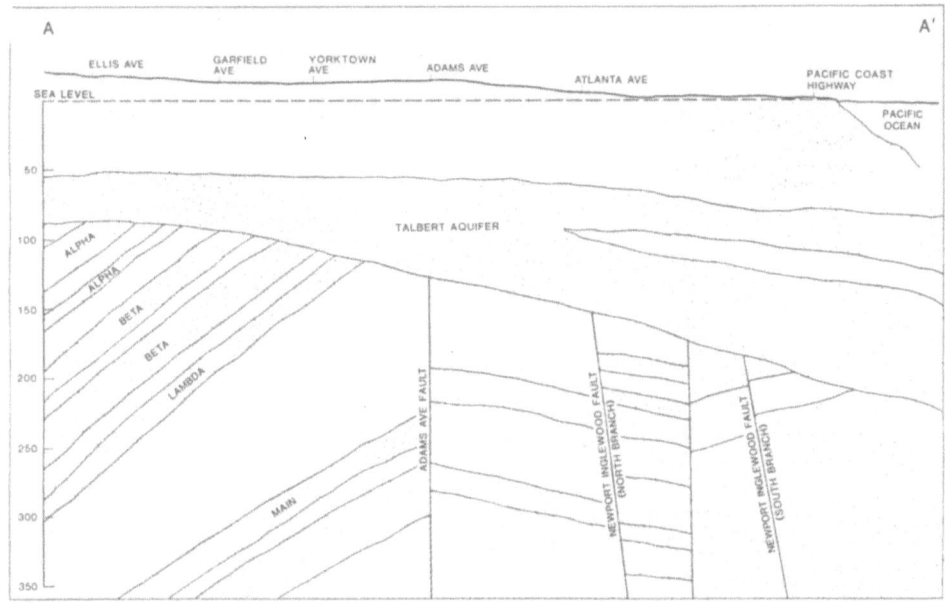

Figure 3. Generalized Cross Section A-A

wells draw this water for domestic use, the quality of the injected supply must be at potable standards. To produce such high-quality water the district has constructed an advanced water treatment facility, called Water Factory 21.

3. WATER FACTORY 21

The Water Factory 21 is designed to produce potable water from an activated sludge secondary effluent. The plant treats 15 million gallons per day of treated wastewater effluent from the Orange County Sanitation Distruct. The advanced waste treatment process includes lime clarification, ammonia stripping, recarbo- nation, mixed-media filtration, activated carbon adsorption, demineralization and chlorination. The reclaimed water is pumped under low pressure to 23 injection wells, which have multiple casing that permit controlled injection into the four aquifers subject to seawater intrusion. The wells are located in a line parallel to the coast at 600-foot intervals in precast 6' x 10' concrete vaults located in a city street 3 1/2 miles inland from the coast. Figure 4 shows the configuration of the wells, along with the location of extraction wells. The wells vary in depth from 90 to 430 feet. Each well has the capacity to inject 450 gallons per minute. Figure 4 is a drawing of a typical injection

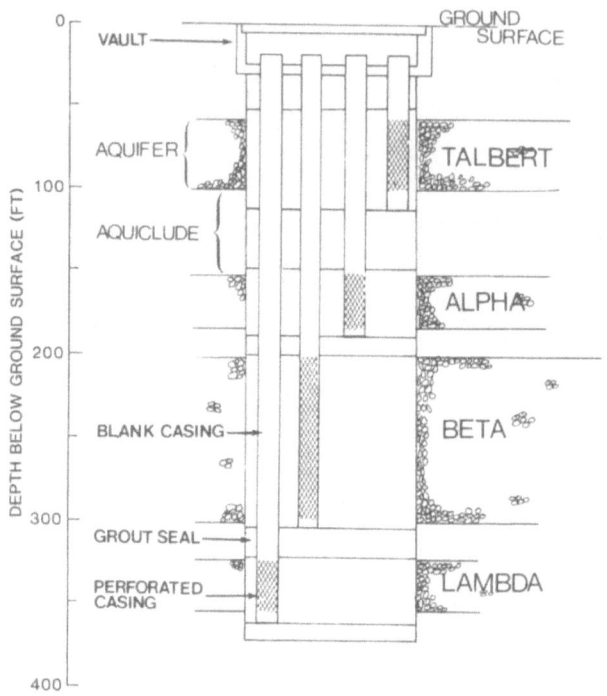

Figure 4.   Typical Injection Well and Observation Well Section

well scheme.   There are 31 monitoring wells distributed landward
and seaward from the injection line.   The monitoring wells are
systematically sampled to determine the hydraulic conditions of
the barrier, and to evaluate water quality characteristics of the
injected water as it passes through the subsurface.   The majority
of the monitoring wells are multiple-casing facilities mirroring
the injection well installations.   To provide the supplementary
water supply necessary to maintian the hydraulic barrier, five
conventional wells were drilled to produce from a deeper aquifer
not subject to seawater contamination.   These wells, ranging from
850 to 1150 feet deep and capable of furnishing 3500 gpm each.
Their proximity in relation to Water Factory 21 are seen also in
Figure 4.

To assure a seaward gradient from the injection mound, and to
regulate the amount of water that will move toward the ocean,
seven extraction wells have been constructed about two miles
inland, as seen in Figure 4 may produce 600 gpm each.

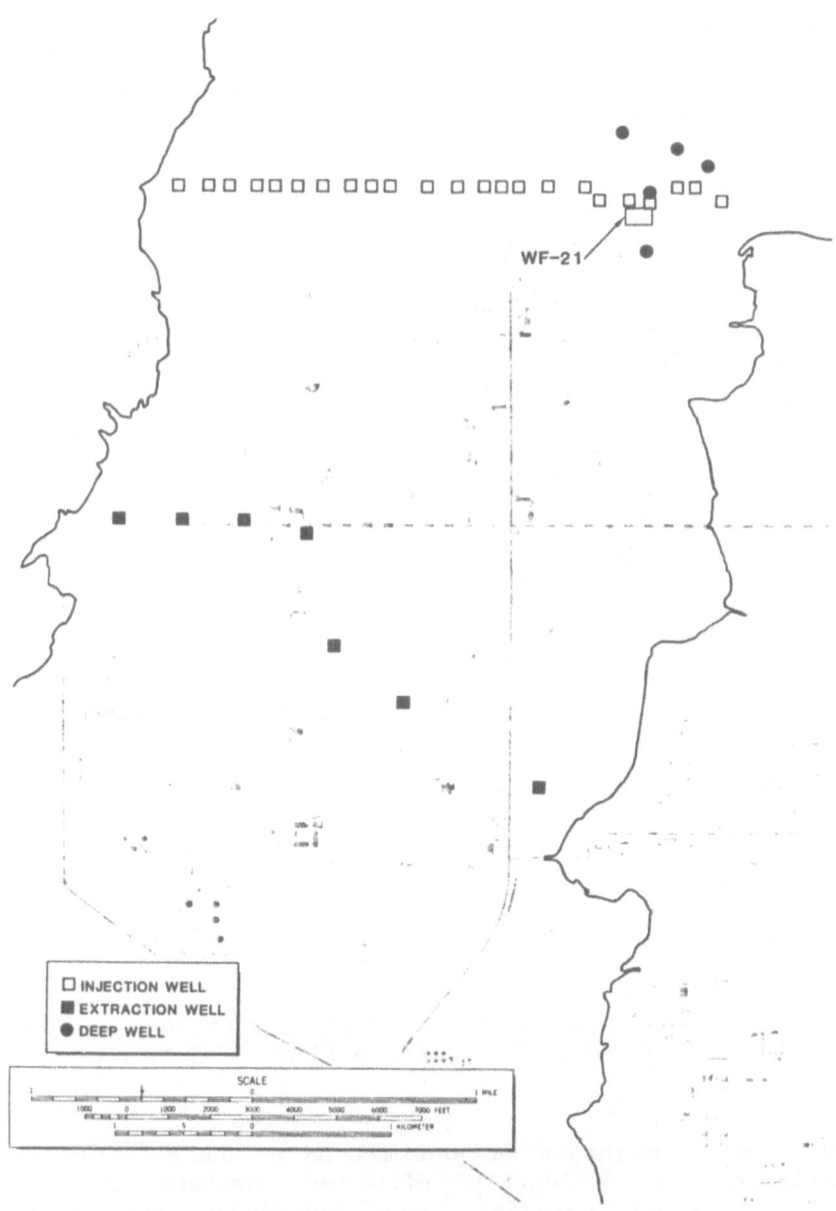

Figure 5. Talbert Gap Barrier Facilities

Figure 6 is a flow diagram showing the liquid processing of the salvaged wastewater. Lime clarification is accomplished by rapid-mix, flocculation and sedimentation. Lime is added in

LIQUID PROCESSING

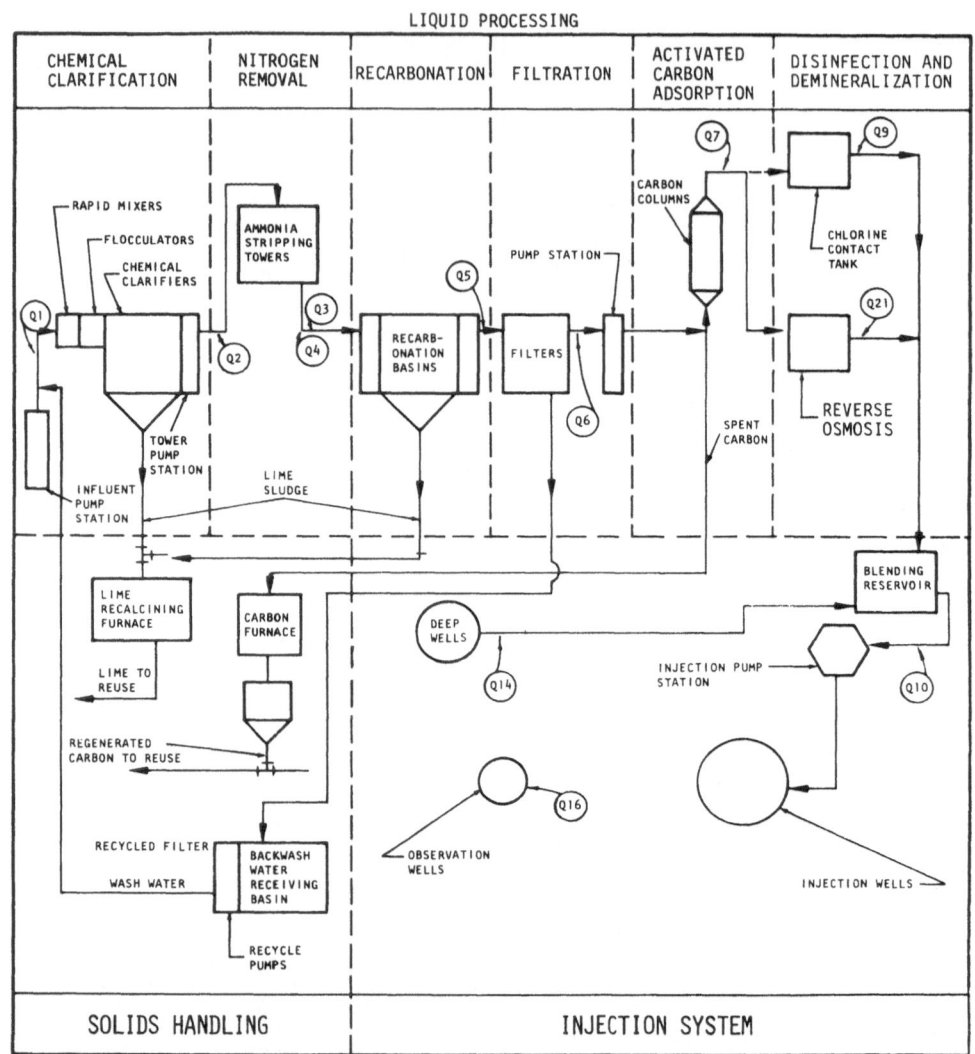

Figure 6.   Flow Schematic and Sampling Locations for 15 mgd
Wastewater Reclamation Plant

slurry form to maintain an optimal pH of 11.3.   This chemical
clarification is a highly effective process for reducing
turbidity, organic materials, and phosphate, and it aids in
ammonia removal.   A multi-hearth lime recalcining furnace has been
installed that accomodates recovery and reuse of about 90 percent
of the lime.

Ammonia stripping is achieved in two large cooling-type towers operated in series. Wastewater is pumped from the clarifiers to the top of the towers and then sprayed through the towers. Ammonia occurs mostly as NH_3 at the elevated pH and consequently it escapes to atompsphere with about 90 percent removal. Recarbonation with carbon dioxide is used to neutralize the effect of previous lime treatment. The carbon dioxide gas is bubbled through the process water with the effect of lowering the pH to about 7.5. The water then flows through a mixed-media filter, composed of coal, silica and garnet sand. Next it is pumped through columns of activated carbon to remove dissolved organic materials such as detergents, taste and odors and color-causing compounds, and other potentially harmful materials. A carbon regeneration furnace is used to recover approximately 93 percent of the spent carbon. Because the average wastewater in Orange County contains approximately 1,000 milligrams per liter total dissolved solids (TDS), it is necessary to provide demin-eralization for a part of the reclaimed effluent in order to satisfy injection quality standards. To accomplish the needed demineralization, the district has been operating a 5 mgd reverse osmosis demineralization plant since July 1977, which was designed to provide 90 percent removal of all salts while achieving an overall product water recovery of 85 percent. The schematic flow of the reverse osmosis system is shown on Figure 7. The system is composed of pretreatment of the activated carbon effluent by adding chlorination and filtration, followed by reverse osmosis membrane processing and post treatment. Table 1 is a summary of the averagy RO unit performance. The operation of the reverse osmosis unit requires pressure of 550 psi, produced by two 900 HP pumps. The basic element of the process consists of spiral-wound cellulose acetate membranes. By regulation, the finished water of the plant must be blended at least 50 percent

Table 1. Typical Performace of Reverse Osmosis Plant

Constituent	Inflow Quality (Activated carbon effluent)	Product Quality
Sodium	150 mg/l	18.5 mg/l
Total hardness	260 mg/l	20.0 mg/l
Sulfate	325 mg/l	15.0 mg/l
Chloride	190 mg/l	24.0 mg/l
Ammonium	20-25 mg/l	4-5 mg/l
COD	20-25 mg/l	1-3 mg/l
EC	1105 μmhos/cm	140 μmhos/cm

464

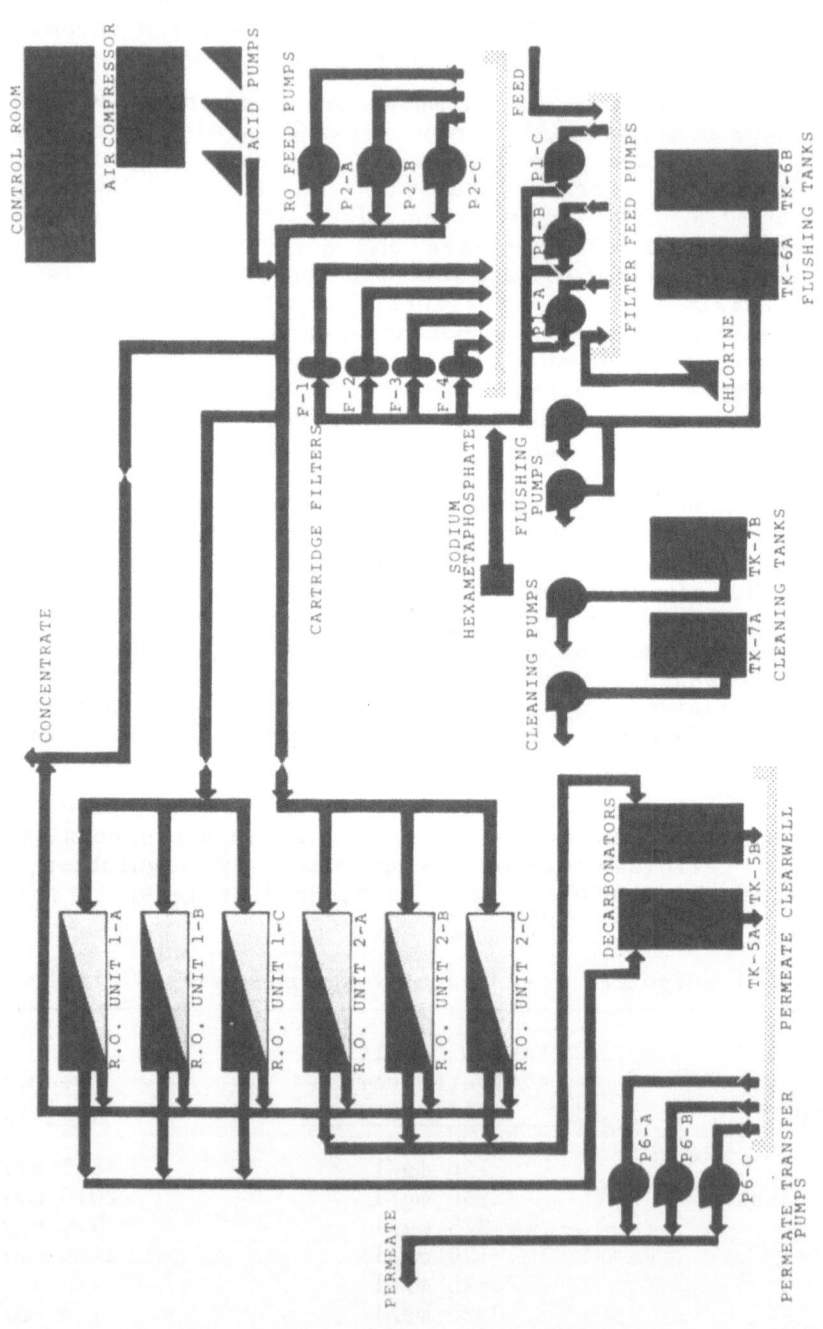

Figure 7. Five mgd Reverse Osmosis Flow Schematic

with demineralized or deep well water prior to injection. Because
the reclaimed water produced by the plant will commingle with
native groundwater and eventually be used for domestic,
agricultural and industrial purposes, very stringent quality
standards have been imposed by the California Regional Water
Quality Control Board and the State Department of Health. These

Table 2.   Water Quality Requirements of California Regional
           Water Quality Control Board for Injection to
           Groundwater

CONSTITUENT	MAXIMUM CONCENTRATION (mg/l)
Ammonium	1.0
Sodium	110.0
Total hardness ($CaCo_3$)	220.0
Sulfate	125.0
Chloride	120.0
Total nitrogen (N)	10.0
Fluoride	0.8
Boron	0.5
MBAS	0.5
Hexavalent chromium	0.05
Cadmium	0.01
Selenium	0.01
Phenol	0.001
Copper	1.0
Lead	0.05
Mercury	0.005
Arsenic	0.05
Iron	0.3
Manganese	0.05
Barium	1.0
Silver	0.05
Cyanide	0.02
Electrical conductivity	900 μmhos/cm
pH	6.5 - 8.0
Taste	None
Odor	None
Foam	None
Color	None
Filter effluent turbidity	1.0 JTU
Carbon adsorption column effluent COD	30 mg/l
Chlorine contact basin effluent	Free chlorine residual

requirements are tabulated in Table 2. In addition, the district conducts a virus monitoring program to ascertain the efficiency of virus removal through the advanced wastewater treatment process. To date only two virus in four years of operation have been determined to have survived the treatment procedure. Because the water is introduced into a groundwater environment the two remote occurrences were not considered a health hazard. The district maintains its own State-approved laboratory to monitor the chemical and biological performance of the plant. Virus work is performed by contract with a local firm.

4. BARRIER OPERATIONS

4.1 Water Quality Monitoring

The product water from the treatment process is chlorinated by breakpoint chlorination and pumped to the 23 injection wells for injection into the subsurface aquifer. The maximum injection pressure is 20 psi, which is to prevent damage to gravel pack and the well seals of the injection units. The injected water and receiving groundwater is measured quantitatively and qualitatively on a daily basis, following a fixed schedule that provides for daily, weekly and monthly measurements. The injection water quality is continuously monitored and analyzed. Twice daily, grab samples are measured for chlorine residual, and every two hours the pH, conductivity and turbidity are analyzed. Each 24 hours a composite sample is tested for ammonia nitrogen, organic nitrogen, total kjeldahl nitrogen, calcium, alkalinity, fecal coliform, and total coliform. Once a week composite samples are tested for TDS, hardness, magnesium, sodium, chloride, flouride, boron, sulfates, foaming agents, color, cyanide, and phenol.

The groundwater quality monitoring program is conducted on samples collected from the four aquifers exposed to seawater intrusion. At weekly intervals four preselected wells are tested for chemical oxygen demand (COD), conductivity, ammonia nitrogen, and chloride, color and odor. Monthly and semiannually all the observation wells are checked for conductivity and the presence of chloride. Due to the special nature of the injection water, there is particular interest concerning the measurements of specific electrical conductance, coliform density, and the potential presence of virus.

The conductivity of the receiving water varies from 1,175 μmhos/cm, to as low as 375 μmhos/cm; the average is about 600 μmhos/cm, which is similar to the injection water. Thus the displacement of native water by injection supply is not discernable.

Tests for total coliform bacteria and fecal coliform are conducted at least every 48 hours on the final blended supply. No fecal coliform has ever been detected in the injection water. Table 3 is a summary of the plant performance during the 1979-1980

Table 3.   Performance of Water Factory 21 during
Period July 1, 1979 - June 30, 1980

Constituent	Units	Regulatory Agency Requirement	Injection (Q10) Mean	Number of Samples	Number of Times Requirements Exceeded
Electrical Conductivity	μmhos	900	534	344	2/344
Ammonium	mg/l	4	0.26	178	0/178
Sodium	mg/l	110	70	55	3/55
TH (as CaCo₃)	mg/l	220	81	54	0/54
Sulfate	mg/l	125	35	84	1/84
Chloride	mg/l	120	60	81	1/81
Fluoride	mg/l	0.8	0.41	55	0/55
Boron	mg/l	0.5	0.39	57	1/57
MBAS	mg/l	0.5	0.02	52	0.52
Hexavalent Chromium	ppb	50	1.70	47	3/47
Cadium	ppb	10	0.98	51	4/51
Selenium	ppb	10	5.00	51	0/51
Phenol	ppb	1	1.92	44	31/44
Copper	ppb	1,000	7.47	47	0/47
Lead	ppb	50	0.90	47	0/47
Mercury	ppb	5	0.54	47	0/47
Arsenic	ppb	50	4.62	47	0/47
Iron	ppb	300	26.00	51	0/51
Manganese	ppb	50	1.88	51	2/51
Barium	ppb	1,000	5.04	47	0/47
Silver	ppb	50	0.39	47	0/47
Cyanide	ppb	200	4.5	51	0/51
TKN	mg/l	10	1.21	156	0/156
pH		6.5 - 8.0	6.90	344	17/344
T Coli	per 100 ml	2.2	0.028	164	0/164
F Coli	per 100 ml	2.2	0.00	164	0/164
Turbidity		<1.0 JTU (Q6)	0.10	339	0/339
		<1.0 JTU (Q10)	0.28	345	0/345
COD	mg/l	(Q8)<30.0	8.83	126	0/126

year. An analysis of coliform occurrence indicates that bacteria were present only when the plant was operating on low flows or coming back on line following a shutdown, which implies some limited bacteria growing during shutdown periods.

4.2 Operational Costs

A summary of the advanced wastewater treatment costs and the reverse osmosis and barrier costs is shown in Tables 4 and 5. Table 4 is a compilation of the annual operation and maintence costs of the entire system, and Table 5 is a summation of total costs including capital and annual expenses. The costs are a

Table 4. Annual Cost of Operations and Maintenance,
Water Factory 21, 15 MGD Plant - 5 MGD RO Plant

Cost Categories	Unit Expenditures	Totals
Salaries		$767,026
Operations	$509,318	
Maintenance	257,708	
Utilities		976,465
Electricity	813,204	
Gas	163,261	
Chemicals		293,843
Lime	113,367	
Polymer	5,473	
Sulfuric Acid	42,078	
Chlorine	46,989	
SHMP	35,936	
Carbon	50,000	
Maintenance		375,813
Parts & Materials	161,921	
Equipment Rentals	5,500	
RO Maint. Contrace	184,822*	
Tools	5,020	
Clothing, Training misc.	18,550	
TOTAL ANNUAL O & M COST		$2,413,147

*RO maintenance contract includes all membrane replacement costs,
cartridge filter replacement, membrane cleaning, and routine
preventive and emergency maintenance.

Table 5. Summary of Operating and Capital Costs,
Water Factory 21

Process	Costs dollars/million gallons		
	Capital	O&M	Total
Inf. Pump & Lime Clarif.	47	104	151
Solids Handling	73	205	278
Ammonia Stripping	123	32	155
Recarbonation	16	61	77
Filtration	37	21	58
Act. Carbon Adsorption	93	56	149
Act. Carbon Regeneration	34	92	126
Chlorination	21	78	99
AWT Subtotal	444	649	1093
RO Demineralization	209	610	819
Deep Well Water	59	147	206
Injection	46	21	67
BLENDED WATER COST	427	692	1119

tabulation of actual expenditures during one-year operation of Water Factory 21. Detailed cost records are maintained to account for treatment processes including personnel expenses, utility costs, chemicals, and all maintenance expenses. All capital costs are amortized over service life of 20 years at 7 percent interest rate with a plant factor of 90 percent.

5. SUMMARY

There has been and remains widespread interest in the effectiveness of using reclaimed water as a supplemental source of water supply in an area of shortage. The experience of the Orange County Water district in this field has led the district to conclude the following:

1. Water Factory 21 has demonstrated the ability to satis- factorily treat secondary effluent from municipal and industrial sources to a level necessary for groundwater recharge and subsequent reuse.

2. With known technology, it is possible through advance waste treatment processes to remove a wide variety of contaminants, including trace organics and viruses.

3. The energy costs of wastewater reclamation are competi- tive with alternative water sources in Southern California where alternatives involve massive interbasin transfer programs.

4. Large-scale wastewater demineralization is an effecitve method of extending water supplies. The results at Water Factory 21 indicate the reverse osmosis processes will efficiently remove a wide range of contaminants from wastewater, including heavy metals, viruses, and organics.

Tables 6 and 7 are summaries of the efficiencies achieved at the District's Water Factory 21 facility.

Table 6. Plant Efficiency of Advanced Wastewater Plant, Water Factory 21.

Parameter	Plant Influent	Plant Effluent	% Removed
TOC (mg/l)	13.1	4.9	63
COD (mg/l)	45.6	6.5	86
MBAS	0.20	0.02	90
Phosphorus (mg/l)	5.5	0.1	98
TKN (mg/l)	30	4.0	86
Turbidity (NTV)	5.4	0.29	95
Coliform (per 100 ml)	7.1×10^5	0	100
Virus	5.0	0	100

Table 7. Plant Efficiency of Reverse Osmosis Plant,
Water Factory 21

Parameter	Plant Influent	Plant Effluent	% Removed
EC (μmhos/cm)	1104	140	87
COD (mg/l)	12.5	0.8	94
TOC (mg/l)	6.0	1.1	82
Sodium (mg/l)	158	18.5	88
Chloride (mg/l)	191	24.2	87

A number of questions have arisen out of the operation of the plant that indicate a need for additional research. These are enumerated as follows:

1. The experience with the reverse osmosis systems indicates a wide variety of complex substances deposit on the surface of the membranes. Additional quantitative fouling studies are needed to alleviate this troublesome and energy-intensive burden.

2. Alternative methods of disinfection appear to be necessary before a widespread use of reclaimed water is practical.

3. Less energy-intensive membranes, which are currently being tested, will lessen the costs of operation and subsequently stimulate broader application of reuse sources.

6. ADDITIONAL REUSE OPPORTUNITIES

Not all water utilized in modern society need to be of potable quality. Recent studies in Southern California have evaluated four general water reuse applications: 1) agricutlutre and landscape irrigation; 2) industrial process and cooling uses; 3) groundwater recharge by spreading; 4) injection to groundwater, as being applied in Orange County, for seawater barriers. In Southern California, the use of 85,000 acre-feet of reclaimed water is being proposed by state and local governments, to be developed in the next ten years.

The reclamation program proposed for Orange County in this regional study is called "Green Acres." The Green Acres program plans to deliver tertiary treated water to 150 landscape irrigation customers within a five-mile radius of Water Factory 21. Approximately 90 miles of pipelines would be constructed in adjacent cities. The uses comtemplated for the project include parks, golf courses, colleges, school yards, freeway landscaping and cemeteries. It is estimated that this program would require 12,000 acre-feet per year and would be substituted in place using the present potable water for these uses. Water costs are estimated to range form $360 to $460 per million gallons, a sum substantially higher than competitive fresh water sources. To encourage a broader application of marginal quality water, the district is contemplating incentive programs, including under-writing the costs of construction by subsidy, or seeking more flexible power rates for treatment and distribution costs. During a prolonged drought several years ago many golf courses in California were enjoined from irrigation and are therefore interested in alternatives that would provide firm water supplies. As a result, there is optimism that recognition of the reality of drought condition will encourage golf course operators to commit to the higher water charges with the assurance that their operations will not be curtailed during times of shortage. There is no doubt that in the next twenty years the costs of imported water in Southern California will surpass local reclamation costs, and therefore broad use of public landscaping will be done by reused water - or not at all.

30. INTERNATIONAL WATER QUALITY MANAGEMENT OF THE RHINE RIVER

Vladimir Mandl and H. Vrijhof
Commission of the European Communities, Brussels

The Rhine River is an international water course. Because of this fact, its management requires international cooperation. Such cooperation is indeed imperative if the economic and social development of the Rhine basin is to be sustained. At the same time, the dense industrial and urban settlements, which represent this development, make the Rhine potentially one of the most polluted rivers in the world. International management must coordinate the uses and interests within the different countries by allocating withdrawals and other uses of the Rhine water, and by protecting its quality. The latter must be done by defining water quality objectives and regulating discharges.

In this chapter, the pollution problems of the Rhine River will be described. Such problems are caused by discharges of oxygen consuming substances, nutrients, and industrial chemical pollutants. Many of the latter are toxic. Also, they are not biodegradable, and they are bio-accumulable. Chloride ion, it should be noted, has become a special problem area because of generally pervasive high concentration levels. Thermal discharges are yet another problem category. At the present time, discharges of redioactive wastes are not significant.

To address some of these problems, international water quality management began in 1950 under auspices of the International Rhine Commission. In 1976 the Commission adopted two conventions to protect the Rhine River against pollution by chloride and other chemicals. One was the Rhine Chemical Agreement, which came into force in 1979. The agreement is reviewed in this chapter within the general discussion of international management activities.

1. THE PROBLEM CONTEXT

1.1 Hydrology

The Rhine River drainage basin covers an area of 185,3000 km^2; it's length is about 700 km, and its width varies between 70 and 500 km. The total length of the main river from its alpine sources to the North Sea is 1,320 km; it ranks only 72nd on the list of the world's longest rivers. The Rhine has a relatively small flow, ranking 38th in the list of the world's largest rivers, but the drainage regime is favorable. Unlike rain-fed rivers, such as the Meuse, the Rhine is a so-called "mixed river." This means that apart from rainwater it is also fed by meltwater from high snow and ice. Thus, during the summer season a considerable mimimum flow is sustained by the melting of "high snow," i.e., 700-3,000 m elevation in the Alps. The mean annual flow at Basle is about 50 percent of that at Lobith at the beginning of the delta on the Dutch border, where it equals 2,300 m^3.s^{-1}. The ratio between the highest and lowest observed flows varies from about 12 to 16 along its course. With rainfed rivers, this ratio can easily reach values of over 100, as the flow during summer can approach zero. This, and the fact that the river is fed by tributaries along the upper and middle reaches make the Rhine navigable the whole year from the North Sea to Basle, which is more than 1,000 km.

1.2 Uses The favorable natural drainage regime of the Rhine River supports such diverse uses such as: inland shipping, disposal of sewage and wastewaters, fisheries, potable and agricultural and industrial water supplies, cooling water, maintaining the level of shipping canals and sluices, flow-through for prevention of seawater intrusion and quality improvement, bathing and boating recreation, and nature conservation. These various uses have been, without doubt, the basic reasons for the high level of economic and social development within the Rhine River basin. The last two are more recent considerations now being given formal attention by the international community of users.

1.3 Development Large areas of the basin are highly industrialized and urbanized. Examples include the Rhur River basin in the Federal Republic of Germany, and the "Randstad Holland" in the Netherlands. The population of the Rhine River basin is 60 million people, and the average population density is 324 per km^2. This makes it the most densely populated river basin in the world. The Rhur basin is the major industrial tributary and has a population density as high as 875 per km^2.

The city of Duisburg in the Ruhr is the world's largest inland port, while Rotterdam has become the world's largest seaport in terms of trans-ocean shipment. In 1977 one Rhine ship passed the Dutch-German border every four minutes and a total of 127 million tons of goods were transported along its course.

Among the industries, chemical industries are important. About 10 percent of all chemical companies in the western world are situated in the Rhine River basin.

1.4 Conflicting Uses

The dense urbanization and industrialization throughout the Rhine River basin, with all the varigated human activities, are the obvious origin of many environmental problems. Pollution is the most pervasive and prominent and the one which will diminish the utility of the river for its many uses.

The uses of the Rhine water are in some cases conflicting or even contradictory. The aggregate requirements cannot be met at the same time at all places.

The interests of the riparian states of the Rhine River are all different. The Federal Republic of Germany is predominantly interested in using Rhine water as industrial process water and cooling water. These uses amount to 900×10^6 m^3 year^{-1}, and $6000 \ 10^6$ m^3 year^{-1}, respectively. The dependance on the Rhine for drinking water supply is similar in the Federal Republic and the Netherlands; each takes about 300×10^6 m^3 year -1. Switzerland also abstracts drinking water from the Rhine, but France does not because groundwater of sufficient quality is available in large quantities.

Abstractions from the Rhine for agriculture and to maintain the level of shipping canals, to prevent salt intrusion from the sea and from groundwater is mainly a Dutch problem which has high priority. Owing to its geographical postition and because the lands of the Rhine River delta are partly below sea level, the water management of the Netherlands is highly complex and sensitive to alterations in the quality and quantity of the Rhine water. Abstractions in Holland amount to approximately 1000×10^6 m^3 year^{-1}.

1.5 Management Framework

From this brief overview it can be seen that coordination of the different uses and national interests is a necessity. Common international objectives must be agreen upon to safeguard water quality and to assure sufficient quantity. The international water management of the Rhine River therefore has three functions:

1. optimization of quality regulation and allocations of quantities of the Rhine River and its tributaries, while taking into account the range of conflictng uses,

2. definition of water quality objectives for each of the uses, i.e., specify the physical, chemical and micro-biological parameters and their recommended or required values corresponding to each of these uses,

3. regulation of wastewater discharges into the watercourses throughout the drainage basin; effluent discharge standards must be imposed according to the water quality objectives, taking into account the self-purification potential of the river and its tributaries.

Negotiations are taking place within an international commission, created in 1963, to manage the pollution of the Rhine and protect its different uses, as outlined within these priciples.

2. POLLUTION PROBLEMS

It is often heard that the Rhine is the "open sewer" of Western Europe. The discharge of untreated sewage and wastewaters has been practiced for many decades. The aggregate amount of treated wastewater discharged into the Rhine has increased, however, in recent years. These pollutants vary in kind over a wide spectrum. They are reviewed here in order to suggest priorities for action, and to outline a strategy for pollution abatement.

2.1 Oxygen Consuming Substances

Since 1946 the biological oxygen demand (BOD) in the river has increased steadily to a peak in the summer of 1971, when over 100 km of its middle reaches were devoid of oxygen. In the last few years, however, both ammonia and BOD have shown distinct decreases. This improvement is due to a few large sewage treatment works, recently installed. The BOD load has been reduced and ammonia has been oxidized to nitrate. As a result there has been an increase in oxygen concentration and in nitrate over the last decade. Average oxygen concentrations at Lobith for 1977, 1978 and 1979 were: 7.0; 7.6 and 7.7 mg/l, respectvely; the latter concentration is the highest since 1953, when measurements were first made. The value of 4.3 mg/l BOD in 1979 is the lowest value ever measured at this spot. The average ammonia concentrations over 1978 and 1979 were the lowest ever measured, e.g., 0.95 and 0.78 mg/l NH_4-N respectively.

A further decrease of the net load of oxygen consuming substances still must be achieved. The means for doing this include the following:

1. more wastewater treatment plants must be constructed for the treatment of domestic and industrial wastes,

2. introducing more complete biological treatment, purification and nitrification for existing wastewater treatment plants,

3. implementation of internal measures within each industry to reduce or eliminate effluents containing substances which cannot be treated.

2.2 Nutrients

Another important group of pollutants are the nutrients, i.e., nitrate and phosphate. These substances increase the primary production of algae and rooted aquatics to make the water unsuitable for some uses. During the nineteen sixties the concentration of phosphates in the Rhine River increased considerably, due mainly to the rising use of detergents. Dissolved ortho-phosphate concentrations increased from 0.06 mg/l P in 1956 to 0.3-0.4 mg/l over the period 1972-1978 as annual averages at Lobith, on the Dutch border. Total phosphates have been measured recently at Lobith in concentrations of 0.7-0.8 mg/l P for years with normal flow, and 0.9-1.0 mg/l for dry years, which are annual averages.

The sewage treatment works did not reduce these concentrations because there has been not tertiary treatment. Due to these heavy nutrient loads, problems of eutrophication are occuring in stagnant parts of the Rhine basin, especially in the Netherlands. As pointed out above, nitrate concentrations have been increasing over the last decade. In 1979 an annual average of 4 mg/l NO_3-N was recorded at Lobith.

2.3 Industrial Pollutants

The concentrations of sulphate, iron, and chloride have shown a steady increase since 1874, when measurements were first made. The chloride load of the Rhine is of major concern even though chloride strictly speaking can hardly be defined as a pollutant.

The natural chloride content of the Rhine is about 10 mg/l^{-1}. In 1880 the concentration measured was still 20 mg/l^{-1}, but from then on it continued to rise along with the industrial development in Western Europe. In 1979 the average chloride concentration was 167 mgl^{-1} corresponding to a mass load of 375 kgs^{-1}. The maximum load in the same year was 900 kg/s^{-1}. About 40 percent of the chloride load comes from Alsace. A similar amount comes from the Federal Republic of Germany, principally from the soda industry and coal mining. The remaining load comes mainly from Switzerland.

2.4 Other Chemicals

Although the oxygen concentration levels in the Rhine have improved recently, this is no reason for being less concerned about pollution in general. Most of this improvement has been obtained as a result of the removal of a limited number of substances from discharges. Many refractory substances continue to be discharged into the river. Among these substances there are toxic, nonbiodegradable and sometimes bio-accumulable compounds.

2.5 Thermal Pollution

Waste heat discharges are extensive also and are due mainly to cooling water from electric power stations and industrial cooling water. Other sources are wastewater discharges from municipalities and heating due to shipping. Until the end of the nineteen sixties the thermal pollution of the Rhine River was limited. At that time a number of electric power stations were being planned and it was thought that they might cause a serious thermal impact.

To investigate the thermal absorption capacity of the Rhine, both Switzerland and the Federal Republic of Germany undertook independent investigations. Based upon the principle that the thermal absorption capacity depends on the flow and the natural temperature of the river, the outcome of both investigations were essentially noncontradictory, and showed that the Rhine should never be heated by more than 3°C above its natural temperature. Only under exceptional circumstances could an increase of 5°C be allowed.

In 1970 a mathematical simulation model made predictions possible for 1975 and 1985. This model was developed by a German group of specialists. One of the assumptions was that all power stations planned would be built with a flow-through cooling system, without recycling of cooling water. It was concluded that the maximum permissible increase in temperature of the Rhine would be exceeded at several stretches of the river, e.g., in the autumn upstream of Mainz-Wiesbaden. Based on this simulation the Federal Republic of Germany decided that power plants to be built must incorporate cooling towers. It may be necessary to include further cooling capacity to enable power stations to adjust their thermal load to the capacity of the river to absorb it.

Since 1970 the number of giant electric power stations along the Rhine has increased, and the total installed electric capacity increased accordingly. In 1973 thirteen power plants were in operation, representing a total installed capacity of approximately 6000 MW. At present the number has more than doubled to 28 in 1980, with a capacity of 25000 MW. Half of this capacity is nuclear.

The total load of waste heat has been approximately 2100 $MJ.s^{-1}$ (5000 $Mcal.s^{-1}$) and it is expected that this value can be maintained by application of optimal (wet) cooling systems. Model calculations by the International Rhine Commission indicate a maximum acceptable waste heat load of 33500 $MJ.s^{-1}$ (8000 $Mcal.s^{-1}$), which is an increase of only 60 percent under optimal conditions.

2.6 Radioactivity

The most important potential sources of radioactive pollution are the nuclear power plants. Based on data from existing installations in the "long-term working program" a typical discharge rate of 20 Ci per 1000 $MW/year^{-1}$ is assumed for beta and gamma emitters, and 20-2000 Ci per 1000 $MW/year^{-1}$ for tritium, depending on the type of reactor. With the present installed capacity of nuclear power plants in the Rhine River basin, this does not present serious problems, except for possible accidents. The total gamma activity of Rhine River water is said to be due to natural radioactivity.

2.7 Pollution Originating from Shipping and Storage

There are various reasons for substances which are transported, transshipped or stored, getting into water. These include: flooding of river banks in periods of high flow, losses during the process of transshipment, disposal of residues, disposal of solid waste, wastewater and waste oil from ships, and accidents. Data on the actual quantities of such wastes and losses are scarce however.

Oil pollution is particularly important. The 16,000 ships using the Rhine River annually discharge 20,000 tons of used oil (bilge). By systematic collection 12,000 tons of this oil is recovered annually (1973), but the average quantity of oil which is transported across the German-Dutch border still amounts to 20,000-24,000 tons per year.

Shipping accidents due to heavy traffic on the river and on the road along the river also contribute to the pollution of the Rhine. Also, an extensive system of pipelines with diameters up to 36 inches is used to transport oil, oil products and ethylene. In recent years and average of about ten serious accidents in the basin are reported every year and it is likely that the risk of accidents will increase.

3. INTERNATIONAL WATER QUALITY MANAGEMENT OF THE RHINE

3.1 Historical Background

The start of the present international consultation and negotiation on the Rhine goes back to the end of the eighteenth century. At that time principles of free navigation were formulated and set forth in the Treaty of Paris in 1814, the Final Act of the Congress of Vienna in 1815, and the Act of Mainz in 1830. These regulations have provided the basis for the "Mannheimer Revidierte Rheinschiffahrtsakte" of October 15, 1868. This Act of Mannheim is still valid and stipulates freedom of navigation on the Rhine for all nationalities. Its executive body is the Central Rhine Navigation Commission. It also aims at the prevention and abatement of pollution originated by shipping transport.

During the nineteenth century the Rhine and Lake Constance were more important as fisheries than for navigation. International cooperation was mainly directed towards maintaining the migration of the salmon. This gave rise to several salmon treaties and commissions, which around 1920 also dealt with industrial pollution from Germany. In spite of all efforts, results were meager. Ten years later the yield of salmon fisheries had decreased considerable and by 1950 the Rhine salmon had become extinct. Deteriorating water quality is said to be the main cause of this.

The first problems of taste of drinking water occurred around 1920. Then in 1935 the Netherlands drew attention to the consequences of increased salinity by a planned increase in the production by the French potash works in Alsace.

After World War II, the Netherlands again drew attention to the deteriorating water quality of the Rhine. The Central Rhine Navigation Commission proved to be an excellent forum to expose the Dutch concern. Discussions on water quality began in 1950, when the International Commission for the protection of the Rhine against pollution was established semiofficially. This International Rhine Commission was provided with a legal basis, when it was institutionalized officially in 1963 by the Agreement of

Berne. Members of the International Rhine Commission are:
Switzerland, France, the Federal Republic of Germany, Luxembourg
(because of Moselle), the Netherlands and, since 1976, also the
European Economic Community. Its field of work is the Rhine
downstream of Lake Constance; thus there is no overlap with the
already existing commission for the protection of Lake Constance
against pollution. The International Rhine Commission cooperates
with this commission, as it does with similar commissions on the
Saar and the Moselle, and the tripartite Frnech-Belgian-Luxembourg
Commission. In this context a number of regional and local international treaties could also be mentioned, e.g., those for the
erection of common wastewater plants.

Other organizations must be mentioned as well, such as the
International Corsortium of Waterworks in the Rhine catchment area
(Internationale Arbeitsgemeinschaft der Wasserwerke im Rheinein-
zugsgebiet, i.e., IAWR). The IAWR has, since its foundation in
1970, been able to mobilize public opinion and has appealed to the
governments to combat the pollution of the Rhine, having
considerable influence.

3.2 The Present Situation

Although the International Rhine Commission met diligently
and dealt extensively with the measures to be taken, no decision
was reached during the first twenty years of its existence. Apart
from compiling a reliable record of the deteriorating water
quality, almost no progress was made. Because of the
International Rhine limited powers of that Commission, nothing
could be achieved unless there was strong support by the governments and public opinion. This support was achieved in 1970.

Because of several important events, such as the
"endosulphane incident" of 1969, and the general rise of environmental awareness, the "illness of Father Rhine" suddenly became a
subject of national and international public concern. The
proposal for the first environmental program by the Commission of
the European Communities contained a first set of practical
proposals aimed to render the work of the Rhine Commission
efficient and deliberations more rapid. As a result of these
developments the Netherlands took the initiative to call the first
conference of ministers, which was held in The Hague in 1972. The
pollution of the Rhine by chloride, chemicals and wasted heat, and
the presentation of a long-term working program were discussed.
The ministers mandated the International Rhine Commission to draft
lists of substances, similar to those of the 1972 Treaty of Oslo
concerning the dumping of wastes into the North Sea. The
discharge of these substances should be banned, limited or
restricted. In the International Rhine Commission special working
groups were formed to examine these topics.

Since their first conference in The Hague in 1972, the ministers have met several times, and have given an important political impetus to the solution of the problem of pollution of the Rhine. They directed the work of the International Rhine Commission, raised efficiency, and added a political element to the decision making process. They also agreed, in 1976, upon two conventions, on chloride and chemicals. A convention on thermal pollution still remains in preparation. Notwithstanding this, a resolution on the load of wasted heat of the Rhine was agreed upon in 1972. This resolution stated that all new electric power stations must have closed cooling systems, or equivalent systems. More recently in 1979, the ministers declared additionally that these systems should be capable of reducing the discharge of waste heat by 10 percent at least, compared to flow-through cooling. They also appealed for the Riparian States to keep in touch with the thermal convention in preparation.

3.3 The Rhine Chloride Agreement

Since the nineteen thirties the water supplies in the big cities along the Rhine have had taste problems. Both chlorophenols and chlorides (salt) were said to be responsible. Later, problems of increased corrosion of water distribution pipes appeared and were ascribed to chloride. Studies at the same time also mention damage to horticulture and the growing of flowers, especially in the Netherlands. The Federal Republic of Germany experienced similar problems. In both countries about four million people depend on the Rhine for their water supply.

The Rhine Chloride Agreement of 1976 aims at a reduction of the salt content of the Rhine. Since 40 percent of the salt load comes from Alsace, France would retain 60 kg $Cl-.s^{-1}$ by 1980; 10 $kg.s^{-1}$ of this quantity would be injected into deep underground layers, the cost of which would be shared by France, the Federal Republic of Germany, the Netherlands and Switzerland. The remaining 40 $kg.s^{-1}$ would be held back pending the agreement on technical modes of this project and its financing, before 1 January 1980. In spite of the quick ratification by other member states, the French government was not in a position to ratify the agreement in parliament. During the last conference of ministers in January 1981 the solution of the salt problem was discussed again. Alternative approaches were discussed such as transport of salt and discharge into the North Sea and making available the chlorides to the soda factories of Lorraine.

The problem of salt is certainly not the most important aspect of the pollution of the Rhine, although publicity may give this impression. It might be the consequence of Dutch policy which gives priority to the salt reduction. Recent studies do not indicate the negative impact on taste of drinking water to be due to salt contents. The cause of bad taste of Rhine water is to be found in organic substances that, whenever discharged, appear in drinking water.

3.4 The Rhine Chemical Agreement

One of the International Rhine Commission's mandates from the first conference of ministers of 1972 was to prepare a convention on chemical pollution. The adoption by the Council of the European Community of the Directive on Pollution caused by certain dangerous substances discharged into the aquatic environment of the community in May 1976, facilitated the preparatory work for the convention. It was signed on 3 December 1976 at Bonn. On the same date, the European Economic Community entered the Agreement of Bern and has taken an active part in the work of the International Rhine Commission.

Like the council directive, the Rhine Chemical Agreement aims at eliminating the pollution of surface water by List I (black list) substances, and reducing the pollution by List II (grey list) substances. Implementation would preserve the different uses of the Rhine water.

List I contains certain individual substances which belong to seven families and groups of substances, selected mainly on the basis of their toxicity, persistence and bio-accumulation. Those which are biologically harmless or which are rapidly converted into substances which are biologically harmless are expected.

List II of the agreement contains:

1. substances belonging to the families and groups of substances in List I for which the limit values referred to in Article 5 of the agreement have not been determined.

2. certain individual substances and categories of substances belonging to the families and groups of substances listed in the agreement and which have a deletorious effect on the aquatic envrionment, which can, however, be confined to a given area and which depend on the characteristics and location of the water into which they are discharged.

Both the Directive and the Chemical Agreement are framework documents, which need to be applied case by case. The Rhine Chloride Agreement might well be considered as a special case for one particular substance.

In applying the Rhine Chemical Agreement, which came into force on 1 February 1979, several phases must be passed before purification of a substance is formally regulated. The actual benefit, however, is realized much earlier than the moment at which the limit values of the convention come into force. Very often the mere decision to study a substance is sufficient to call forth appropriate countermeasures. Obvious examples of this effect are endosulphane and mercury. The application of the Rhine Chemical Agreement has an influence that is further reaching than might be expected.

In detailed application of the procedures for substances in the black and grey lists, many questions have to be answered. For the grey list substances, the procedure is now being studied on the metal chromium. For the black list substances the following phases can be distinguished.

1. On the basis of preliminary data, one or more contracting parties may propose that a substance be dealt with priority, in working group B of the International Rhine Commission. This group is charged to propose the implementation of the convention.

2. Evaluation of the proposed substance is accomplished according to the criteria: toxicity (including carcino-genity), persistence, bio-accumulation, occurrence in the Rhine water, quantity that might enter the Rhine water during production and application or use.

3. For selected substances there will be an investigation into: (1) the nature of the industrial sector involved, (2) the emission per category of polluting activity, and (3) existing national legislation on manufacture and application.

4. The next phase is a study of possible means of reducing or eliminating the emissions.

5. Limit values for substances and industrial sectors are formulated; dates for these limits to come into force are proposed and procedures for surveillance and control agreed upon.

6. These agreements have to be approved by the contracting parties and come into force when they have been accepted unanimously.

7. The Permanent Working Group of the International Rhine Commission, which is responsible for measuring the water quality, will also measure the new black list substances.

A recent survey of black list substances indicates that the procedure for mercury discharged by the chlor alkali electrolysis industry is now in the last phase of the decision making process. Another twelve substances of List I had already been studied by the departments of the Commission of the European Communities. Additionally 66 substances were selected and are currently being studied.

Priority was given to substances which, apart from being toxic, persistent or accumulable in organisms or foodchains, are produced or used in large quantities, which could lead to potentially dangerous concentrations in the Rhine. Most of these 66 substances have been detected in the Rhine.

4. CONCLUSION

In recent years substantial progress in combating pollution of the Rhine has been achieved. The oxygen saturation has been improved, the efficiency of the Internatinal Rhine Commission has increased and the first effects of the Rhine Chemical Agreement are becoming evident.

Much still needs to be done to make the Rhine clean and its waters suitable for the different used they are intended to serve. The Commission of the European Communities will continue to play an active part in this work. The case of the Rhine River illustrates the idea of operation and management of a highly developed complex water system from both technical and institutional points of view. Modeling of the river, political decision making, and development of administrative organizations have become institutionalized to work toward more effective management to achieve common goals.

31. INSTITUTIONAL PROBLEMS IN THE OPERATION OF IRRIGATION SYSTEMS

Luis S. Pereira
High Institute of Agronomy,
Technical University of Lisbon

1. INTRODUCTION

The success of the irrigation systems and, consequently, of the water resource systems from which they are a part, depends on the efficiency of agricultural uses. Therefore, the management of such systems must take into account not only good design but also the institutional problems.

These institutional questions concern both the planning and operation phases. During the planning phase the concerns are policies in the area of soil and water management, crop management, agrarian structures, and the effectiveness of the research and extension related to appropriate agricultural technologies. During the operation phase emphasis is on participation and involvement of farmers, their training and ability to use modern technologies, and the agrarian policies that influence the decisions of farmers about soil and water uses.

This chapter outlines the role of institutions in the operation of an irrigation system. Further it relates irrigation water use to overall project planning.

1.1 Characteristics of Irrigation Systems

The use of water for irrigation may be characterized as follows: water is dispersed over large land areas; a large number of users, the farmers, often have different criteria for water use; water is a fundamental factor of production, and consequently its value depends on the agricultural output; in arid and semiarid regions, such as the Mediterranean, irrigation consumptive use is

the major water uses within the water system. These
characteristics indicate that management of the irrigation system
is related to the success of the irrigated agriculture, and therefore to the factors that are controlled to optimize production.
These include water management, farming improvement and farmer
education and involvement. At the same time, the success of a
multipurpose water resource system requires good performance in
improvement of agricultural output.

Thus it is important to identify the factors that influence
the success of an irrigation system. From this it is possible to
devise institutional measures to improve water use and the
efficiency of the water resource system as a whole.

2. PHASES OF AN IRRIGATION PROJECT

The development of an irrigation system has three basic
phases: planning, construction and operation. Figure 1 illustrates in the format of a flowchart, with further disaggregation
to show important steps.

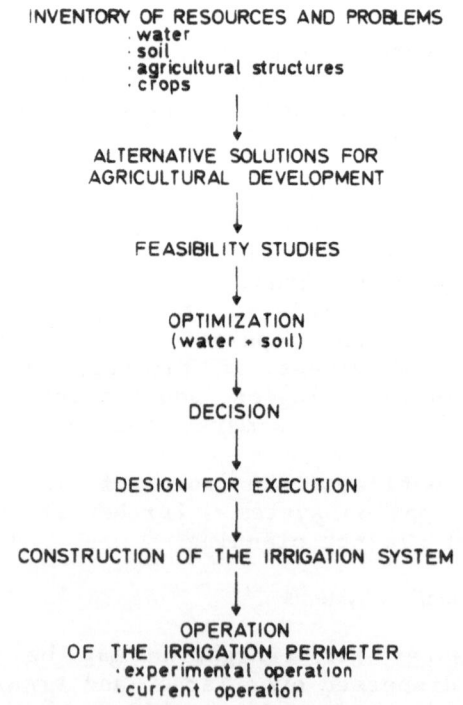

Figure 1. Phases in development of an irrigation system.

All these phases have similar importance and are interrelated. In particular, the tasks associated with operation of an irrigation system, such as reservoir operation, irrigation and drainage network operation and maintenance, on-farm water management, crop production, and farm management, depends on the planning and design of the project.

Planning for successful irrigation begins at the inventory of resources to be utilized and of problems to be solved. It is necessary to identify all questions related with water and soil uses, with crops to be produced, and with agrarian structures such as human and social, farm sizes and parcellation, farm equipment and infrastructures, that condition the agricultural development.

This initial step of problem identification and resource inventory is the basis for devising alternative solutions for agricultural development and land improvement. These solutions must coordinate both the engineering and agricultural aspects of the development, and also provide enough information in the feasibility analysis for optimization.

The optimization within the alternative solutions selected must take into account the scarce natural resources, e.g., water and soil. It must be done so that the decision-making is clear with respect to the social and economic costs and benefits of the optimized solutions. The decisions must consider not only the quantifiable aspects of the system but also the less tangible management factors that influence the irrigation system operation. These are related to overall water management and with on-farm water use. These are key factors that affect the project's output.

The last step in the planning phase is design, based on the alternative solution chosen. This is the basis for the next phase, which is construction.

The construction of an irrigation system includes reservoirs, canals, pumping stations, irrigation and drainage networks, hydraulic structures, rural roads, electricity distribution systems, land leveling, etc. This must be done along with the farm structures improvement, and land consolidation, in order to transform the agriculture from rainfed to irrigation. It is evident that, just as the planning and design are requisite to the construction phase, the technical conditions of execution are important for the system operation and maintenance.

In the third phase, there are two principal periods. First, is the testing period, during which the structures are tested, the operation rules are implemented, the soil and water uses are improved, and the irrigation agriculture technologies are

488

introduced. Second, there is day to day operation, where the
irrigation system and the irrigated agriculture will run according
to modes established during the testing period.

The day to day operation leads to the development of
operating rules, contingency water management planning such as
during droughts, of irrigation district perimeters, and on-farm
soil, water, and crop management. Further, there must be reconcil-
iation with the problems from the antecedent phases.

3. RELATIONSHIPS BETWEEN AN IRRIGATION AND A COMPLEX WATER SYSTEM

An irrigation system is no longer isolated from the larger
multiple use complex water system. Therefore, to understand the
impacts of agricultural institutional problems on a multipurpose
complex water system, it is important to recognize the relationships between the development phases of both systems. Figure 2

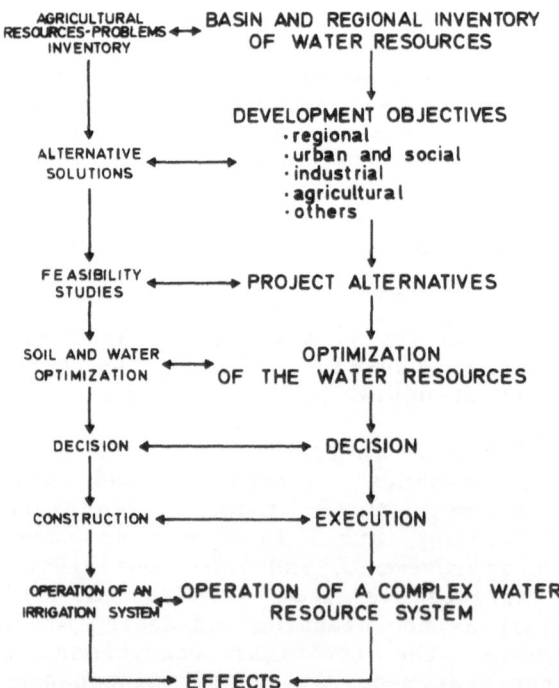

Figure 2.   Corresponding development phases between an irrigation
project and the overall multipurpose water project.

summarizes graphically. Two flow tracks are shown, the irrigation development track on the left and the overall water system track is on the right. The relationships at corresponding steps are seen by the horizontal lines.

These relationships are quite evident at the beginning inventory step where the inventory of agricultural resources and problems are related to the water resources inventory at the regional level. The second phase in the overall water system planning is the definition of development objectives. This is necessary in order to formulate the alternative solutions for agricultural development. Such coordination permits integration between agricultural development, regional development, and the development of nonagricultural sectors. Therefore, the feasibility studies concerning the alternative solutions for agriculture must take into account the complex multipurpose project alternatives. Also, the optimization of the irrigation system must be integrated with the water resource optimization at basin or regional level. This leads to an interrelated decision process. The decision-making process must provide the rules for water allocation between agricultural and nonagricultural uses based upon a broad, unified vision of the overall regional development.

The water management institutions are faced with more difficult tasks in these multipurpose projects. They must be able to develop operating rules which integrate the system operation with resources availability and the needs of various users. Also they must be able to solve problems of water allocation during periods of water shortages. All these relationships require coordination between regional interests and those of individuals with the objective of effective system operation and efficient use of water.

In general, in multipurpose projects in arid and semiarid conditions, the efficiency of irrigation is a decisive factor in the success of the overall water system. Institutional problems are often critical in this question of irrigation efficiency, even causing inadequate engineering and agricultural solutions.

4. FACTORS INFLUENCING THE SUCCESS OF IRRIGATION

A multitude of factors influence the successful operation of an irrigation project. Figure 3 identifies some of them and shows where they should be considered in planning and development. The phases are discussed below with respect to the influences of these influencing factors. They are:

490

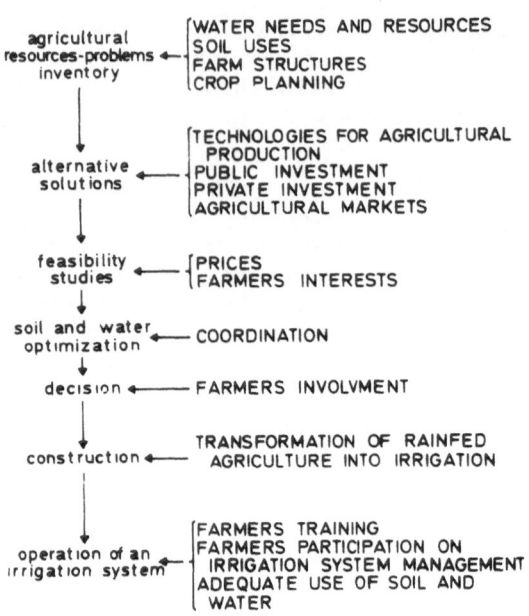

Figure 3. Factors influencing the planning and development
of an irrigation system.

a.   The inventory of agricultural resources and problems
must identify questions that constitute fundamental factors to be
considered in the following stages. They include available water
resources and the water needs, for agricultural production;
through the land evaluation, the soil uses capability and the
suitable soil uses; agrarian structures, mainly the existing farm
structures, as farm sizes and parcellation, farm ownership, farm
equipment and farm management problems; crops to be planted under
irrigation.

b.   The alternative solutions for agricultural development
are based upon the inventory information and according with the
development objectives. Factors to be considered at this stage
include technologies to be used for the agricultural production,
for irrigation and drainage, and for the operation of the irrigation system; the alternative engineering solutions, must correspond to the different levels of public investment; the agricultural and irrigation solutions must consider the availability
of private investments to transform rainfed farming into irrigation; considering the irrigation will change crops and
quantities of products, the alternative solutions must consider
the markets for the agricultural products.

c. The feasibility studies will analyze the alternative solutions. Therefore their results depend on the quality of data obtained during the antecedent planning stages. But they are also influenced, among others, by two main questions, i.e., the quality of information and forecast on prices of agricultural products and production factors, and the interests of the farmers in the project, because the production decisions at farm level are done by the farmers themselves.

d. The optimization process, with regard to soil and water, is based upon the feasibility analysis and on the allocation rules at basin or regional level.

e. The decision-making process must take into consideration the resources allocation, the social and economical aspects of the project, and the technical solutions. But above all, the farmers must be involved. They will be the users of the irrigation system and the success of it depends on the farmers decision about soil and water use, about crops, about farming, about almost all questions that condition the irrigation system operation.

f. Following the selection of an alternative solution, design and construction is scheduled. The main focus during this phase engineering and changing rainfed agriculture to irrigation.

g. The operation of an irrigation system utilizes the facilities constructed. This means that the effective consideration of all factors during the antecedent phases leads to better conditions for operation. In addition, some particular questions dealing with operation are: training of farmers toward their utilization of the technologies and the production of the suitable crops; participation of the farmers in the organization dealing with operation and maintenance of the irrigation system and in the overall water management; adequate use of the soil and water resources to provide not only the best mix of production factors but also the highest incomes to farmers, and therefore, the highest project output.

5. INSTITUTIONAL PROBLEMS INFLUENCING OPERATION OF IRRIGATION SYSTEMS

As shown above, the operating conditions depend not only on the factors directly associated with this phase, but also on conditions during the antecedent stages. It is then important to recognize the institutional problems which shape the planning, the construction and the operation conditions. Figure 4 is a diagram which identifies the various kinds of irrigation institutions associated with project operating factors. How these various

492

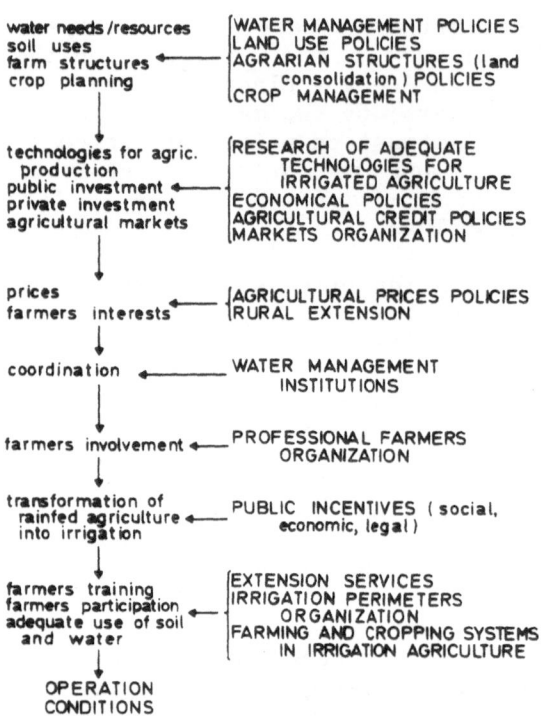

water needs/resources   WATER MANAGEMENT POLICIES
soil uses   LAND USE POLICIES
farm structures ← AGRARIAN STRUCTURES (land
crop planning   consolidation) POLICIES
     CROP MANAGEMENT

technologies for agric.   RESEARCH OF ADEQUATE
   production   TECHNOLOGIES FOR
public investment ←   IRRIGATED AGRICULTURE
private investment   ECONOMICAL POLICIES
agricultural markets   AGRICULTURAL CREDIT POLICIES
     MARKETS ORGANIZATION

prices   AGRICULTURAL PRICES POLICIES
farmers interests ← RURAL EXTENSION

coordination ←   WATER MANAGEMENT
     INSTITUTIONS

farmers involvement ← PROFESSIONAL FARMERS
     ORGANIZATION

transformation of
  rainfed agriculture ← PUBLIC INCENTIVES (social,
  into irrigation   economic, legal)

farmers training   EXTENSION SERVICES
farmers participation   IRRIGATION PERIMETERS
adequate use of soil   ORGANIZATION
  and water   FARMING AND CROPPING SYSTEMS
     IN IRRIGATION AGRICULTURE

OPERATION
CONDITIONS

Figure 4.   Irrigation institutions and their association
with project operating factors.

institutions are related to practical operating questions is
enumerated in terms of the points following:

     a.   The information given from the inventory of agricultural
resources and problems must support the application of agricultural policies which result in the appropriate technical
solutions. These include crop planning information which depends
on a general crop management policy. Very often this policy is
not defined and consequently the technical decisions on crops can
cause failure in subsequent stages. Water resources must be
allocated according to the policies on water management, otherwise
there are no bases to build up balances between resources and
needs, and the project alternatives have no strong basis, thus
influencing the development of the system. The ways to achieve
good soil conservation depends on land use policies and on the

implementation measures. If such policies are not defined, the project is developed on mere hypothesis. The existing policies on agrarian institutions such as land consolidation, condition the adaptation of farmers to new production conditions and, consequently, to the irrigation and cropping solutions through the irrigation project.

b. The formulation of alternative solutions depends not only upon the above mentioned institutional forms, but a host of others. They include agricultural credit policies, from which depends the private investment to the farming development; organization of agricultural markets and the farmer's participation; and the agricultural research policies concerning technological solutions, irrigation and drainage, and water management.

c. During the feasibility studies two main institutional questions are to be considered. These are: The agricultural prices policies, because their knowledge is essential to forecast price trends and their analysis give us a picture of the impact of prices on farm management decision; and the institutional organization of rural extension services which help to understand the interest and motivations of farmers about the irrigation project, and induce farmers to participate in the transformation of rainfed agriculture into irrigation.

d. The most important institutional problem concerns intersectorial coordination: the water management institutions that influence water allocation rules at basin or regional level. Without the effectiveness of these institutions the optimization process is insufficient and its consequences on the decision-making processes are less efficient, generating problems on operation of water resource systems.

e. Concerning the decision-making process one must emphasize the questions related with farmers involvement in this process, which is fundamental for their participation in investments and in the agricultural development process. The commitments of farmers depend on their having an organization through which farmer's representatives can take part in the decision-making process.

f. As said earlier under section 4f, during construction phases it is necessary to prepare the transformation from rainfed agriculture to irrigated agriculture. Therefore it is important to prepare and to provide public incentives, regarding political, social, economic and legal measures that influence the desired transformation.

g. The above mentioned institutions problems have both positive or negative impacts on the operation of irrigation projects and multipurpose water resource systems. The institutional problems which influence directly the operation of an irrigation system are described in the following: The farmers' training on use of modern irrigation technologies, good cropping patterns, and efficient farming management, depends upon the effectiveness of the agricultural extension services. The farmers' participation in operation and maintenance of the irrigation system is related to the institutional forms concerning the water management services for operation of the water resources system. The adequacy of soil and water use and then the maximization of the agricultural output is a result of the technical solutions chosen for the irrigation project and of the institutional forms leading to the suitable farming and cropping systems in irrigated agriculture.

6. CONCLUSIONS

It has been shown that the operation conditions of an irrigation system depend very much on the quality of the technical solutions arising in the antecedent phases, e.g., planning and construction, which gives the background for establishing the most effective operation rules and water management decision. It has been recognized also that effective operation of an irrigation system requires support by appropriate institutional forms. The institutional problems have then an important role for the success of an irrigation system. They should be considered along with technical problems in importance. Therefore, when looking for the improvement of the operation of complex water resource systems it is fundamental ro recognize the institutional problems and to implement their solutions.

RÉFERENCES

1. Biswas, A. K., A. H. Samaha, M. H. Amer and M. A. Abu-Zeid (editors). Water Management for Arid Lands in Developing Countries. Pergamon Press, Oxford, 1980.

2. Carruthers, I. Contentious Issues in Planning Irrigation Schemes, in The Social and Ecological Effects of Water Development in Developing Countries. Pergamon Press, Oxford, pp. 301-308, 1978.

3. Papadopoulos, G. E. Basic Prerequisites for and Adequate Management of Reclamation Projects. Proceedings 10th Congr. Irrig. Drain., ICID, Athens, Vol. 5, pp. 1-21, 1978.

4. Pereira, L. S. Aménagements Hydroagricoles. Considérations sur leur Project et leur Réhabilitation. Proceedings, 11th Congress on Irrigation and Drainage, ICID, Grenoble, Vol. 1, pp. 41-62, 1981.

5. Vlachos, E. C., G. E. Radosevich and G. V. Skogerboe. Operational and Organizational Characteristics for Effective Irrigation Systems. Proceedings 10th Congr. Irrig. Drain., ICID, Athens, Vol. 5, pp. 33-54, 1978.

6. Widstrand, C. (editor). Water Conflicts and Research Priorities. Pergamon Press, Oxford, 1980.

32. THE WATER RESOURCES OF TURKEY

Ibrahim Gürer
Hacettepe University, Beytepe-Ankara, Turkey

1. GENERAL

Annual precipitation in Turkey varies between 220 mm. to 2500 mm., wIth an average annual precipitation of 679 mm. This corresponds to a total water potential of 518×10^9 m^3/year. About 245 mm., or approximately 180×10^9 m^3 per year occurs as flow in rivers. According to recent feasibility studies with the construction of 450 to 500 dams, it will be possible to regulate all the rivers and make use of 95×10^9 m^3/year of this amount. In addition, the available groundwater reserve of Turkey is 9.4×10^9 m^3/year. Therefore the net sum of available water is 105×10^9 m^3/year.

With the completion of the planned water works, it will be possible to protect 213,760 hectares of land from floods, to drain and make use of 18,990 hectares of swampy lands, to irrigate 4,600,000 hectares of lands, and to supply 1.1×10^9 m^3 of water to community centers for domestic and industrial use.

When the Turkish republic was found in 1923, the electric energy production was 33,000 kW per year, which is one fourth of today's need of capital Ankara. Just after the second world war, the electric energy production was 246,000 kW per year, and with 10 years, it increased to 611,000 kW per year. Between 1956 and 1962 it became 1,332,000 kW per year. Average rate of increase is about 13 percent per year. With the construction of 300 to 350 hydroelectric power plants the associated electric energy production will be about 100×10^9 kWh per annum from 25,000 MW installed capacity. At present, only 9-10 percent of the total water resources potential of the country is utilized. Utilization of the whole capacity will be possible only through large investments, covering a long period of construction.

Even the first tribes living in Anatolia were successful in constructing simple water structures. It is possible to find small dams and water storage units built mostly for drinking water supplies. Ruins of small dams 8 to 10 meters high, dating from Roman Era, were also found in Anatolia. These were destroyed by historical floods and so only the ruins remain.

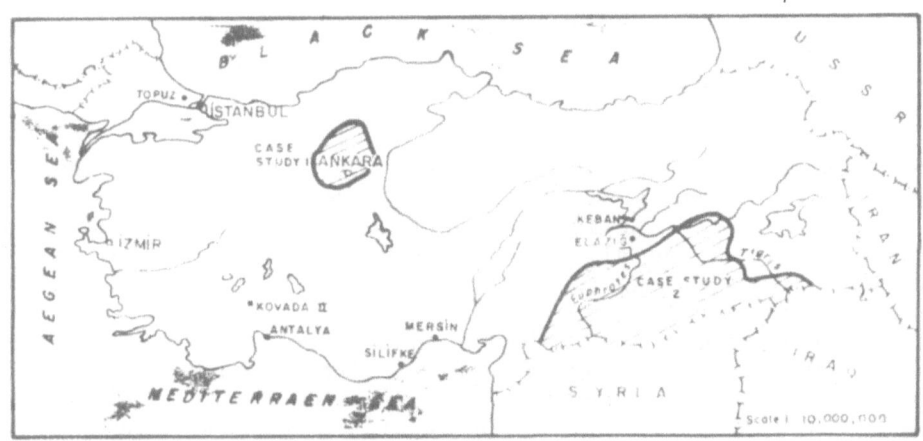

Figure 1. General Map of Turkey with the Areas of Case Studies

The number of small dams built before the Republican Era is about fifteen; the dates of construction of the oldest five are not known. The oldest recorded dam is Topuz, near Istanbul, was built about 1620. It had 70,000 m^3 of lake volume and was 8.6 m in height. There are at present 66 dams and hydroelectric power plans in operation. The largest one is Keban; Figure 1 shows its location. Keban dam was built in 1974 on the Euprates River and controls 64,000 km^2 of drainage area. It is the 12th highest, 13th in water volume stored behind the dam, and 24th in energy production in the world list of large dams. Keban is a composite dam of mass concrete and rock fill with 848 m of crest elevation,

207 m of height on talweg, 845 m of average water level, 30.6 x 10^9 m^3 of water storage, 675 km^2 lake area, 1240 MW installed capacity, and 5800 x 10^6 kWh annual energy production. It is a multi-purpose dam for irrigation, flood protection and electric energy production. The average inflow to dam is about 635 m^3/sec; 70 percent comes from snow-melt floods of eastern Anatolian highlands.

There are also 16 river-run and canal hydro power plants in operation with capacity varying from 0.38 MW at Silifke, to 53 MW at Kovada, Antalya. At present, there are 50 dams and hydropower plants under construction, 24 with final designs completed, and 34 under design.

2. OPERATION OF RESERVOIRS

The operation process starts with the completion of the project. Usually a satisfactory operation program of a multi-purpose project depends on the coordination among the owner, the user, and the plant. The Turkish Electric Authority (TEK) decides on the energy use, and allocates the sources to answer the energy demand in different sectors of country. The Turkish State Hydraulics Works (DSI) prepares the water budget programs of 64 reservoirs under operation. Almost all the operation programs are annual and are based on long term forecasting of annual as well as seasonal inflows to reservoirs by classic linear step wise, multiple regression analysis. The analysis is based on a long period of observations of inflow as the dependent variable. The independent variables are winter flow, monthly precipitation, and snow water equivalents. With the pattern of the past years, the total annual flow is distributed by month. At the beginning of each month the hydrologic forecasting office of the operation and maintenance division dictates the maximum and minimums of reservoir elevations. This is not the most flexible way to operate reservoirs, especially in high water seasons of snow melt floods of April and May. Therefore the snow surveys become more and more important for long term and short term forecasts especially in the Central as well as Eastern Anatolian reservoirs. Here a substantial part of the annual inflows comes from the snow pack.

3. CASE STUDY 1: WATER SUPPLY SYSTEM TO ANKARA

Due to a sudden increase in population, coinciding with industrialization, and the consequent immigration of people from rural areas to big cities, the domestic and industrial water demand of the city of Ankara has increased sharply in recent years. The DSI prepared a water supply project for Ankara. Figure 2 shows the water resources and the planned water works for the project. There are four reservoirs to supply water to the city. These are Cubuk I built in 1936, Cubuk II built in 1964 to provide 28×10^6 m^3/year, Bayindir built in 1965 adds 5×10^6 m^3/year, and Kurtbogazi built in 1967, provides 83×10^6 m^3/year. All of this water is for domestic and industrial use in Ankara. Cubuk I dam is the first dam of the Republican era. It has almost completed its economic life due to sediment deposition. It is now only a weekend resort serving the capital. Therefore Cubuk II dam was built to store spring floods and work in conjunction with Cubuk I, and other units of the project.

At present, Camlidere dam and tunnel, the †vedik water purification plant, and the Ankara city water pumps and transportation network are under construction. Figure 2 shows their layout.

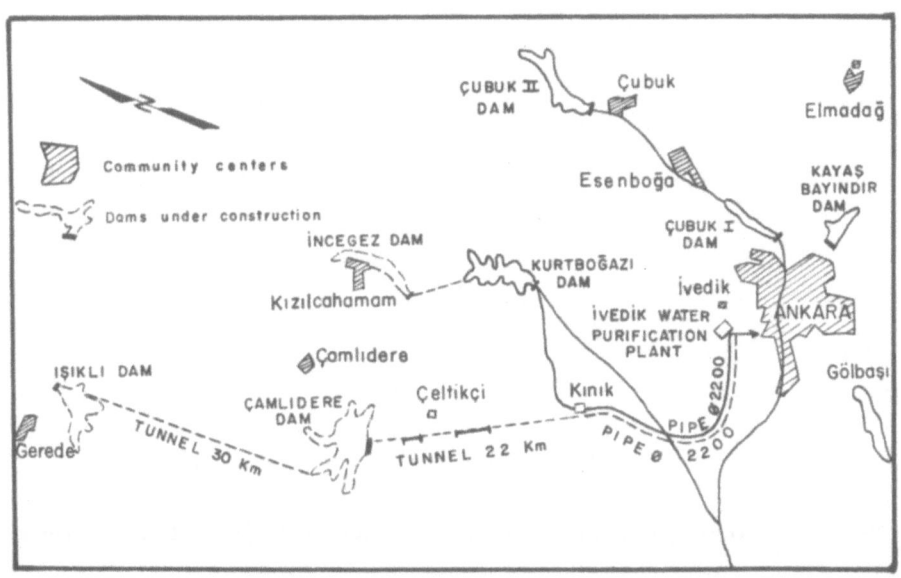

Figure 2.   Area of Case Study 1.   The Water Supply System to Ankara City

It is anticipated that by constructing three more dams and two more tunnels, water demand of Ankara will be satisfied until the year 2020.

4. CASE STUDY 2: SOUTHEAST ANATOLIA PROJECT

The Southeast Anatolia Project (SAP) will be the largest project of Turkey. It is a complex water resources system as well as a development project. It will cover a large area between the Euprates and Tigris Rivers as shown in Figure 3. The system consists of 15 dams, 18 hydroelectric plants, and will irrigate 18 x 10^4 hectares of land covering 12 large projects. The project will be completed in 30 years. The electric energy which will be produced with this complex system is 22 x 10^9 kWh/year based upon 7620 MV installed capacity. It is anticipated that the people living in project areas will improve their present living conditions by a significant margin.

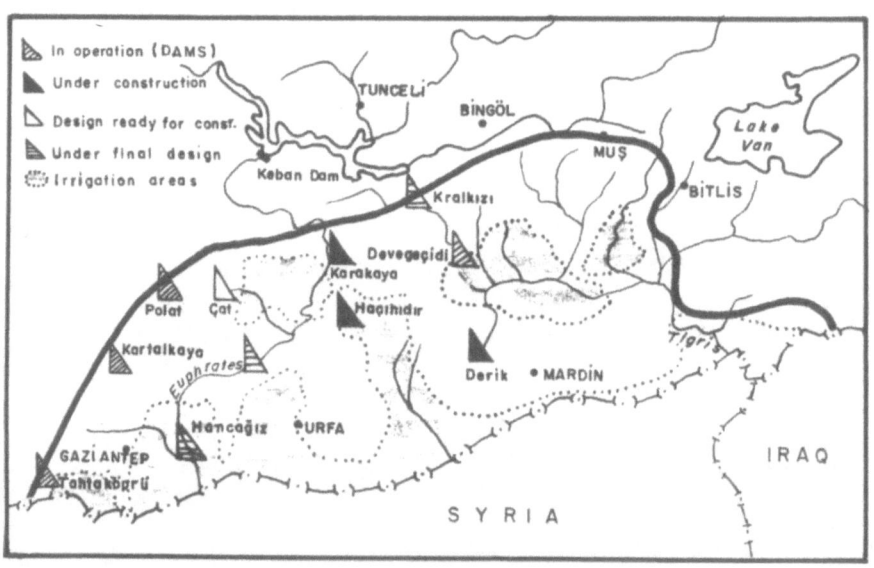

Figure 3.   Area of Case Study 2.   Southeast Anatolia Project

5. CONCLUSION

With twenty-six large river basins, the total available water of Turkey is 105 x 10^9 m^3 per annum. At present about 10 percent is utilized, made possible by 450 dams and 350 hydropower plants. This collection of dams, reservoirs, and ancillary structures comprised a nationwide complex water system. Its operation will be supplemented and enlarged greatly by the new projects being planned and under construction.

REFERENCES

1. DSI, 1980. Türkiyede yapilmis ve Projelendirilen Baraj ve Hidroelektrik santraller. DSI, Ankara.

2. DSI, 1969. Keban Projesi, G. Yayin No: 636, Grup No: VII, Ozel No: 116, DSI, Ankara.

3. DSI, 1979. DSI 5. Bölge, Ankara.

4. DSI, 1981. Güneydogu Anadolu Projesi, DSI, Ankara. 1981.

33. GENERATION OF RUNOFF DATA FOR UNGAUGED STREAMS

G. A. Fuller
University of Regina, Canada

1. INTRODUCTION

Streamflow records are one of the bases used in the design and operation of complex water resource systems. Unfortunately records available from many streams are of insufficient length and no streamflow data have been recorded for some streams. Consequently estimates of unrecorded flows are often required for hydrolgic studies. To estimate data at a site for which some records exist, relationships may be established between the existing record and records from nearby sites, (3). The more difficult problem of estimating streamflow data at sites for which no records exist is the topic of this chapter.

2. UNGAUGED STREAMFLOW MODEL

A least squares regression model has been developed to estimate ungauged streamflow data, (1). This model uses streamflow data from nearby sites as predictor variables as given in Equation 1.

$$y = a + b_1x_1 + b_2x_2 + \ldots + b_{nxn} + ZS_y (1-R^2)^{\frac{1}{2}} \qquad (1)$$

where
y = estimated streamflow data
x_i = ith streamflow predictor
a = regression constant
b_i = regression coefficient on ith streamflow predictor
R = multiple correlation coefficient
Z = random standardized normal deviate
S_y = standard deviation of y

The least squares regression coefficients used in Equation 1 are calculated from estimates of correlation coefficients between the flows at the ungauged site and the records from nearby hydrometric sites. Physiographic characteristics were used to estimate these correlation coefficients, the means and standard deviations of the flow at the ungauged site. The regression coefficients were derived from the relationship given in Equation 2 from Richardson (4).

$$b_i = \frac{S_y}{S_{x_i}} \cdot \frac{M_{yi}}{M_{yy}} \cdot (-1)^{i+1} \tag{2}$$

where b_i = regression coefficient for ith streamflow

S = standard deviation

M_{yi} = minor of correlation matrix between y and x's obtained by deleting the row associated with y and the column associated with the ith streamflow record.

In estimating ungauged streamflow data from estimated correlation coefficients, care must be taken to insure that the matrix of correlation coefficients between gauged and ungauged streamflows is positive semidefinite, since all correlation matrices are positive semidefinite. To insure that this occurs, an algorithm based on the Rosenbrock hill climbing method has been developed to adjust the estimated correlation coefficients to make the matrix positive semidefinite when necessary (2).

3. RESULTS

To investigate the suitability of the least squares model for estimating ungauged streamflow, estimates were made of flows which were assumed to be unrecorded. Pertinent statistics of the estimated data were compared with those of the actual records (1). The results indicated that even a simple single-season model was capable of giving reliable streamflow estimates, especially if the model parameters were accurately estimated. As more comprehensive data banks are develped, better estimates of the model parameters will be possible. The errors in estimating the model parameters for daily data are greater than for monthly data. Therefore daily data cannot be estimated as accurately as monthly data.

504

REFERENCES

1.   Fuller, G. A., 1978, Generation of Ungauged Streamflow Data, J. Hydraul. Div., Am. Soc. Civil Engr., Vol. 104, No. HY3.

2.   Fuller, G. A. and N. E. Fuller, 1974, Inconsistent Matrices in Estimating Ungauged Streamflow Data, Water Resour. Res., Vol. 10, No. 6, pp. 1949-1950.

3.   Fuller, G. A. and H. M. Hill, 1970, Data Preparation for Great Lakes Studies, Proc. 1970 Conf. of Intl. Assoc. for Great Lakes Research, Buffalo, N.Y., Mar. 31 to Apr. 3, pp. 997-1003.

4.   Richardson, C. H., 1944, An Introduction to Statistical Analysis, Harcourt, Brace, and Co., N.Y., pp. 277-305.